CliffsStudySolver™
Algebra I

By Mary Jane Sterling

WILEY

Wiley Publishing, Inc.

Published by:
Wiley Publishing, Inc.
909 Third Avenue
New York, NY 10022
www.wiley.com

Published by Wiley Publishing, Inc., New York, NY
Published simultaneously in Canada

Library of Congress Control Number available from the Library of Congress

ISBN: 0-7645-3763-6

Printed in the United States of America

10 9 8 7 6 5

1B/RS/QW/QT/IN

For general information on our other products and services or to obtain technical support, please contact our Customer Care Department within the U.S. at 800-762-2974, outside the U.S. at 317-572-3993, or fax 317-572-4002.

Wiley also publishes its books in a variety of electronic formats. Some content that appears in print may not be available in electronic books.

Publisher's Acknowledgments

Editorial

Project Editor: Kelly D. Dobbs

Acquisitions Editor: Greg Tubach

Technical Editor: Anthony Bedenikovic

Editorial Assistant: April Yoder

Composition

Indexer: Tom Dinse

Proofreader: Vicki Broyles

Wiley Publishing Inc. Composition Services

Table of Contents

Introduction

Algebra is a typical student's first introduction into the world of symbols and rules for symbolic representations and manipulations. With a good arithmetic base, a student can move through algebra with relative ease and understanding. That's the ideal situation. Algebra can sometimes be more of a challenge for those not as comfortable with numbers or symbols. This book is designed to fill in the gaps where necessary to help you succeed in algebra. Use this book to get practice in the specific areas you need. Explanations given in this book are followed by generous examples, and it's all tied together with pretest, customized full-length exam, problems, and more problems.

One of the keys to being successful with an algebra experience is to understand the "language." The language of algebra consists of symbols representing numbers and operations. Another important part of the language is the actual language—the **words**. Algebraic symbols are very precise. They mean a specific thing, and you have to know what that thing is. The words are also very precise. The word "term" is used to describe "that bunch of numbers and letters that are bound together by multiplication and division and separated from others of its kind by addition and subtraction." You can see why the four-letter word is preferred, but its meaning has to be understood exactly.

Use these first few pages as a reference when working through a new or unfamiliar section of the book. The terms and symbols are described here for that purpose.

Types of Numbers

Numbers are classified into several types depending on what they're being used for in a particular situation. A bank will credit you with $45.02 in interest, even though the exact figure might have been $45.01935. These are decimal or **rational** numbers. If you're ignoring the cents and just want the dollar amount, $45, then you're dealing with a **whole** number. Here are the different types of numbers used in algebra and throughout this book.

Natural numbers (also known as counting numbers)

These are the numbers 1, 2, 3, 4, ...

The smallest is the number 1, and natural numbers increase by 1. No fractions or decimals are considered natural numbers unless they can be reduced or rounded to that form.

Whole numbers

These are the numbers 0, 1, 2, 3, ...

Whole numbers are different from the natural numbers only because they contain the number 0. No fractions or decimals are included here, either.

Integers

These are the numbers . . . –3, –2, –1, 0, 1, 2, 3 . . ..

They are the positive and negative whole numbers and 0. The number 0 is not positive or negative—it separates those two groups. No fractions or decimals are included here either.

Rational numbers

These are any numbers that can be written as a fraction, $\frac{p}{q}$, where p and q are integers and q is not equal to 0. Any number that can be written as a fraction is a rational number, and it has a decimal that "behaves." The decimal equivalent of any fraction either **terminates** (comes to an end, eventually) or **repeats** (goes on forever in a pattern of digits such as .123123123123 . . .).

Irrational numbers

These are numbers that have decimal values that never repeat and never end. They're "irrational" because there isn't a pattern of the same numbers repeating over and over in their decimal. Examples of irrational numbers are $\sqrt{2}, \sqrt{3}, \sqrt{10}, \ldots$ The square roots of any numbers that aren't perfect squares are irrational numbers. The exact value of $\sqrt{2}$ is just that, the value written with the radical. Any decimal representation is a rounded-off version and not exactly equal to the number. The same goes for π; its decimal value never repeats or terminates.

Real numbers

These numbers consist of all the types of numbers previously mentioned. Natural numbers, whole numbers, integers, rational and irrational numbers are all real numbers. They are the types that are dealt with in this book. An example of a number that is **not** real is $\sqrt{-5}$. The negative under the radical makes this a complex number, or one that is not real. These types of numbers would be included in a more advanced discussion of algebra, but we are not delving into them within this text.

Notation

Algebra involves a language with lots of symbols. The symbols represent numbers, operations, or processes. The symbols are necessary to keep things simple in computation and explanations. When the same process is repeated, you don't want to write out a wordy description. A "shorthand" notation is used, instead. You have to become familiar with what this notation means in order to proceed with any particular problems. Notation is divided into groupings: operations, relations, and other.

Operations

$+, -$ The operations of addition and subtraction are fairly standard. The subtraction symbol, $-$, is also read opposite, minus, or negative, depending on the situation.

$\times, \cdot, *$ The multiplication operation usually is designated by a dot, $\cdot$, in algebra. The $\times$ isn't used very much, because it can get confused with the variable, x. The $*$ is seen more in terms of graphing calculators and computer language. Multiplication is often silent and not shown at all, because parenthesis, ( ); brackets, [ ]; or braces, { } next to a number indicate multiplication. Variables or letters written next to each other are multiplied. The expression xy means x times y.

$\div, /$ The division operation can also be designated by the fraction line that is horizontal, with a numerator and denominator indicating what is being divided by what.

$\sqrt{}, \sqrt[3]{}$ The radical indicates that a root is being taken. If no number is above the left shelf, then it's assumed that the root is a 2, or square root. Any integer can be inserted for the root. The section in Chapter 1 on radicals offers more of an explanation.

$^{2}, ^{3}$ The superscripts written to the right and above a number indicate powers. They're called exponents. See the section in Chapter 1 on exponents for more information.

| | Absolute value is an operation performed on a single number put inside the lines. This operation produces positive values or zero when it's performed. See "Basic Math Operations," in Chapter 1 for more on absolute values.

Relations

=   This symbol represents the basic algebra relation. One thing equals another. The two things are exactly the same. The equal sign is used to establish equations that will need to be solved.

≤, ≥   These symbols are less than or equal to and greater than or equal to. These relations belong with < and > in showing the relationship between two values. They are discussed more in Chapter 2, "Number Lines."

≠   This symbol means not equal to.

≈   This symbol is read as "about" or "approximately." It's used when rounding the decimal value of a number to fewer than all of its digits, making the value not exact any more.

Other Symbols

∞   This is the infinity symbol. It represents a value that's so large (or small, $-\infty$) that it can't be expressed. You cannot assign a value or approximation to this symbol.

π This symbol is the Greek letter pi. It represents the ratio between the circumference (distance around the outside) and diameter (distance across) a circle. It has an approximate value of 3.14 or $\frac{22}{7}$. These two numbers are approximates of π and are used most frequently, but you can do more accurate computations if more decimal places are used. The value π isn't a rational number—no fraction is exactly equal to it—but approximate values are used in the different applications.

. . .   The three dots following a list of numbers or letters is called an ellipsis. It's the mathematical version of et cetera.

Vocabulary

Explaining processes in algebra is made much simpler if the precise mathematical vocabulary is used. The words described here are in a standard dictionary with other definitions. The math definition is very exact, however. It's important to know what is meant, mathematically. It helps to avoid vagueness or confusion, and it is a big savings in wording and the written/spoken language.

Factor, Term, Coefficient

A **factor** is a value that multiplies another value. The main distinction is the multiply part. In the expression or **term** $2xy$, the 2 is a factor; the x is a factor; and the y is a factor. They're all multiplied together.

A **term** is an expression that's all put together by multiplication and division. It's separated from other terms by addition and subtraction. The expression $4xy^2 + \frac{2xy}{3}$ has two terms. The first is $4xy^2$, and the second is $\frac{2xy}{3}$. Each of these terms has several **factors**.

The numerical part of a term usually is referred to as a **coefficient**. The coefficient of the term $2x^2$ is the 2. Coefficients tell you how many of that kind of term there are. Terms that are alike can be combined by addition or subtraction by adding and subtracting their coefficients.

Exponent, Power

An **exponent** is a small-case number written to the right and slightly above a number or letter. The exponent in x^3 is the 3. The **base** is the x. The exponent tells how many times to multiply the base times itself.

Power is another word for **exponent**.

Sum, Difference, Product, Quotient

Sum is the result of adding.

Difference is the result of subtracting.

Product is the result of multiplying.

Quotient is the result of dividing.

Numerator, Denominator

The **numerator** is the value(s) in the top of a fraction—those above the fraction line. In the fraction $\frac{4x+2}{y-1}$, the numerator is $4x + 2$.

The **denominator** is the value(s) in the bottom of a fraction. In the previous fraction, $y - 1$ is the denominator. A common reminder in algebra is that the denominator cannot be equal to 0.

Opposite, Reciprocal

The **opposite** is having a different sign than something else. The numbers -3 and 3 are opposites of one another.

A **reciprocal** is a fraction with 1 in the numerator and the original number in the denominator. Two numbers are reciprocals if their product is 1. The reciprocal of 7 is $\frac{1}{7}$. The reciprocal of $\frac{3}{4}$ is $\frac{1}{\frac{3}{4}}$ or $\frac{4}{3}$

See Chapter 2's section, "Complex Fractions," for more on this topic.

Variable, Constant

A **variable** is a symbol, usually x, y, z, w, and so on, which can represent a number. It varies, because its value isn't always known. A variable can represent any number in a situation and can be solved for in equations and inequalities.

A **constant** is a definite number—something that doesn't vary or change. When a constant is a number, such as the 2 in $2x$, then it's easy to spot. Constants also can be represented by letters. Usually, the first letters in the alphabet, a, b, c, are used to indicate constants.

Proper Fraction, Improper Fraction, Mixed Number, Complex Fraction

A **proper fraction** has an absolute value between 0 and 1. The numerator is smaller than the denominator. Some proper fractions are $\frac{2}{3}, \frac{89}{1234}$, and $\frac{9999}{10,000}$.

An **improper fraction** has a numerator that is greater in magnitude than the denominator. Its absolute value is greater than 1. Some improper fractions are $\frac{4}{3}, \frac{19}{18}$, and $\frac{123}{5}$.

A **mixed number** is a value that contains both a whole number and a fraction. Improper fractions can be rewritten as mixed numbers. For example, $\frac{4}{3} = 1\frac{1}{3}$ and $\frac{123}{5} = 24\frac{3}{5}$. A **complex fraction** has a fraction in its numerator, its denominator, or both. The fraction $\frac{\frac{2}{3}}{\frac{5}{9}}$ is a complex fraction.

Pretest

Pretest Questions

In problems 1 through 57, circle the letter corresponding to the correct answer.

1. Simplify $\frac{12}{3} + 10 \div 2$.

   A. 5

   B. 7

   C. 9

2. Simplify $\frac{\sqrt{10-1}}{3} + 2(5) - 3^2$.

   A. 2

   B. 5

   C. 7

3. Simplify $2x + 3y + x - 2y$.

   A. $3x + y$

   B. $4xy$

   C. $2x^2 - y^2$

4. Simplify $\sqrt{100a^2} - |-2a|$. Assume that a is a positive number.

   A. $8a$

   B. $10a$

   C. $12a$

5. Rewrite using exponents $6 \cdot 6 \cdot x \cdot x \cdot x \cdot x \cdot x \cdot x \cdot y \cdot y \cdot y$.

   A. $12x^6 y^3$

   B. $36x^6 y^3$

   C. $(6xy)^{11}$

6. Simplify $7a^2 + 2a + 1 - 5a^2 + 3a + 2$.

   A. $10a^3$

   B. $2a^2 + 5a + 3$

   C. $7a^2 + 3$

7. Simplify $3(x^2 + 2x^2) + 3(4a + a) + 5 - 1$.

   A. $5x^2 + 13a + 4$

   B. $5x^2 + 12a + 4$

   C. $9x^2 + 15a + 4$

8. Complete the statement to make it true: $5 + (a+3) = ($_____$) + 3$.

   A. $5a$

   B. $8 + a$

   C. $5 + a$

9. Complete the statement to make it true: $-4 +$ ____ $= 0$.

   A. 4

   B. 0

   C. −4

10. Complete the statement to make it true: $\frac{6}{5} \cdot$ ____ $= 1$.

    A. $\frac{5}{6}$

    B. $-\frac{6}{5}$

    C. $-\frac{5}{6}$

11. Simplify $(2x^2)^3$.

    A. $6x^5$

    B. $6x^6$

    C. $8x^6$

12. Simplify $m^2 m^4 + (m^2)^4$.

    A. $2m^8$

    B. $2m^6$

    C. $m^6 + m^8$

13. Simplify $\dfrac{12(x^2 y^3)^4}{3x^3 y^2}$.

    A. $4x^3 y^5$

    B. $4x^5 y^{10}$

    C. $4x^5 y^6$

14. $\sqrt{196} =$

    A. 14

    B. 16

    C. 18

15. $\sqrt[3]{216} =$

   A. 4

   B. 6

   C. 8

16. Estimate the square root to the nearer tenth: $\sqrt{21}$.

   A. 4.3

   B. 4.4

   C. 4.6

17. Combine the terms after simplifying: $\sqrt{40} + \sqrt{90}$.

   A. $\sqrt{130}$

   B. $5\sqrt{10}$

   C. $13\sqrt{10}$

18. Rationalize $\dfrac{6}{\sqrt{15}}$.

   A. $\dfrac{2\sqrt{15}}{5}$

   B. $2\sqrt{5}$

   C. 6

19. Simplify $\sqrt{x^4}\,\sqrt[3]{x^9}$. Assume that x is positive.

   A. $\sqrt[5]{x^{13}}$

   B. x^8

   C. x^5

20. Simplify $\dfrac{8^{2/3}}{8^{1/3}}$.

   A. 2

   B. $\dfrac{1}{3}$

   C. $\dfrac{8}{3}$

21. Which of the following is divisible by 3?

   A. 501,204

   B. 369,022

   C. 905,030

22. Which of the following is divisible by 12?

   A. 142,030

   B. 180,012

   C. 234,234

23. Which of the following is divisible by 11?

    A. 410,014

    B. 11,112

    C. 10,001

24. Which of the following is prime?

    A. 21

    B. 51

    C. 73

25. Which of the following is prime?

    A. 19

    B. 91

    C. 169

26. Find the prime factorization for 12.

    A. $2 \cdot 3^2$

    B. $2^2 \cdot 3$

    C. $2 \cdot 3 \cdot 4$

27. Find the prime factorization for 100.

    A. $2^2 \cdot 5^2$

    B. $2^5 \cdot 5$

    C. $2 \cdot 5^3$

28. The interval notation for $-4 \leq x < 8$ is

    A. $(-4, 8]$

    B. $(-4, 8)$

    C. $[-4, 8)$

29. The inequality notation for $(2, \infty)$ is

    A. $x < 2$

    B. $x > 2$

    C. $x < -2$

30. The interval notation for $-2 \leq x < 13$ is

    A. $[-2, 13)$

    B. $(-2, 13]$

    C. $(-2, 13)$

31. The inequality notation for $[-3, \infty)$ is

    A. $x \leq -3$

    B. $x \geq -3$

    C. $x < -3$

32. Add $4 + (-8)$.

    A. -4

    B. 4

    C. -12

33. Add $-5 + (-11)$.

    A. 16

    B. -16

    C. -6

34. Add $6 + (-6)$.

    A. 12

    B. -12

    C. 0

35. Subtract $4 - (-9)$.

    A. 13

    B. -5

    C. 5

36. Subtract $-11 - (-5)$.

    A. -16

    B. -6

    C. 6

37. Subtract $-2 - 3$.

    A. -1

    B. -5

    C. 1

38. Divide $\dfrac{-24}{-6}$.

    A. 4

    B. -4

    C. $\dfrac{1}{4}$

39. Simplify $\dfrac{3(-8)}{-4}$.

    A. -6

    B. 6

    C. -5

40. Distribute $-4(3-x+2x^2)$.

    A. $-12+4x-8x^2$

    B. $-12-4x+8x^2$

    C. $-12-x+2x^2$

41. Simplify $\dfrac{x^3}{x^{-2}}$.

    A. x^5

    B. x^1

    C. x^{-1}

42. Simplify $\left(\dfrac{a^{-4}}{a}\right)^{-3}$.

    A. a^6

    B. a^9

    C. a^{15}

43. Simplify $\left(\dfrac{6m^4 n^{-2}}{2m^{-4} n}\right)^{-2}$.

    A. $\dfrac{n^6}{9m^{16}}$

    B. $-\dfrac{3n^5}{m^{10}}$

    C. $\dfrac{6n^5}{m^{10}}$

44. Simplify $\dfrac{2a-2b}{4a+2b}$.

    A. $\dfrac{1}{2a}-1$

    B. $\dfrac{a-b}{2a+b}$

    C. $\dfrac{a-b}{a+b}$

45. Add $\dfrac{5}{2z} + \dfrac{1}{z^2}$.

   A. $\dfrac{6}{2z^3}$

   B. $\dfrac{5z+2}{2z^3}$

   C. $\dfrac{5z+2}{2z^2}$

46. Subtract $\dfrac{3}{xy} - \dfrac{2}{5y}$.

   A. $\dfrac{15-2x}{5xy}$

   B. $\dfrac{1}{5xy^2}$

   C. $\dfrac{15-2x}{5xy^2}$

47. Multiply $\dfrac{4x^2}{9y^3} \cdot \dfrac{27y}{28x^5}$.

   A. $\dfrac{3}{7x^3y^2}$

   B. $\dfrac{3x^3y^2}{7}$

   C. $\dfrac{7y^2}{3x^3}$

48. Divide $\dfrac{50m^2n^3}{77rt} \div \dfrac{30mn^4}{49rt^2}$.

   A. $\dfrac{15mt}{77n}$

   B. $\dfrac{12mt}{14n}$

   C. $\dfrac{35mt}{33n}$

49. Simplify $\dfrac{\frac{4}{7}}{\frac{8}{9}}$.

   A. $\dfrac{9}{14}$

   B. $\dfrac{14}{9}$

   C. $\dfrac{32}{63}$

50. Simplify $\dfrac{\frac{7x^2}{m}}{\frac{21x}{4m^2}}$.

   A. $\dfrac{3}{4mx}$

   B. $\dfrac{4mx}{3}$

   C. $\dfrac{147x^3}{4m^3}$

51. Change the decimal to a fraction: .45.

    A. $\frac{4}{5}$

    B. $\frac{9}{20}$

    C. $\frac{5}{11}$

52. Change the fraction to a decimal: $\frac{9}{16}$.

    A. 0.5625

    B. 0.6915

    C. 0.6525

53. Change the decimal to a fraction: .8181....

    A. $\frac{81}{100}$

    B. $\frac{8}{9}$

    C. $\frac{9}{11}$

54. Change the percent to a fraction: 16%.

    A. $\frac{1}{6}$

    B. $\frac{6}{25}$

    C. $\frac{4}{25}$

55. Change the scientific notation to its equivalent value: 4.3×10^{10}.

    A. 43,000,000,000

    B. 4,300

    C. 430,000,000,000

56. Change the number to its equivalent in scientific notation: 0.00000326.

    A. 3.26×10^{6}

    B. 3.26×10^{-6}

    C. 3.26×10^{-5}

57. Simplify and write the answer in scientific notation: $(4 \times 10^{-9})(8 \times 10^{15})$.

    A. 3.2×10^{5}

    B. 3.2×10^{6}

    C. 3.2×10^{7}

In problems 58 through 64, solve for the value of the variable.

58. $x + 4 = 47$

A. $\frac{47}{4}$

B. 43

C. 51

59. $y - 6 = -2$

A. 4

B. −4

C. −8

60. $\frac{z}{9} = -2$

A. −18

B. 18

C. 7

61. $7w = 217$

A. 147

B. 31

C. $\frac{1}{3}$

62. $4y + 7 = y - 2$

A. 1

B. 3

C. −3

63. $5(x + 4) = 3x + 8$

A. −14

B. 14

C. −6

64. $\frac{z - 3}{2} = -4$

A. 5

B. −5

C. −11

65. Solve for h: $V = \pi r^2 h$.

A. $h = V - \pi r^2$

B. $h = \dfrac{Vr^2}{\pi}$

C. $h = \dfrac{V}{\pi r^2}$

66. Solve for l: $P = 2l + 2w$.

A. $l = \dfrac{P}{2} - 2w$

B. $l = \dfrac{P - 2w}{2}$

C. $l = P - w$

67. Solve for z: $M = 3(z + 4t)$.

A. $z = \dfrac{M - 12t}{3}$

B. $z = M - 4t$

C. $z = \dfrac{M}{3} - 12t$

68. Solve for x: $\dfrac{x}{20} = \dfrac{24}{30}$.

A. 2

B. 8

C. 16

69. Solve for x: $\dfrac{18}{x} = -\dfrac{27}{33}$.

A. $-\dfrac{1}{22}$

B. -22

C. $-\dfrac{11}{22}$

70. Multiply $\dfrac{a(a+1)(a-1)}{x^2(x+2)^2} \cdot \dfrac{x(x+2)(x-1)}{a^3(a+1)^3}$.

A. $\dfrac{(a-1)(x-1)}{a^2 x(x+2)(a+1)^2}$

B. $\dfrac{a^2 x(a-1)(x-1)}{(x+2)(a+1)^2}$

C. $\dfrac{1}{a^2 x(a+1)}$

71. Divide $\dfrac{4z(z-2)^2}{9a^2(a+2)^3} \div \dfrac{12z^2(z-2)(x+2)^2}{81a^4(a+2)^5}$.

A. $\dfrac{3a^2 z(z-2)(a+2)}{(x+2)^2}$

B. $\dfrac{3(a+2)^4(z-2)}{z(x+2)^2}$

C. $\dfrac{3a^2(z-2)(a+2)^2}{z(x+2)^2}$

72. Add $\dfrac{3}{2ab} + \dfrac{4b}{a^2}$.

A. $\dfrac{3a+4b^2}{2a^2 b}$

B. $\dfrac{3a+8b^2}{2a^2 b}$

C. $\dfrac{3+4b}{2a^2 b}$

73. Subtract $\dfrac{4+y}{3y} - \dfrac{6}{5y^2}$.

A. $\dfrac{5y^2+20y-18}{15y^2}$

B. $\dfrac{10+y}{15y^2}$

C. $\dfrac{2+5y^2}{15y^2}$

74. Solve for y: $\dfrac{y}{5} - \dfrac{7}{30} = \dfrac{7y}{30} + \dfrac{1}{10}$.

A. -6

B. -8

C. -10

75. Solve for y: $\dfrac{6y+1}{20} - \dfrac{2y}{5} = \dfrac{y}{5} - \dfrac{1+7y}{20}$.

A. -2

B. 2

C. -4

76. Distribute: $-3y(4y^3 - 2y^2 + x - 1)$.

A. $-12y^4 + 6y^3 - 3xy + 3y$

B. $-12y^4 - 6y^3 + 3xy - 3y$

C. $-12y^4 + 6y^3 - 3xy - 1$

77. Distribute: $4z^3(z^2 - 2z + 5z^{-1} - z^{-3})$.

   A. $4z^5 - 8z^4 + 20z^2$

   B. $4z^5 - 8z^4 + 20z^2 - 4$

   C. $4z^6 - 8z^3 + 20z^{-1} - 4z^{-9}$

78. Multiply: $(x + 4)(x - 5)$.

   A. $x^2 + x - 20$

   B. $x^2 - x - 20$

   C. $x^2 - x - 9$

79. Multiply: $(2x - 1)(5x - 8)$.

   A. $10x^2 - 11x + 8$

   B. $10x^2 - 15x + 8$

   C. $10x^2 - 21x + 8$

80. Multiply: $(x + 3)(x^2 - 5x - 6)$.

   A. $x^3 - 8x^2 + 9x + 18$

   B. $x^3 - 2x^2 - 9x + 18$

   C. $x^3 - 2x^2 - 21x - 18$

81. Multiply: $(2x - 9)(2x + 9)$.

   A. $4x^2 - 81$

   B. $4x^2 - 36x + 81$

   C. $4x^2 - 18x - 81$

82. Multiply: $(1 + 3z)^2$.

   A. $1 + 9z^2$

   B. $1 + 3z + 6z^2$

   C. $1 + 6z + 9z^2$

83. Multiply: $(n + 1)^3$.

   A. $n^3 + 1$

   B. $n^3 + 3n^2 + 3n + 1$

   C. $n^3 + 3n^2 + 3n + 3$

84. Divide: $(4y^3 - 3y^2 + y - 1) \div y$.

 A. $4y^2 - 3y + 1 - \dfrac{1}{y}$

 B. $4y^3 - 3y^2 + y - 1$

 C. $4y^2 - 3y - 1$

85. Divide: $(3x^4 - 2x^2 + 3) \div (x - 1)$.

 A. $3x^3 - 3x^2 + x - 1 + \dfrac{4}{x-1}$

 B. $3x^3 + x^2 + \dfrac{4}{x-1}$

 C. $3x^3 + 3x^2 + x + 1 + \dfrac{4}{x-1}$

86. Factor out the Greatest Common Factor: $5x^2 y^2 - 10x^2 y^3 + 20xy^4$.

 A. $5x^2 y^4 (y^2 - 2y + 4x)$

 B. $5xy^2 (x - 2xy + 4y^2)$

 C. $5xy(xy - 10xy^2 + 4xy^3)$

87. Factor out the Greatest Common Factor: $4(m+1)^2 - 6m(m+1)^3 + 12m^2(m+1)^4$.

 A. $2m(m+1)^2 [2 - 3(m+1) + 6m(m+1)^2]$

 B. $2(m+1)^2 [2 - 3m(m+1) + 6m^2(m+1)^2]$

 C. $2(m+1)^4 [2(m+1)^2 - 3m(m+1) + 6m^2]$

For problems 88 through 96, factor completely.

88. $4 - z^2 =$

 A. $(2-z)(2+z)$

 B. $(2+z)(2+z)$

 C. $z(2-z)$

89. $a^3 + 27 =$

 A. $(a+3)(a^2 - 9)$

 B. $(a+3)(a^2 + 9)$

 C. $(a+3)(a^2 - 3a + 9)$

90. $m^4 - 64 =$

 A. $(m^2 + 8)(m^2 - 8)$

 B. $(m-2)(m+2)(m^2 + 16)$

 C. $(m^2 - 8)(m^2 - 8)$

91. $x^2 - 2x - 24 =$

A. $(x - 4)(x + 6)$

B. $(x - 6)(x + 4)$

C. $(x - 12)(x + 2)$

92. $16x^2 - 22x - 3 =$

A. $(4x - 3)(4x + 1)$

B. $(16x + 3)(x - 1)$

C. $(8x + 1)(2x - 3)$

93. $2x^2 - 9x + 7 =$

A. $(2x - 7)(x - 1)$

B. $(2x + 7)(x + 1)$

C. $(2x - 1)(x - 7)$

94. $y^4 + 9y^2 + 8 =$

A. $(y + 1)(y + 1)(y^2 + 9)$

B. $(y^2 + 9)(y^2 + 1)$

C. $(y^2 + 1)(y^2 + 8)$

95. $a^2 x + x - 2a^2 - 2 =$

A. $(a^2 + 1)(x - 2)$

B. $(a^2 + 2)(x - 1)$

C. $(a - 1)(a + 1)(x + 2)$

96. $5y^2 z + 15y^2 - 20z - 60 =$

A. $5(z - 4)(y^2 + 3)$

B. $5(z + 3)(y - 2)(y + 2)$

C. $5(z - 2)(y + 2)(y + 32)$

97. Which of the choices is in the solution of $5x - 2 > x - 1$?

A. -1

B. 0

C. 1

98. Which of the choices is in the solution of $-4 < 1 - 2x < 5$?

A. -8

B. -3

C. -1

99. Which of the choices is in the solution of $\frac{3}{z} > \frac{4}{z-1}$?

    A. −4

    B. 0

    C. 2

100. Find the common solution of $x > 4$ and $x \le 11$.

    A. $4 < x \le 11$

    B. $-4 < x \le 11$

    C. $x < 4$ or $x \ge 11$

101. Find the common solution of $-3 < z < 5$ and $z - 2 > 1$.

    A. $-3 < z < 3$

    B. $-5 < z < -3$

    C. $3 < z < 5$

102. Solve for x: $|2x + 1| = 7$.

    A. 3 or −3

    B. 3 or −4

    C. 4 or −3

103. Solve for y: $|6y - 11| + 3 = 8$.

    A. 1 or $\frac{8}{3}$

    B. 1 or 0

    C. 0 or $\frac{8}{3}$

104. Solve for z: $|4 - 3z| \le 6$.

    A. $-\frac{10}{3} \le z \le \frac{2}{3}$

    B. $-\frac{2}{3} \le z \le \frac{10}{3}$

    C. $z \le -\frac{2}{3}$ or $z \ge \frac{10}{3}$

105. Solve for w: $2|w - 4| + 3 > 9$.

    A. $w < 1$ or $w > 7$

    B. $1 < w < 7$

    C. $-7 < w < 1$

For problems 106 through 110, assume that the variables are positive.

106. Simplify $\sqrt{900a^3 b^4 c^8}$.

   A. $30a^2 b^2 c^6$

   B. $30ab^2 c^4 \sqrt{a}$

   C. $30ab^2 c^6 \sqrt{a}$

107. Simplify $\sqrt{8x^2 y^3} \sqrt{6xy^4}$.

   A. $4x^2 y^6 \sqrt{3xy}$

   B. $16xy^3 \sqrt{3xy}$

   C. $4xy^3 \sqrt{3xy}$

108. Simplify and add $2\sqrt{45ax^2} + 3x\sqrt{20a}$.

   A. $6x\sqrt{65ax^2}$

   B. $36x^2 \sqrt{5a}$

   C. $12x\sqrt{5a}$

109. Simplify $\sqrt[3]{16m^4 n^6}$.

   A. $2mn^2 \sqrt[3]{2m}$

   B. $2m^2 n^3 \sqrt[3]{2m}$

   C. $8mn^3 \sqrt[3]{2mn^2}$

110. Simplify $\sqrt[4]{9x^8 y^9} \sqrt[4]{27xy^6}$.

   A. $81x^8 y^{12}$

   B. $3x^2 y^3 \sqrt[4]{3xy^3}$

   C. $3xy^2 \sqrt[4]{3xy}$

111. Solve for x: $\sqrt{5x+1} = 9$.

   A. 4

   B. 8

   C. 16

112. Solve for y: $6 - \sqrt{8-y} = 3$.

   A. 1

   B. −1

   C. −3

In problems 113 through 116, solve for the value(s) of the variable.

113. $x^2 + 13x + 40 = 0$

    A. 5 or 8

    B. −5 or −8

    C. −4 or −10

114. $3x^2 + 11x - 4 = 0$

    A. $\frac{1}{3}$ or −4

    B. $\frac{4}{3}$ or −1

    C. $-\frac{4}{3}$ or 1

115. $x^2 + 3x - 2 = 0$

    A. $\dfrac{3 \pm \sqrt{15}}{2}$

    B. $\dfrac{3 \pm 1}{2}$

    C. $\dfrac{-3 \pm \sqrt{17}}{2}$

116. $3y^2 + y - 5 = 0$

    A. $\dfrac{-1 \pm \sqrt{61}}{6}$

    B. $\dfrac{1 \pm \sqrt{61}}{2}$

    C. $\dfrac{-1 \pm \sqrt{18}}{6}$

117. Solve for z by completing the square: $z^2 - 12z + 32 = 0$.

    A. 4 or 8

    B. −16 or −2

    C. 16 or 2

For problems 118 through 122, solve for the value(s) of the variable.

118. $x^{1/2} + 4x^{1/4} - 21 = 0$

    A. 81

    B. 3

    C. 3 or −7

119. $y^4 - 40y^2 + 144 = 0$

    A. 12 or −12

    B. 2 or −2 or 72 or −72

    C. 2 or −2 or 6 or −6

120. $(x - 8)(3x + 1) \le 0$

    A. $-\frac{1}{3} \le x \le 8$

    B. $-8 \le x \le \frac{1}{3}$

    C. $x \le -\frac{1}{3}$ or $x \ge 8$

121. $\frac{z}{z-3} \ge 0$

    A. $0 \le z < 3$

    B. $-3 \le z < 0$

    C. $z \le 0$ or $z > 3$

122. $\sqrt{1 - 16x} = x + 10$

    A. −3 or −33

    B. −3 only

    C. 3 or 33

123. Match the points on the graph with their coordinates.

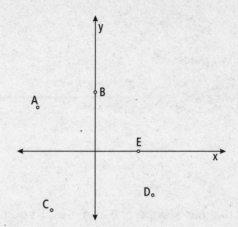

The choices are (0,4), (3,0), (−4,3), (−3,−4), (4,−3).

A:

B:

C:

D:

E:

124. The graph of $y = -\frac{1}{3}x + 4$

A. crosses the x-axis at 4.

B. crosses the y-axis at 4.

C. crosses the x-axis at $\left(-\frac{1}{3}, 4\right)$.

125. What is the vertex of the parabola $y = 3(x-1)^2 + 2$?

A. (1,2)

B. (−1,2)

C. (1,−2)

126. The graph of the parabola $y = 3(x - 1)^2 + 2$

  A. opens upward.

  B. opens downward.

  C. crosses the x-axis in two places.

127. The graph of $y = x(x - 1)(x + 3)$ crosses the x-axis at

  A. 1 and −3

  B. 0 and −1 and 3

  C. 0 and 1 and −3

128. The equation of the line that has a slope of 6 and goes through the point (3,−2) is

  A. $y = 6x - 2$

  B. $y = 6x + 3$

  C. $y = 6x - 20$

129. A point in the solution of $3x - 2y \geq 6$ is

  A. (0,2)

  B. (2,0)

  C. (3,4)

130. The common solution of the lines $y = -4x + 3$ and $x + y = 6$ is

  A. (−1,7)

  B. (1,−1)

  C. (7,−1)

131. Which of the following is a function of x?

  A. $x^2 + y^2 = 7$

  B. $y = 3x + 2$

  C. $y^2 = -5x - 1$

132. What is the domain of the function $y = \sqrt{x - 7}$?

  A. $x \geq 7$

  B. $x \leq 7$

  C. all real numbers

133. What is the range of the function $y = x^2 + 3$?

  A. $y \geq 0$

  B. $y \geq 3$

  C. all real numbers

134. Which is the inverse of the function $f(x) = 3x - 1$?

    A. $f^{-1}(x) = \dfrac{1}{3x - 1}$

    B. $f^{-1}(x) = \dfrac{1}{3}x + 1$

    C. $f^{-1}(x) = \dfrac{x + 1}{3}$

135. One number is 1 more than 3 times another. Their sum is 21. What is their product?

    A. 16

    B. 80

    C. 100

136. The sum of 3 consecutive odd integers is 99. What is the largest of them?

    A. 33

    B. 35

    C. 41

137. Ken is three times as old as Karen. In 10 years, he'll be twice as old as Karen. How old is Karen?

    A. 5

    B. 10

    C. 20

138. A rectangle has a length that is 2 inches more than twice the width. The area is 60. What is the length?

    A. 10

    B. 12

    C. 15

139. Two trains are 360 miles apart at 7:00 AM and are traveling on the same track moving toward one another. One train is traveling at 40 mph, and the other is traveling at 50 mph. At what time will they meet?

    A. 8:00 AM

    B. 10:00 AM

    C. 11:00 AM

140. How many quarts of 80% solution need to be added to 5 quarts of 50% solution to produce a mixture that's 70% solution?

    A. 6

    B. 8

    C. 10

141. Marissa has 24 coins in quarters and nickels. She has $3. How many of the coins are quarters?

    A. 9

    B. 10

    C. 11

142. Arlene can do a job in 3 hours, and Ben can do the same job in 12 hours. How long will it take the two of them to do the job working together?

    A. 1 hour

    B. $1\frac{1}{4}$ hours

    C. $2\frac{2}{5}$ hours

Key to Pretest Questions

1. C

2. A

If you missed 1 or 2, go to "Order of Operations," page 31.

3. A

4. A

5. B

If you missed 3, 4, or 5, go to "Basic Math Operations," page 35.

6. B

7. C

If you missed 6 or 7, go to "Combining 'Like' Terms," page 38.

8. C

9. A

10. A

If you missed 8, 9, or 10, go to "Properties of Algebraic Expressions," page 41.

11. C

12. C

13. B

If you missed 11, 12, or 13, go to "Integer Powers (Exponents)," page 44.

14. A

15. B

If you missed 14 or 15, go to "Square Roots and Cube Roots," page 48.

16. C

If you missed 16, go to "Approximating Square Roots," page 50.

17. B

18. A

If you missed 17 or 18, go to "Simplifying Square Roots," page 52.

19. C

20. A

If you missed 19 or 20, go to "Exponents for Roots," page 56.

21. A

22. B

23. A

If you missed 21, 22, or 23, go to "Divisibility Rules," page 58.

24. C

25. A

If you missed 24 or 25, go to "Prime Numbers," page 61.

26. B

27. A

If you missed 26 or 27, go to "Prime Factorization," page 63.

28. C

29. B

30. A

31. B

If you missed 28, 29, 30, or 31, go to "Number Lines," page 73.

32. A

33. B

34. C

If you missed 32, 33, or 34, go to "Addition of Signed Numbers," page 75.

35. A

36. B

37. B

If you missed 35, 36, or 37, go to" Subtraction of Signed Numbers," page 78.

38. A

39. B

40. A

If you missed 38, 39, or 40, go to "Multiplication and Division of Signed Numbers," page 80.

41. A

42. C

43. A

If you missed 41, 42, or 43, go to "Exponents That Are Signed Numbers," page 82.

44. B

45. C

46. A

If you missed 44, 45, or 46, go to "Lowest Terms and Equivalent Fractions," page 86.

47. A

48. C

If you missed 47 or 48, go to "Multiplying and Dividing Fractions," page 89.

49. A

50. B

If you missed 49 or 50, go to "Mixed Numbers and Complex Fractions," page 91.

51. B

52. A

53. C

If you missed 51, 52, or 53, go to "Decimals," page 94.

54. C

If you missed 54, go to "Percents," page 97.

55. A

56. B

57. C

If you missed 55, 56, or 57, go to "Scientific Notation," page 99.

58. B

59. A

60. A

61. B

If you missed 58, 59, 60, or 61, go to "Solving Linear Equations," page 107.

62. C

63. C

64. B

If you missed 62, 63, or 64, go to "Solving Linear Equations with More Than One Operation," page 111.

65. C

66. B

67. A

If you missed 65, 66, or 67, go to "Solving Linear Formulas," page 116.

68. C

69. B

If you missed 68 or 69, go to "Ratios and Proportions," page 120.

70. A

71. C

If you missed 70 or 71, go to "Multiplying and Dividing Algebraic Fractions," page 125.

72. B

73. A

If you missed 72 or 73, go to "Adding and Subtracting Algebraic Fractions," page 128.

74. C

75. A

If you missed 74 or 75, go to "Equations with Fractions," page 131.

76. A

77. B

If you missed 76 or 77, go to "Multiplying Monomials," page 145.

78. B

79. C

80. C

If you missed 78, 79, or 80, go to "Multiplying Polynomials," page 148.

81. A

82. C

83. B

If you missed 81, 82, or 83, go to "Special Products," page 151.

84. A

85. C

If you missed 84 or 85, go to "Dividing Polynomials," page 156.

86. B

87. B

If you missed 86 or 87, go to "Greatest Common Factor," page 163.

88. A

89. C

90. A

If you missed 88, 89, or 90, go to "Factoring Binomials," page 165.

91. B

92. C

93. A

If you missed 91, 92, or 93, go to "Factoring Trinomials," page 168.

94. C

95. A

96. B

If you missed 94, 95, or 96, go to "Factoring Other Polynomials," page 172.

97. C

98. C

99. A

If you missed 97, 98, or 99, go to "Inequalities," page 185.

100. A

101. C

If you missed 100 or 101, go to "Solving Inequalities by Graphing on a Number Line," page 189.

102. B

103. A

If you missed 102 or 103, go to "Absolute Value Equations," page 192.

104. B

105. A

If you missed 104 or 105, go to "Absolute Value Inequalities," page 199.

106. B

107. C

108. C

If you missed 106, 107, or 108, go to "Simplifying Square Roots," page 203.

109. A

110. B

If you missed 109 or 110, go to "Simplifying Other Roots," page 206.

111. C

112. B

If you missed 111 or 112, go to "Radical Equations," page 208.

113. B

114. A

If you missed 113 or 114, go to "Solving Quadratic Equations by Factoring," page 224.

115. C

116. A

If you missed 115 or 116, go to "Solving Quadratic Equations with the Quadratic Formula," page 228.

117. A

If you missed 117, go to "Solving Quadratic Equations by Completing the Square," page 231.

118. A

119. C

If you missed 118 or 119, go to "Solving Quadratic-Like Equations," page 237.

120. A

121. C

If you missed 120 or 121, go to "Quadratic and Other Inequalities," page 241.

122. B

If you missed 122, go to "Radical Equations with Quadratics," page 245.

123. A (−4,3); B (0,4); C (−3,−4); D (4,−3); E (3,0)

If you missed 123, go to "Graphing Points and Quadrants," page 260.

124. B

If you missed 124, go to "Graphing Lines," page 263.

125. A

126. A

127. C

If you missed 125, 126, or 127, go to "Graphs of Other Curves," page 273.

128. C

If you missed 128, go to "Finding the Equation of a Line," page 283.

129. B

If you missed 129, go to "Graphing Inequalities," page 286.

130. A

If you missed 130, go to "Solving Systems of Equations by Graphing," page 290.

131. B

If you missed 131, go to "Functions and Relations," page 313.

132. A

133. B

If you missed 132 or 133, go to "Domain and Range," page 316.

134. C

If you missed 134, go to "Inverse Functions and Function Notation," page 325.

135. B

If you missed 135, go to "Number Problems," page 333.

136. B

If you missed 136, go to "Consecutive Integer Problems," page 335.

137. B

If you missed 137, go to "Age Problems," page 337.

138. B

If you missed 138, go to "Geometric Problems," page 339.

139. C

If you missed 139, go to "Distance Problems," page 344.

140. C

If you missed 140, go to "Mixture Problems," page 347.

141. A

If you missed 141, go to "Coin and Interest Problems," page 351.

142. C

If you missed 142, go to "Working Together Problems," page 353.

Chapter 1

The Basics

Think of this first chapter as a resource or reference for much of what follows. You can come back to review this material if something puzzles you in a later chapter. These underlying principles make up much of every kind of mathematics.

Order of Operations

Mathematics deals in so many symbols. This is a good thing. The use of symbols is efficient; it saves time and writing. It also makes the language of mathematics universal—understood worldwide. Likewise, a universal agreement exists about how operation symbols such as add, subtract, and so on are to be handled in an equation. The order in which things are done makes a difference. Think of putting the cap on a bottle of soda pop and then shaking the bottle. The result is a lot different than if you first shake the bottle and **then** put the cap on. The same is true here; order makes a difference. Many operations are used in mathematics, and, accompanying them, some rules and conventions need to be followed. These rules or procedures were established so that anyone reading a mathematical statement written by someone else would know exactly what was intended. Mathematicians throughout the world use the same rules.

Basic Order of Operations

When more than one operation is indicated in an algebraic expression, the operations are done in the following order, except when grouping symbols, such as parentheses, interrupt:

> First: Powers and roots
>
> Second: Multiplication and division
>
> Third: Addition and subtraction

If more than one of the same level of operation appears in the expression, do them in order, moving from left to right.

Example Problems

These problems show the answers and solutions.

1. Simplify $50 - 2^2 \cdot 6$.

   answer: 26

By the Order of Operations, calculate the power first, then the multiplication, and then the subtraction.

First do the power, $50 - 2^2 \cdot 6 = 50 - 4 \cdot 6$.

Then multiply, $50 - 4 \cdot 6 = 50 - 24$.

Finally subtract, $50 - 24 = 26$.

2. Simplify $\sqrt{625} - 2 \cdot 6 + 7^2$.

answer: 62

By the Order of Operations, the root of 625 and the power of 7 are done first.

$$\sqrt{625} - 2 \cdot 6 + 7^2 = 25 - 2 \cdot 6 + 49$$

Next, multiply the 2 and 6, $25 - 2 \cdot 6 + 49 = 25 - 12 + 49$.

Then add and subtract. Since addition and subtraction are on the same level, perform the operations moving from left to right.

$$25 - 12 + 49 = 13 + 49 = 62$$

Grouping Symbols

Grouping symbols can "interrupt" the Order of Operations. The most commonly used grouping symbols are parentheses ( ), brackets [ ], and braces { }. Also, fraction lines and radicals (root symbols) act to group expressions above, below, and inside them. The rule is that you perform the operations within the grouping symbols first and then go to the Order of Operations. Grouping symbols more often than not help clarify what is meant in a mathematical statement. Think of them as being like punctuation in a written statement—helping you to understand the meaning.

Example Problems

These problems show the answers and solutions.

1. Simplify $6(14 - 3) + 8$.

   answer: 74

   Perform what's in the parentheses first. $6(14 - 3) + 8 = 6(11) + 8$

   Then multiply and, finally, add. $6(11) + 8 = 66 + 8 = 74$

2. Simplify $3\sqrt{16 - 7} + (5 - 3) \cdot 7 - \dfrac{14}{9 - 2}$.

   answer: 21

Three grouping symbols are here: radical, parentheses, and fraction line. Perform the operations within, above, or below them first.

$$3\sqrt{16-7}+(5-3)\cdot 7-\frac{14}{9-2}=3\sqrt{9}+(2)\cdot 7-\frac{14}{7}$$

The root has to be calculated next, because powers and roots are on the first level.

$$3\sqrt{9}+(2)\cdot 7-\frac{14}{7}=3\cdot 3+(2)\cdot 7-\frac{14}{7}$$

Now, do the two multiplications and the division.

$$3\cdot 3+(2)\cdot 7-\frac{14}{7}=9+14-2$$

Now add and subtract, moving from left to right.

$$9+14-2=23-2=21$$

Work Problems

Use these problems to give yourself additional practice.

1. $4^2+3\cdot 6-2$

2. $\frac{15-3}{4}-\sqrt{9}+8$

3. $\frac{6^2+9}{9}+5(8-2^2)-1$

4. $10+3^2-4\left(\sqrt{36}-5\right)$

5. $\frac{3(6-2)+7^2(5-1)}{5+3}$

Worked Solutions

1. **32** First raise 4 to the second power.

$$4^2+3\cdot 6-2=16+3\cdot 6-2$$

Next, multiply the 3 and 6.

$$16+3\cdot 6-2=16+18-2$$

Last, add and subtract from left to right.

$$16+18-2=34-2=32$$

2. **8**   First, subtract the 3 from 15, because they're "grouped."

$$\frac{15-3}{4} - \sqrt{9} + 8 = \frac{12}{4} - \sqrt{9} + 8$$

Next, find the square root of 9.

$$\frac{12}{4} - \sqrt{9} + 8 = \frac{12}{4} - 3 + 8$$

Now, divide 12 by 4 and combine the terms from left to right.

$$\frac{12}{4} - 3 + 8 = 3 - 3 + 8 = 0 + 8 = 8$$

3. **24**   First, raise the 6 in the numerator of the fraction to the second power and raise the 2 in the parentheses to the second power.

$$\frac{6^2 + 9}{9} + 5(8 - 2^2) - 1 = \frac{36 + 9}{9} + 5(8 - 4) - 1$$

Now add the two numbers in the numerator of the fraction and subtract the two numbers in the parentheses.

$$\frac{36 + 9}{9} + 5(8 - 4) - 1 = \frac{45}{9} + 5(4) - 1$$

Next, do the division and multiplication.

$$\frac{45}{9} + 5(4) - 1 = 5 + 20 - 1$$

Last, add and subtract in order from left to right.

$$5 + 20 - 1 = 25 - 1 = 24$$

4. **15**   First, raise the 3 to the second power and find the square root of 36.

$$10 + 3^2 - 4\left(\sqrt{36} - 5\right) = 10 + 9 - 4(6 - 5)$$

Now combine the two numbers in the parentheses and then multiply the result by 4.

$$10 + 9 - 4(6 - 5) = 10 + 9 - 4(1) = 10 + 9 - 4$$

Last, add and subtract in order.

$$10 + 9 - 4 = 19 - 4 = 15$$

5. **26** First perform the operations in the parentheses and in the denominator of the fraction.

$$\frac{3(6-2)+7^2(5-1)}{5+3}=\frac{3(4)+7^2(4)}{8}$$

Now square the 7.

$$\frac{3(4)+7^2(4)}{8}=\frac{3(4)+49(4)}{8}$$

Now do the multiplications in the numerator. The two terms can't be added until those multiplications are first performed.

$$\frac{3(4)+49(4)}{8}=\frac{12+196}{8}$$

Add the two numbers in the numerator and then divide the result by 8.

$$\frac{12+196}{8}=\frac{208}{8}=26$$

Basic Math Operations

The basic math operations that apply to numbers also apply to variables, which are represented by letters. The main difference between dealing with numbers and variables is that with numbers you can see, directly, what the operation does to them. When dealing with variables, you sometimes don't know what the variable represents, and difficulties could arise depending on whether the variable represents a positive or negative number, a fraction or whole number, an even or odd number, and so on. Following is a discussion of the basic operations and how variables are handled in each situation.

Addition and Subtraction

When you add and subtract terms with the same variable in them, the coefficient (number multiplier) indicates how many of that variable there are. So just combine the coefficients. For instance,

$3a + 2a = 5a$ (Think: "Three apples plus two apples equals five apples. You don't change the things that are being added to 'apple-apples'.")

$7b - 5b = 2b$

$8c + c = 9c$ (No coefficient shows on the second term, so you can assume that it's "1.")

Multiplication and Division

When multiplying or dividing expressions with variables by numbers, just multiply or divide the coefficients by the multiplying or dividing number. For example, if z stands for a number of zebras at the zoo, then $2z$ means two times or double that number. Multiplying by 5, $5 \cdot 2z$

represents five times that doubled number, or $5 \cdot 2z = 10z$. If p is the number of octopuses in a tank, then $8p$ is the number of legs in the tank. If the $8p$ number of legs is to be divided into four groups, then there would be $\frac{8p}{4} = 2p$ or $2p$ legs in each of the four groups. Here are some more examples.

$$(5m) \cdot 6 = 30m$$

$$8(3z) = 24z$$

$$\frac{16w}{8} = 2w$$

$$\frac{96t}{16} = 6t$$

Powers (Exponents)

Before mathematicians agreed on superscripts as the notation for powers (in 8^2, the 2 is the *power* or *exponent*), multiplying a variable times itself repeatedly was shown by repeating that letter. What you write as y^5 was once written as $y \cdot y \cdot y \cdot y \cdot y$. The power or superscript tells you how many times the variable multiplies itself. The expression $2xxxxxx$ is more conveniently written as $2x^6$. The 2 is written in front. By the Order of Operations, you compute the power first and then multiply the result by 2. This is, of course, if you know what number the x represents. Also, a convenient way of writing repeated multiples of a variable when it occurs in the denominator of a fraction is to use a **negative** exponent: $\frac{1}{wwww} = \frac{1}{w^4} = w^{-4}$.

Example Problems

These problems show the answers and solutions.

1. Write $3aaaaaaaaaa$ using exponents.

    answer: $3a^{10}$

    Simply count the number of identical variables (that is, the number of a's used) and make that number the exponent.

2. Write $2 \cdot 2 \cdot 2xxyyzzzz$ using exponents.

    answer: $2^3 x^2 y^2 z^4$

    The letters can't be combined, because they can stand for different numbers. Also, notice that the multiplication dot ($\cdot$) is shown between the 2s but not between the letters or variables. That's because when the letters are written next to one another, multiplication is assumed. If I didn't put the multiplication dot between the 2s, you would think I meant the number 222.

Roots

A root of a number is the value that has to be multiplied over and over to get the original number. The symbol for taking a root is a radical, $\sqrt{}$. If there's no number above the "shelf" on the left, then it's assumed that you mean the square root. A "square" root indicates that you want to find the value whose square (multiplied twice) gives you the number under the radical. If there is a 3 above the shelf, then you want the cube root, or the value whose third power gives you the value under the radical. Assume, in each case here, that the variable under the radical represents a positive number. If you want more information on what happens when you don't know whether the variable is positive or not see "Absolute Value," which follows the Example Problems.

Example Problems

These problems show the answers and solutions.

Simplify each root.

1. $\sqrt{9}$

   answer: 3

   $$\sqrt{9} = \sqrt{3 \cdot 3} = 3$$

2. $\sqrt{400}$

   answer: 20

   $$\sqrt{400} = \sqrt{20 \cdot 20} = 20$$

3. $\sqrt[6]{5^6}$

   answer: 5

   $$\sqrt[6]{5^6} = \sqrt[6]{5 \cdot 5 \cdot 5 \cdot 5 \cdot 5 \cdot 5} = 5$$

4. $\sqrt[3]{m^3}$

   answer: m

   $$\sqrt[3]{m^3} = \sqrt[3]{mmm} = m$$

5. $\sqrt{4b^4}$

   answer: $2b^2$

   $$\sqrt{4b^4} = \sqrt{2 \cdot 2 \cdot b^2 \cdot b^2} = 2b^2$$

Absolute Value

This operation appears to change negative numbers to positive numbers and leave positive numbers as they are. Although this is what happens, practically, the operation is really telling you how far a number is from 0. Absolute value is written with two vertical bars: $|$some number$|$ = the number if it's positive and the opposite of a number if it's negative. (The opposite of a negative is a positive.) For example,

$$|-2| = 2$$

$$|4| = 4$$

$|6 - 5| = |1| = 1$ Absolute value can be a grouping symbol. Do what's inside first.

$|m| = m$ if the variable represents a positive number and $-m$ if the variable represents a negative number. The $-m$ is read, "The *opposite* of m."

$\sqrt{x^2} = |x|$ when you don't know whether the variable x is positive or negative.

Work Problems

Use these problems to give yourself additional practice.

1. Simplify: $14a - 10a + 3(2b)$

2. Simplify: $3(5a - a)$

3. Simplify: $2 \cdot 3 \cdot n \cdot n \cdot n \cdot n \cdot p$

4. Simplify: $\sqrt{a^2 b^6}$

5. Simplify: $\sqrt{49x^2 y^{10}}$

Worked Solutions

1. **$4a + 6b$**   First multiply the 3 and $2b$.

$$14a - 10a + 3(2b) = 14a - 10a + 6b$$

Now combine the two terms with the variable a.

$$14a - 10a + 6b = 4a + 6b$$

2. **$12a$**   First combine the terms in the parentheses. Then multiply the result by 3.

$$3(5a - a) = 3(4a) = 12a$$

3. **$6n^4 p$**   Multiply the 2 and 3 and rewrite the repeated multiplication of n as a factor with an exponent.

$$2 \cdot 3 \cdot n \cdot n \cdot n \cdot n \cdot p = 6n^4 p$$

4. **ab^3**   Rewrite the factors under the radical as two repeated variables before taking the square root.

$$\sqrt{a^2 b^6} = \sqrt{a \cdot a \cdot b^3 \cdot b^3} = ab^3$$

5. **$7xy^5$**   Rewrite the factors under the radical as two repeated variables before taking the square root.

$$\sqrt{49x^2 y^{10}} = \sqrt{7 \cdot 7 \cdot x \cdot x \cdot y^5 \cdot y^5} = 7xy^5$$

Combining "Like" Terms

One major objective of working with algebraic expressions is to write them as simply as possible and in a logical, generally accepted arrangement. When there is more than one term (a term consists of one or more factors multiplied together and separated from other terms by + or −), then you check to see whether they can be combined with other terms that are "like" them. Numbers, by themselves without letters or variables, are "like" terms. You can combine 14 and 8 because you know what they are and know the rules. For instance, $14 + 8 = 22$, $14 - 8 = 6$, $14(8) = 112$, and so on. Most numbers can be written so that they can combine with one another. Fractions can be added if they have a common denominator. Decimals can be subtracted if you line up the decimal points in the two decimal numbers so that the tens place is under the tens place, the hundredths place is under the hundredths place, etc. The exception to this is that some numbers, written under a radical, can't be combined. These numbers are called "irrational." This is a good name for them, because they sometimes are difficult to manipulate.

Algebraic expressions involving variables or letters have to be dealt with carefully. Because the numbers that the letters represent aren't usually known, you can't add or subtract terms with different letters. The expression $2a + 3b$ has to stay that way. That's as simple as you can write it. But, the expression $4c + 3c$ can be simplified. You don't know what c represents, but you can combine the terms to tell how many of them you have (even though you don't know what they are!): $4c + 3c = 7c$. Here are some other examples.

$$5ab + 9ab = 14ab$$

$$5x^2 y - x^2 y + 6xy^2 + 2xy^2 = 4x^2 y + 8xy^2$$

Notice that two different kinds of terms are shown, one with the x squared and the other with the y squared. Only those that have the letters exactly alike with the exact same powers can be combined. The only thing affected by adding and subtracting these terms is the coefficient.

Example Problems

These problems show the answers and solutions.

1. Simplify $4a + 2b + 3ac + 3a - 2b + ac$.

    answer: $7a + 4ac$

    Three different kinds of terms are shown here. First rearrange the terms so that the like terms are together. Then combine the like terms.

    $$4a + 2b + 3ac + 3a - 2b + ac = 4a + 3a + 2b - 2b + 3ac + ac$$
    $$= 7a + 0 + 4ac$$
    $$= 7a + 4ac$$

2. Simplify $9x^2 + 5x^3 - 8x + 4 + 4x^3 - 7x^2 + 8x + 9$.

    answer: $9x^3 + 2x^2 + 13$

    Again, rearrange the terms so that the like ones are together. By convention, you write terms that have different powers of the same variable in either decreasing or increasing order of their powers.

    $$9x^2 + 5x^3 - 8x + 4 + 4x^3 - 7x^2 + 8x + 9 = 5x^3 + 4x^3 + 9x^2 - 7x^2 - 8x + 8x + 4 + 9$$
    $$= 9x^3 + 2x^2 + 0 + 13$$
    $$= 9x^3 + 2x^2 + 13$$

3. Simplify $6x^2y^2 + 2x^2y + 3xy^2 + 4x^2y^2 - 2xy^2 + 9x^2y^2 + 8xy + 1$.

    answer: $19x^2y^2 + 2x^2y + xy^2 + 8xy + 1$

    Rearrange and then combine like terms:

    $6x^2y^2 + 2x^2y + 3xy^2 + 4x^2y^2 - 2xy^2 + 9x^2y^2 + 8xy + 1 =$
    $6x^2y^2 + 4x^2y^2 + 9x^2y^2 + 2x^2y + 3xy^2 - 2xy^2 + 8xy + 1 = 19x^2y^2 + 2x^2y + xy^2 + 8xy + 1$

Work Problems

Use these problems to give yourself additional practice.

1. Simplify by combining like terms. $2m + 3n + 4p + 5m - 2n - p$

2. Simplify by combining like terms. $3x^2 + 5x^4 + 2x + 4x^3 - 2x^2 + x^3 + 3$

3. Simplify by combining like terms. $3ab + 4ac + 5bd + 6ad + 4ac + 9ab + 6de$

4. Simplify by combining like terms. $x^3 + 2x^2y + 2xy^2 + xy^2 + x^2y + y^3$

5. Simplify by combining like terms. $2(8a^3b^2) + 7ab^2 + 3(5ab) - 3a^3b^2 - ab$

Worked Solutions

1. **$7m + n + 3p$**   Rearrange and combine:

    $$2m + 3n + 4p + 5m - 2n - p = 2m + 5m + 3n - 2n + 4p - p$$
    $$= 7m + n + 3p$$

2. **$5x^4 + 5x^3 + x^2 + 2x + 3$**   Rearrange and combine:

    $$3x^2 + 5x^4 + 2x + 4x^3 - 2x^2 + x^3 + 3 = 5x^4 + 4x^3 + x^3 + 3x^2 - 2x^2 + 2x + 3$$
    $$= 5x^4 + 5x^3 + x^2 + 2x + 3$$

    Notice that the terms are listed in decreasing powers of the variable x.

3. **$12ab + 8ac + 5bd + 6ad + 6de$**   In this expression, there are only two sets of terms that are alike—the two sets with ab and ac in them. You have to read carefully to be sure that they match in variables exactly.

    $$3ab + 4ac + 5bd + 6ad + 4ac + 9ab + 6de = 3ab + 9ab + 4ac + 4ac + 5bd + 6ad + 6de$$
    $$= 12ab + 8ac + 5bd + 6ad + 6de$$

4. **$x^3 + 3x^2y + 3xy^2 + y^3$**   Rearrange and combine the terms:

    $$x^3 + 2x^2y + x^2y + 2xy^2 + xy^2 + y^3 = x^3 + 3x^2y + 3xy^2 + y^3$$

5. $13a^3b^2 + 7ab^2 + 14ab$ First, simplify the two terms with multiplication problems.

$$2(8a^3b^2) + 7ab^2 + 3(5ab) - 3a^3b^2 - ab = 16a^3b^2 + 7ab^2 + 15ab - 3a^3b^2 - ab$$

Now, the like terms can be combined after rearranging.

$$16a^3b^2 + 7ab^2 + 15ab - 3a^3b^2 - ab = 16a^3b^2 - 3a^3b^2 + 7ab^2 + 15ab - ab$$
$$= 13a^3b^2 + 7ab^2 + 14ab$$

Properties of Algebraic Expressions

Just as mathematics and algebra are governed by rules and conventions regarding the operations, some rules allow for manipulating expressions so that the changed version of the expression has the same meaning and "worth" as the original. The changed expression is usually something more convenient or usable in the particular situation. Many of these rules are common sense as far as you're concerned. They're written here, formally, so you can refer to them when needed in more complicated situations and so you can feel confident that you aren't violating a rule when making a particular adjustment or move in simplifying.

The rules listed in the following section involve what adjustments can be made to an expression and still keep it equivalent or equal to the original.

Commutative Property of Addition and Multiplication

The commutative property says that the order doesn't matter. You get the same thing when you add $2 + 3$ as you do when you add $3 + 2$. The same thing happens in multiplication. Multiplying 4×7 is the same as multiplying 7×4. This is **not** the case in subtraction or division.

Commutative Property of Addition: $a + b = b + a$
Commutative Property of Multiplication: $a \cdot b = b \cdot a$

Associative Property of Addition and Multiplication

The associative property assures you that you can group or associate terms differently and get the same result. When adding three numbers together such as $15 + 9 + 1$, it's easier to group the 9 and 1 together to get 10 and add it to the 15 rather than add the 15 and 9 and then add the 1. Using the associative property, it looks like this: $(15 + 9) + 1 = 15 + (9 + 1)$. Notice that it's all addition, not a mixture of addition and subtraction. The same is true for multiplication.

Associative Property of Addition: $a + (b + c) = (a + b) + c$
Associative Property of Multiplication: $a \cdot (b \cdot c) = (a \cdot b) \cdot c$

The Associative Property does **not** hold for subtraction or division.

Distributive Property of Multiplication over Addition (or Subtraction)

This property allows you to multiply a factor times several terms within a grouping symbol. The opposite of distributing is factoring. Each of these processes has its place in algebraic manipulations. The use of one or the other depends on the particular application.

Distribution Property of Multiplication over Addition: $a(b + c) = ab + ac$

Distribution Property of Multiplication over Subtraction: $a(b - c) = ab - ac$

Identities

If you keep your identity, then you don't change. Two numbers are considered to be identities in algebra. The number "0" is the *additive identity*. If you add 0 to a number, that number doesn't change—it keeps its identity. The number 1 is the *multiplicative identity*. If you multiply 1 times a number, that number doesn't change.

Additive Identity: $0 + a = a$ and $a + 0 = a$

Multiplicative Identity: $1 \cdot a = a$ and $a \cdot 1 = a$

It's tempting to include subtraction and division here, but you have to be careful. It's true that $4 - 0 = 4$, but, because subtraction isn't commutative, you can't change the order and expect to get the same answer. $0 - 4 \neq 4$. The same with division, $8 \div 1 = 8$ but $1 \div 8 \neq 8$.

Inverses

Inverses or opposites are used extensively when solving equations. An inverse can reverse the effect of a number that's operating on a variable. Inverses are classified as either additive inverse or multiplicative inverse. When two additive inverses are combined with addition, the result is 0. When two multiplicative inverses are combined with multiplication, the result is 1.

Additive Inverses: $a + (-a) = 0$ or $(-a) + a = 0$

Multiplicative Inverses: $a \cdot \left(\dfrac{1}{a}\right) = 1$ and $\left(\dfrac{1}{a}\right) \cdot a = 1$

Technically, one number doesn't have either inverse. It's the number 0. The additive inverse of a number has the opposite sign of the original number, so 0 doesn't have an additive inverse, because 0 doesn't have a sign. It isn't positive or negative, so it can't have an opposite. Also, 0 doesn't have a multiplicative inverse, because you can't divide by 0. Writing $\dfrac{1}{0}$ doesn't make any sense, because that fraction doesn't have an answer. Other than 0, all the real numbers have additive and multiplicative inverses.

Multiplication Property of Zero

Zero is a special number in many respects. It doesn't have an inverse, but it has a property that sets it aside and makes it valuable in algebraic manipulations. This is the Multiplication Property of Zero. You may wonder what's so special about that. After all, whenever you multiply something times 0, you always get 0, no matter how big or small the other number. This may not seem important or especially useful, but it is. In fact, this property is the backbone of solving most algebraic equations. Consider the puzzle: "I'm thinking of two numbers, and their product is 0. I'll give you $100 if you can tell me what either one of the numbers is." For the product to be 0, one of the numbers **must** be 0. There's no other way to get the 0 when you're multiplying. That's an easy $100. You just say, "One of your numbers is 0!" Compare that to a similar puzzle: "I'm thinking of two numbers, and their product is 36. I'll give you $100 if you can tell me what either one of the numbers is." Unless you're really lucky, you won't be able to guess what either number is, because the puzzler has an infinite (uncountable) number of choices. Here are a few: $36 \cdot 1, 18 \cdot 2, 9 \cdot 4, 6 \cdot 6, 72 \cdot \dfrac{1}{2}, 108 \cdot \dfrac{1}{3}, \cdots$. This can go on forever, because there's no end to the combinations that can be used. This is what's so special about 0. If the answer is 0, one of the multipliers has to be 0.

Multiplication Property of Zero: $a \cdot 0 = 0$ and $0 \cdot a = 0$

Example Problems

These problems show the answers and solutions.

Name the property illustrated by each statement (give the property that makes the statement true).

1. $6 + 8 = 8 + 6$

    answer: Commutative Property of Addition   It doesn't matter in what order you add numbers.

2. $3 \cdot (4 \cdot x) = (3 \cdot 4) \cdot x$

    answer: Associative Property of Multiplication   It's easier to see that the product is 12x from the right-hand side of the equation.

3. $5 + (-5) = 0$

    answer: Additive Inverses

4. If $a \cdot b = 0$, then $a = 0$ or $b = 0$.

    answer: Multiplication Property of Zero

5. $\frac{1}{5} \cdot 5 = 1$

    answer: Multiplicative Inverses   Their product is 1.

6. $\frac{3}{4} \cdot \frac{4}{3} = 1$

    answer: Multiplicative Inverses   Their product also is 1.

7. $5(10 + 2) = 5 \cdot 10 + 5 \cdot 2 = 50 + 10 = 60$

    answer: Distributive Property

8. $9 \cdot \frac{3}{3} = 9 \cdot 1 = 9$

    answer: Multiplicative Identity   Multiplying by 1 and using the Multiplicative Identity doesn't seem to make much sense until you use it when changing fractions to those with common denominators.

Work Problems

Use these problems to give yourself additional practice.

Use the property indicated with each problem to complete the statement correctly.

1. Commutative Property of Multiplication: $523 \cdot 8 =$

2. Additive Inverse: $23 + (\underline{}) = 0$

3. Distributive Property: $4(7 + 3) =$

4. Multiplication Property of Zero: If $m \cdot n = 0$, then _____

5. Associative Property of Addition: $(5 + 3) + 7 =$

6. Multiplicative Inverse: $\frac{1}{z} \cdot (\_\_) = 1$

7. Commutative Property of Addition: $41 + 11 = 11 + \_\_$

8. Associative Property of Multiplication: $5(2 \cdot 11) = (\_\_) \cdot 11$

Worked Solutions

1. **$523 \cdot 8 = 8 \cdot 523$**   The order doesn't matter in multiplication.

2. **$23 + (-23) = 0$**   The sum of two opposites is 0.

3. **$4(7 + 3) = 4 \cdot 7 + 4 \cdot 3$**   Multiply the 4 over the two numbers in the binomial.

4. **If $m \cdot n = 0$, then either m is 0 or n is 0.**   One or the other of the factors has to be 0.

5. **$(5 + 3) + 7 = 5 + (3 + 7)$**   The grouping doesn't matter in addition.

6. **$\frac{1}{z} \cdot (z) = 1$**   The product of a number and its reciprocal is 1.

7. **$41 + 11 = 11 + 41$**   The order doesn't matter in addition.

8. **$5(2 \cdot 11) = (5 \cdot 2) \cdot 11$**   The grouping doesn't matter in multiplication.

Integer Powers (Exponents)

A power or exponent tells how many times a number multiplies itself. There are many opportunities in algebra for combining and simplifying expressions with two or more of these exponential terms in them. The rules used here to combine numbers and variables work for any expression with exponents. They are found in formulas and applications in science, business, and technology as well as math. The term a^4 has an exponent of 4 and a base of a. The base is what gets multiplied repeatedly. The exponent tells how many times that repeatedly is.

Laws for Using Exponents

$$a^n \cdot a^m = a^{n+m}$$

When multiplying two numbers that have the same base, add their exponents.

$$\frac{a^n}{a^m} = a^{n-m}$$

When dividing two numbers that have the same base, subtract their exponents.

$$\left(a^n\right)^m = a^{n \cdot m}$$

When raising a value that has an exponent to another power, multiply the two exponents.

$$\left(a \cdot b\right)^n = a^n \cdot b^n$$

The product of two numbers raised to a power is equal to raising each of the numbers to that power and then multiplying them together.

$$\left(\frac{a}{b}\right)^n = \frac{a^n}{b^n}$$

The quotient of two numbers raised to a power is equal to the power of each of the numbers in the quotient.

$$a^{-n} = \frac{1}{a^n}$$

A value raised to a negative power can be written as a fraction with the positive power of that number in the denominator.

$$a^0 = 1$$

Any number (except 0) raised to the 0 power is equal to 1.

Example Problems

These problems show the answers and solutions.

Examples involving $a^n \cdot a^m$

1. Multiply $2^2 \cdot 2^3$.

   answer: 2^5

   $2^2 \cdot 2^3 = 2^{2+3} = 2^5$

2. Multiply $x^4 \cdot x^5$.

   answer: x^9

   $x^4 \cdot x^5 = x^{4+5} = x^9$

3. Multiply $y^8 \cdot y$.

   answer: y^9

   $y^8 \cdot y = y^8 \cdot y^1 = y^9$

4. Multiply $z^{-7} \cdot z^7$.

   answer: 1

   $z^{-7} \cdot z^7 = z^{-7+7} = z^0 = 1$

Examples involving $\dfrac{a^n}{a^m}$

5. Divide $\dfrac{3^6}{3^4}$.

    answer: 3^2

    $$\dfrac{3^6}{3^4} = 3^{6-4} = 3^2$$

6. Divide $\dfrac{x^8}{x^5}$.

    answer: x^3

    $$\dfrac{x^8}{x^5} = x^{8-5} = x^3$$

7. Divide $\dfrac{y^3}{y^3}$.

    answer: 1

    $$\dfrac{y^3}{y^3} = y^{3-3} = y^0 = 1$$

Examples involving $\left(a^n\right)^m$

8. Simplify $\left(4^3\right)^2$.

    answer: 4^6

    $$\left(4^3\right)^2 = 4^{3 \cdot 2} = 4^6$$

9. Simplify $\left(z^5\right)^6$.

    answer: z^{30}

    $$\left(z^5\right)^6 = z^{5 \cdot 6} = z^{30}$$

10. Simplify $\left(w^{23}\right)^0$.

    answer: 1

    $$\left(w^{23}\right)^0 = w^{23 \cdot 0} = w^0 = 1$$

Examples involving $(a \cdot b)^n$

11. Simplify $(4 \cdot 5)^2$.

    answer: 400

    $$(4 \cdot 5)^2 = 4^2 \cdot 5^2 = 16 \cdot 25 = 400$$

12.   Simplify $(x \cdot y)^3$.

     answer: $x^3 \cdot y^3$

     $(x \cdot y)^3 = x^3 \cdot y^3$

13.   Simplify $\left(3yz^2\right)^4$.

     answer: $3^4 y^4 z^8$

     $\left(3yz^2\right)^4 = 3^4 \cdot y^4 \cdot (z^2)^4 = 3^4 y^4 z^8$

Examples involving $\left(\dfrac{a}{b}\right)^n$

14.   Simplify $\left(\dfrac{6}{x}\right)^2$.

     answer: $\dfrac{36}{x^2}$

     $\left(\dfrac{6}{x}\right)^2 = \dfrac{6^2}{x^2} = \dfrac{36}{x^2}$

15.   Simplify $\left(\dfrac{2x}{y^3}\right)^4$.

     answer: $\dfrac{16x^4}{y^{12}}$

     $\left(\dfrac{2x}{y^3}\right)^4 = \dfrac{2^4 \cdot x^4}{(y^3)^4} = \dfrac{16x^4}{y^{12}}$

Examples involving a^{-n}

16.   Rewrite 2^{-3}.

     answer: $\dfrac{1}{2^3}$

     $2^{-3} = \dfrac{1}{2^3}$

17.   Rewrite x^{-4}.

     answer: $\dfrac{1}{x^4}$

     $x^{-4} = \dfrac{1}{x^4}$

18.   Rewrite y^{-1}.

     answer: $\dfrac{1}{y}$

     $y^{-1} = \dfrac{1}{y}$

19.   Rewrite $\left(\dfrac{x^2}{4}\right)^{-2}$.

answer: $\dfrac{16}{x^4}$

$$\left(\dfrac{x^2}{4}\right)^{-2} = \dfrac{1}{\left(\dfrac{x^2}{4}\right)^2} = \dfrac{4^2}{(x^2)^2} = \dfrac{16}{x^4}$$

A nice property of fractions is that when you raise them to a negative power, you can rewrite the expression and change the power to a positive if you "flip" the fraction. So $\left(\dfrac{x^2}{4}\right)^{-2} = \left(\dfrac{4}{x^2}\right)^2$.

Work Problems

Use these problems to give yourself additional practice.

Tell whether each statement is true or false. If false, correct the error.

1. $\left(3ab^3\right)^4 = 3^4 a^4 b^7$

2. $\left(\dfrac{4}{3}\right)^2 = \dfrac{4^2}{3^2} = \dfrac{16}{9}$

3. $m^2 \cdot m^8 = m^{16}$

4. $\dfrac{n^{12}}{n^4} = n^8$

5. $\left(3^8\right)^{-1} = \dfrac{1}{3^8}$

Worked Solutions

1.   **False**   The power on b is wrong. $\left(3ab^3\right)^4 = 3^4 a^4 b^{12}$

2.   **True**

3.   **False**   The exponents should be added, not multiplied. $m^2 \cdot m^8 = m^{2+8} = m^{10}$

4.   **True**

5.   **True**

Square Roots and Cube Roots

Finding a square root or cube root is a common technique used when solving quadratic (second degree) and cubic (third degree) equations. Taking a root "undoes" the act of raising to a power. The symbol for finding a root is the radical, $\sqrt{}$. If no number is above the "flag" or left part, then it's assumed that you're looking for the square root or undoing the act of raising to power 2. All other roots require numbers above the flag. $\sqrt[3]{}$ stands for the cube root—undoing raising to the third power. $\sqrt[4]{}$ stands for the fourth root, and so on.

Solving $\sqrt{25}$ means that you find the single number that, when it multiplies itself, gives you 25. That number is 5. Because $5 \cdot 5 = 25$, then $\sqrt{25} = 5$. Likewise, $\sqrt{49} = 7, \sqrt{121} = 11$, and so on. It's helpful to know the perfect squares between 0 and 200 when solving algebraic problems. Being familiar with these first 14 perfect squares makes life easier.

$$1^2 = 1 \quad 2^2 = 4 \quad 3^2 = 9 \quad 4^2 = 16 \quad 5^2 = 25 \quad 6^2 = 36 \quad 7^2 = 49$$
$$8^2 = 64 \quad 9^2 = 81 \quad 10^2 = 100 \quad 11^2 = 121 \quad 12^2 = 144 \quad 13^2 = 169 \quad 14^2 = 196$$

Cube roots are the next most used roots in algebra. Solving $\sqrt[3]{27}$ means that you find the number, which when multiplied times itself three times, gives you 27. Because $3 \cdot 3 \cdot 3 = 27$, then $\sqrt[3]{27} = 3$. Here are some more: $\sqrt[3]{125} = 5, \sqrt[3]{1} = 1, \sqrt[3]{216} = 6$. Again, it helps to know what some of the perfect cubes are. Because the cubes get so big so fast, this list doesn't go as far as the list of squares.

$$1^3 = 1 \quad 2^3 = 8 \quad 3^3 = 27 \quad 4^3 = 64 \quad 5^3 = 125$$
$$6^3 = 216 \quad 7^3 = 343 \quad 8^3 = 512 \quad 9^3 = 729 \quad 10^3 = 1000$$

Other roots aren't used as frequently, but it's nice to have a process to help solve them. For instance, to solve $\sqrt[4]{81}$, you would think of what number times itself four times gives you 81. Because 81 is an odd number, you should try only odd numbers when trying to get the answer. Start with 1, then 3, then 5, and so on, until you find the number for which you're looking.

Example Problems

These problems show the answers and solutions.

1. Find $\sqrt[4]{81}$.

    answer: 3

    Try multiplying odd numbers times themselves four times.

    $1 \cdot 1 \cdot 1 \cdot 1 = 1$ That's not it.

    $3 \cdot 3 \cdot 3 \cdot 3 = 81$. That's the number, so $\sqrt[4]{81} = 3$.

2. Find $\sqrt[3]{2197}$.

    answer: 13

    Try multiplying odd numbers times themselves three times to find the root.

    Don't bother with the 1s. They only give you another 1.

    $$3 \cdot 3 \cdot 3 = 27$$

    Skip the 5s, because the answers will all end in 5, and you need a 7 at the end.

    $$7 \cdot 7 \cdot 7 = 343$$
    $$9 \cdot 9 \cdot 9 = 729$$
    $$11 \cdot 11 \cdot 11 = 1331$$
    $$13 \cdot 13 \cdot 13 = 2197$$
    $$\sqrt[3]{2197} = 13$$

Of course, it would be much easier just to use a calculator, but sometimes they're not handy, sometimes they're not allowed, and usually it's just better to have a good guess in your head and work it out.

Approximating Square Roots

Of the first 100 positive integers, the integers from 1 to 100, only 10 of these are perfect squares. So only 10 of these have nice, integer answers when the radical is applied. $\sqrt{1} = 1$, $\sqrt{4} = 2$, $\sqrt{9} = 3$, and so on. What about the other 90 numbers? What if you want an answer to $\sqrt{40}$, even if it's just an approximate answer? The square roots of numbers that aren't perfect squares are called irrational numbers. Their decimal values go on forever, and they never have a nice pattern to them. The best you can do is to approximate the roots to a predetermined stopping point. Usually, you want an answer to one decimal place or two decimal places, depending on the situation.

For instance, $\sqrt{40}$ is some number between 6 and 7. You can tell this because $\sqrt{40}$ is between $\sqrt{36}$ and $\sqrt{49}$, roots of two consecutive perfect squares. Because $\sqrt{36} = 6$ and $\sqrt{49} = 7$, then $\sqrt{40}$ is between 6 and 7.

One way to find the square root is to use a calculator and round to a particular number of digits.

$$\sqrt{40} = 6.3 \text{ rounded to 1 decimal place}$$
$$\sqrt{40} = 6.32 \text{ rounded to 2 decimal places}$$
$$\sqrt{40} = 6.325 \text{ rounded to 3 decimal places}$$
$$\sqrt{40} = 6.3246 \text{ rounded to 4 decimal places}$$

In algebra, it's sometimes just easier to estimate the root and calculate that root to one place. If you're graphing points or solving practical applications, you don't need any more accuracy than that.

To estimate a root to one decimal place, use the following steps:

A. Determine between which two perfect squares the number lies and select the smaller of the square roots.

B. Make a guess as to the decimal value that's to be added to the root.

C. Divide the number you're finding the root of by your guess and carry the division to two places.

D. Average the guess and the quotient of the division problem.

E. Round the answer to one place.

Example Problems

These problems show the answers and solutions.

1. Estimate $\sqrt{40}$ correct to one decimal place.

   answer: 6.3

A. Determine between which <u>two</u> perfect squares the number lies and select the smaller of the square roots. Because $\sqrt{40}$ is between $\sqrt{36}$ and $\sqrt{49}$, and the roots of 36 and 49 are 6 and 7, choose the smaller, which makes 6 the choice.

B. Make a guess as to the decimal value that's to be added to the root. The number 40 is closer to 36 than 49, so a guess as to the decimal value might be 6.2.

C. Divide the number you're finding the root of by your guess and carry the division to two places. Divide 40 by 6.2, $40 \div 6.2 = 6.45$ and carry the division to two places.

D. Average the guess and the quotient of the division problem:

$$\frac{6.2 + 6.45}{2} = \frac{12.65}{2} = 6.325$$

E. Round the answer to one place.

$$\sqrt{40} \approx 6.3.$$

This is the square root of 40 correct to one decimal place. This means that 6.3 squared is closer to 40 than 6.2 squared, 6.4 squared, or any other number squared.

2. Estimate $\sqrt{200}$ correct to one decimal place.

answer: 14.1

A. Determine between which two perfect squares the number <u>lies</u> and select the <u>smaller of</u> the square roots. Because 200 is between 196 and 225, $\sqrt{200}$ is between $\sqrt{196}$ and $\sqrt{225}$. The roots of 196 and 225 are 14 and 15. Choose the smaller, 14.

B. Make a guess as to the decimal value that's to be added to the root.

Guess 14.3, because 200 is closer to 196 than 225.

C. Divide the number you're finding the root of by your guess and carry the division to two places:

$$200 \div 14.3 = 13.99 \text{ rounded to 2 places}$$

D. Average the guess and the quotient of the division problem.

$$\frac{14.3 + 13.99}{2} = \frac{28.29}{2} = 14.145$$

E. Round the answer to one place. The answer is 14.1, rounded to 1 place.

It really doesn't matter what you guess, although a good guess is better in terms of the arithmetic. If your guess is too low, then the division result will be high on the other side of the actual answer. If it's too high, then the division result will be on the low side of the actual answer. The averaging takes care of these high and low amounts.

Simplifying Square Roots

One way to deal with the square roots of numbers that aren't perfect squares is to estimate their value with a calculator or estimate with the guess-divide-average method discussed in the preceding section. Another choice, however, is to keep the **exact value** of the square root and not estimate. Break the square root into smaller parts, when possible, so that those smaller parts or factors can be combined with other square roots that are exactly alike. One of the rules for dealing with exponents, given in the earlier section, "Integer Powers (Exponents)," allows you to write $\sqrt{a \cdot b} = \sqrt{a}\,\sqrt{b}$. See how this works with specific numbers: $\sqrt{4 \cdot 9} = \sqrt{4}\,\sqrt{9}$. In this case, $\sqrt{36}$ has been broken down into two factors that are both perfect squares. Continuing with the simplifying, $\sqrt{4}\,\sqrt{9} = 2 \cdot 3 = 6$. This same principle is used when simplifying roots that have only one perfect square factor. A root can be simplified if there's a factor of the number that's a perfect square.

Example Problems

These problems show the answers and solutions.

1. Simplify $\sqrt{20}$.

   answer: $2\sqrt{5}$

   You can write 20 as the product of two numbers in several ways, but this is the way that has a perfect square in it.

   $$\sqrt{4 \cdot 5} = \sqrt{4}\,\sqrt{5} = 2\sqrt{5}$$

   The product under the radical is broken up into two factors. The root of the factor that is a perfect square is computed, and the result multiplies that root times the other factor. This is considered to be the simplified form.

2. Simplify $\sqrt{48}$.

   answer: $4\sqrt{3}$

   You can write the number 48 as a product of a perfect square and another factor in two ways. They are $48 = 4 \cdot 12$ and $48 = 16 \cdot 3$. The second choice will be used, because it has the larger perfect square. If the first choice is used, then the resulting expression will need to be simplified again to complete the job.

   $$\sqrt{48} = \sqrt{16 \cdot 3} = \sqrt{16}\,\sqrt{3} = 4\sqrt{3}$$

3. Simplify $\sqrt{66}$.

   answer: $\sqrt{66}$

   The number 66 can be factored into $2 \cdot 33$ or $3 \cdot 22$ or $6 \cdot 11$, but none of these contains a perfect square. This radical can't be simplified.

4. Add $\sqrt{18} + \sqrt{50}$.

   answer: $8\sqrt{2}$

These two radicals can't be combined as they are. They're completely different numbers. By simplifying each and finding a common radical, however, you can compute the addition.

First, simplify each radical.

$$\sqrt{18} + \sqrt{50} = \sqrt{9 \cdot 2} + \sqrt{25 \cdot 2} = \sqrt{9}\sqrt{2} + \sqrt{25}\sqrt{2} = 3\sqrt{2} + 5\sqrt{2}$$

Now the two terms are simplified, and each is a multiple of $\sqrt{2}$.

Three "radical 2s" plus 5 "radical 2s" equals 8 "radical 2s."

$$3\sqrt{2} + 5\sqrt{2} = 8\sqrt{2}$$

5. Combine $\sqrt{72} - \sqrt{32} + \sqrt{12} + \sqrt{98} - \sqrt{3}$.

    answer: $9\sqrt{2} + \sqrt{3}$

    None of the radicals is the same, and it appears that they can't be combined. Simplifying the radicals will help determine whether any adding or subtracting can be done.

    $$\sqrt{72} = \sqrt{36}\sqrt{2}$$
    $$\sqrt{32} = \sqrt{16}\sqrt{2}$$
    $$\sqrt{12} = \sqrt{4}\sqrt{3}$$
    $$\sqrt{98} = \sqrt{49}\sqrt{2}$$

    You can see that there are some common radicals. Now, rewrite the original problem in terms of these radicals:

    $$\sqrt{36}\sqrt{2} - \sqrt{16}\sqrt{2} + \sqrt{4}\sqrt{3} + \sqrt{49}\sqrt{2} - \sqrt{3} = 6\sqrt{2} - 4\sqrt{2} + 2\sqrt{3} + 7\sqrt{2} - \sqrt{3}$$

    Now, rearrange the terms and combine like terms:

    $$6\sqrt{2} - 4\sqrt{2} + 2\sqrt{3} + 7\sqrt{2} - \sqrt{3} = 6\sqrt{2} - 4\sqrt{2} + 7\sqrt{2} + 2\sqrt{3} - \sqrt{3} = 9\sqrt{2} + \sqrt{3}$$

Work Problems

Use these problems to give yourself additional practice.

1. Solve $\sqrt[4]{625}$.

2. Solve $\sqrt[3]{27,000,000}$.

Estimate each of the following correct to one decimal place.

3. $\sqrt{11}$

4. $\sqrt{500}$

5. $\sqrt{901}$

Worked Solutions

1. **5**

2. **300**　When doing problems with lots of zeros, break them into two parts: the zeros part and the rest of the problem. The zeros part has 6 zeros. Since it's a cube root, divide 6 by 3 and get 2. That's the number of zeros in the root. The other part is the 27. The cube root of 27 is 3. Put that together with the two zeros to get 300.

3. **3.3**　Since 11 is between the perfect squares 9 and 16, the root is between the roots of these two numbers, 3 and 4.

   Guess 3.2 and divide. $11 \div 3.2 = 3.44$

   Now, average the guess and the result of division:

   $$\frac{3.2 + 3.44}{2} = \frac{6.64}{2} = 3.32$$

   Round to 1 decimal place. $\sqrt{11} = 3.3$.

4. **22.4**　Since 500 is between the perfect squares 484 and 529, the root is between the roots of these two numbers, 22 and 23.

   Guess 22.5 and divide. $500 \div 22.5 = 22.22$.

   Now, average the guess and the result of division:

   $$\frac{22.5 + 22.22}{2} = \frac{44.72}{2} = 22.36$$

   Round to 1 decimal place. $\sqrt{500} = 22.4$

5. **30**　Since 901 is between the perfect squares 900 and 961, the root is between the roots of these two numbers, 30 and 31.

   Guess 30.1 and divide. $900 \div 30.1 = 29.90$

   Now, average the guess and the result of division:

   $$\frac{30.1 + 29.90}{2} = \frac{60.00}{2} = 30$$

   Round to 1 decimal place, $\sqrt{901} = 30.0$. This is the same answer as the square root of 900. The number 901 is very close to 900, and the perfect squares are farther apart as the numbers get bigger, so rounding to one decimal place will give many duplicate answers. For more accuracy, the division has to be carried out with a number that has more decimal places, and it has to be carried out farther. The more decimal places—the more accuracy and the closer you are to the actual square root.

Rationalizing Fractions

In algebra, it's not considered to be in "good form" to have a radical in the denominator of a fraction. The radical, if it can't be simplified, is an irrational number with no end or pattern to the decimal equivalent. It's hard to divide that kind of number into another. Instead, a procedure called "rationalization" is performed to write a fraction that's equivalent to the first but no longer has a radical in the denominator.

The method used involves multiplying both the numerator and denominator of the fraction by the same radical that's in the denominator. This creates a perfect square under the radical in the denominator so that simplifying gets rid of the radical there. It does leave a radical in the numerator, but that's an okay format.

Example Problems

These problems show the answers and solutions.

1. Rationalize $\dfrac{3}{\sqrt{2}}$.

   answer: $\dfrac{3\sqrt{2}}{2}$

   Multiply both numerator and denominator by $\sqrt{2}$. Basically, what that's doing is multiplying by 1.

   $$\frac{\sqrt{2}}{\sqrt{2}} = 1$$

   Because 1 is the Multiplicative Identity, this isn't changing the value of the original fraction—it's just changing the form.

   $$\frac{3}{\sqrt{2}} \cdot \frac{\sqrt{2}}{\sqrt{2}} = \frac{3\sqrt{2}}{\sqrt{2}\sqrt{2}} = \frac{3\sqrt{2}}{\sqrt{4}} = \frac{3\sqrt{2}}{2}$$

2. Rationalize $\dfrac{6\sqrt{2}}{\sqrt{3}}$.

   answer: $2\sqrt{6}$

   Even though there's a radical in the numerator, it's only the denominator you need be concerned with in rationalizing the denominator. Multiply both numerator and denominator by $\sqrt{3}$.

   $$\frac{6\sqrt{2}}{\sqrt{3}} \cdot \frac{\sqrt{3}}{\sqrt{3}} = \frac{6\sqrt{2}\sqrt{3}}{\sqrt{3}\sqrt{3}} = \frac{6\sqrt{6}}{3} = \frac{\overset{2}{\cancel{6}}\sqrt{6}}{\underset{1}{\cancel{3}}} = 2\sqrt{6}$$

Work Problems

Use these problems to give yourself additional practice.

1. Simplify $\sqrt{24}$.

2. Simplify and combine $\sqrt{75} - \sqrt{27}$.

3. Simplify and combine $\sqrt{99} + \sqrt{125} - \sqrt{44} + \sqrt{80}$.

4. Rationalize $\dfrac{2}{\sqrt{5}}$.

5. Rationalize $\dfrac{4\sqrt{10}}{\sqrt{6}}$.

Worked Solutions

1. $2\sqrt{6}$

$$\sqrt{24} = \sqrt{4 \cdot 6} = \sqrt{4}\sqrt{6} = 2\sqrt{6}$$

2. $2\sqrt{3}$

$$\sqrt{75} - \sqrt{27} = \sqrt{25 \cdot 3} - \sqrt{9 \cdot 3} = \sqrt{25}\sqrt{3} - \sqrt{9}\sqrt{3} = 5\sqrt{3} - 3\sqrt{3} = 2\sqrt{3}$$

3. $\sqrt{11} + 9\sqrt{5}$

$$\sqrt{99} + \sqrt{125} - \sqrt{44} + \sqrt{80} = \sqrt{9}\sqrt{11} + \sqrt{25}\sqrt{5} - \sqrt{4}\sqrt{11} + \sqrt{16}\sqrt{5}$$
$$= 3\sqrt{11} + 5\sqrt{5} - 2\sqrt{11} + 4\sqrt{5}$$
$$= 3\sqrt{11} - 2\sqrt{11} + 5\sqrt{5} + 4\sqrt{5}$$
$$= \sqrt{11} + 9\sqrt{5}$$

4. $\dfrac{2\sqrt{5}}{5}$

$$\frac{2}{\sqrt{5}} \cdot \frac{\sqrt{5}}{\sqrt{5}} = \frac{2\sqrt{5}}{5}$$

5. $\dfrac{4\sqrt{15}}{3}$

$$\frac{4\sqrt{10}}{\sqrt{6}} \cdot \frac{\sqrt{6}}{\sqrt{6}} = \frac{4\sqrt{60}}{6} = \frac{4\sqrt{4}\sqrt{15}}{6} = \frac{4 \cdot 2\sqrt{15}}{6} = \frac{4 \cdot 2\sqrt{15}}{\cancel{6}_3} = \frac{4\sqrt{15}}{3}$$

Exponents for Roots

Expressions involving radicals can be written using exponents instead of radicals. The exponents in this case are fractions. This allows for many types of manipulations and operations to be performed more quickly and easily. The rules involving exponents, found in the earlier section on Integer Powers, still hold. The rule or convention for using exponents instead of radicals is that the root of the radical appears in the denominator of the fractional exponent, and any powers under the radical appear in the numerator of the fractional exponent.

Fractional exponents for radicals:

$$\sqrt[n]{a} = a^{\frac{1}{n}} = a^{1/n}$$

$$\sqrt[n]{a^m} = a^{\frac{m}{n}} = a^{m/n}$$

So $\sqrt{a} = a^{1/2}$, $\sqrt[3]{a} = a^{1/3}$, $\sqrt[7]{a} = a^{1/7}$, and $\sqrt[4]{a^3} = a^{3/4}$, $\sqrt[5]{a^2} = a^{2/5}$, and so on. This even allows for easier simplification of expressions with powers and roots. For instance, assuming that a is a positive number, $\sqrt[4]{a^2} = a^{2/4} = a^{1/2}$. This is the equivalent of writing that $\sqrt[4]{a^2} = \sqrt{a}$. $\sqrt[6]{a^4} = a^{4/6} = a^{2/3}$. This is another way of saying that $\sqrt[6]{a^4} = \sqrt[3]{a^2}$.

Example Problems

These problems show the answers and solutions.

1. Simplify $\sqrt[7]{z^2} \cdot \sqrt[7]{z^3}$.

   answer: $\sqrt[7]{z^5}$

   $$\sqrt[7]{z^2} \cdot \sqrt[7]{z^3} = z^{2/7} \cdot z^{3/7} = z^{2/7 + 3/7} = z^{5/7} = \sqrt[7]{z^5}$$

   As you see, the rule used to multiply numbers with the same base still applies here. You add the exponents.

2. Simplify $\dfrac{4^{4/3}}{4^{1/3}}$.

   answer: 4

   $$\frac{4^{4/3}}{4^{1/3}} = 4^{4/3 - 1/3} = 4^{3/3} = 4^1 = 4$$

3. Simplify $9^{1/4} \cdot 9^{1/4}$.

   answer: 3

   $$9^{1/4} \cdot 9^{1/4} = 9^{1/4 + 1/4} = 9^{2/4} = 9^{1/2} = \sqrt{9} = 3$$

4. Simplify $\left(8^{5/6}\right)^2$.

   answer: 32

   $$\left(8^{5/6}\right)^2 = 8^{10/6} = 8^{5/3}$$

   Again, the rule involving raising a power to a power applies here. Now the result can be simplified further by rewriting the 8 as a power of 2.

   $$(8)^{5/3} = \left(2^3\right)^{5/3} = 2^{3 \cdot \frac{5}{3}} = 2^5 = 32$$

5. Simplify $20^{3/2}$.

   answer: $8 \cdot 5^{3/2}$

   $$20^{3/2} = (4 \cdot 5)^{3/2} = 4^{3/2} \cdot 5^{3/2}$$

   Now that first factor, the 4, and its exponent can be simplified.

   $$4^{3/2} \cdot 5^{3/2} = (2^2)^{3/2} \cdot 5^{3/2} = 2^{2 \cdot \frac{3}{2}} \cdot 5^{3/2} = 2^3 \cdot 5^{3/2} = 8 \cdot 5^{3/2}$$

Work Problems

Use these problems to give yourself additional practice.

1. Simplify $\sqrt[5]{x^2} \cdot \sqrt[5]{x^3}$.

2. Simplify $\dfrac{y^{8/3}}{y^{2/3}}$.

3. Simplify $\dfrac{25^{5/2}}{25^2}$.

4. Simplify $18^{1/3} \cdot 18^{1/6}$.

5. Simplify $\left(27^{1/6}\right)^4$.

Worked Solutions

1. x $\sqrt[5]{x^2} \cdot \sqrt[5]{x^3} = x^{2/5} \cdot x^{3/5} = x^{2/5+3/5} = x^{5/5} = x^1 = x$

2. y^2 $\dfrac{y^{8/3}}{y^{2/3}} = y^{8/3-2/3} = y^{6/3} = y^2$

3. 5 $\dfrac{25^{5/2}}{25^2} = 25^{5/2-2} = 25^{1/2} = \sqrt{25} = 5$

4. $3\sqrt{2}$ $18^{1/3} \cdot 18^{1/6} = 18^{1/3+1/6} = 18^{2/6+1/6} = 18^{3/6} = 18^{1/2}$

This last term can be simplified using either radicals or fractional exponents.

$$18^{1/2} = (9 \cdot 2)^{1/2} = 9^{1/2} \cdot 2^{1/2} = \sqrt{9}\sqrt{2} = 3\sqrt{2}$$

5. 9 $\left(27^{1/6}\right)^4 = \left((3^3)^{1/6}\right)^4$

This time, change the 27 to a power of 3 first, so all the exponents can be multiplied together. The product of the exponents is $3 \cdot \dfrac{1}{6} \cdot 4 = \dfrac{12}{6} = 2$. So, $\left((3^3)^{1/6}\right)^4 = 3^2 = 9$.

Divisibility Rules

Performing algebraic operations frequently requires factoring out multipliers that the different terms have in common. This is what is done when reducing fractions. The fraction $\dfrac{10}{12}$ can be reduced, because both the numerator and denominator have factors of 2; they're both divisible by 2. When reducing fractions or factoring out other algebraic expressions, the process is made easier when you recognize the Greatest Common Factor (GCF) of the numbers. For instance, the GCF of the numbers 48 and 60 is the number 12. True, both 48 and 60 are also evenly divisible by 2 or 3 or 4 or 6, but recognizing the **greatest** possible divisor is more efficient—saving time in the long run. To help determine the Greatest Common Factor for two or more numbers, here are some **divisibility rules**. These are helpful rules or methods for recognizing when a number is divisible by another—or not. You can always just do long division and see whether there's a remainder or not, but these rules are intended to save you time and a lot of unnecessary divisions.

Rules of Divisibility

A number is divisible by...	...if
2	it ends in 0, 2, 4, 6, or 8
4	the last two digits form a number divisible by 4
8	the last three digits form a number divisible by 8
5	it ends in 0 or 5

A number is divisible by...	...if
10	it ends in 0
3	the sum of the digits is divisible by 3
9	the sum of the digits is divisible by 9
6	it's divisible by both 2 and 3
12	it's divisible by both 3 and 4
15	it's divisible by both 3 and 5
11	the sums of the alternating digits have a difference of 0 or 11 or 22...

Example Problems

These problems show the answers and solutions.

1. What is 90 divisible by?

    answer: The number 90 is divisible by 2, 3, 5, 6, 9, 10, 15, and 30.

    It's divisible by 2 and 5 and 10, because it ends in 0.

    It's divisible by 3, because the sum of the digits is $9 + 0 = 9$, and 9 is divisible by 3.

    It's divisible by 9, because the sum of the digits is $9 + 0 = 9$, and 9 is divisible by 9.

    It's divisible by 6, because it's divisible by 2 and 3 both.

    It's divisible by 15, because it's divisible by 3 and 5 both.

2. What is 10,164 divisible by?

    answer: The number 10,164 is divisible by 2, 3, 4, 6, and 11.

    It's divisible by 2, because it ends in a 4.

    It's divisible by 3, because the sum of the digits is 12, and 12 is divisible by 3.

    It's divisible by 4, because the last two digits form the number 64, and 64 is divisible by 4.

    It's divisible by 6, because it's divisible by both 2 and 3.

    It's divisible by 11, because the sum of the first, third, and fifth digits is 6; the sum of the second and fourth digits is 6; and the difference between 6 and 6 is 0.

3. What is 1999 divisible by?

    answer: The number 1999 is prime.

It's not divisible by any of the numbers you have rules of divisibility for, and it's not divisible by any other number other than 1 and itself.

It ends in an odd number, so it isn't divisible by 2, 4, 6, 8, 10, or 12.

The sum of the digits is 28, so it isn't divisible by 3 or 9.

The sum of the first and third digits is 10, and the sum of the second and fourth digits is 18. The difference between 10 and 18 is 8, which means it isn't divisible by 11.

To check for divisibility by other numbers such as 7, 13, 17, 19 and so on, you can find rules for them or, because those rules are usually no easier than just doing long division, it's best just to divide it.

Work Problems

Use these problems to give yourself additional practice.

Determine whether or not the numbers given are divisible by 2, 3, 4, 5, 6, 8, 9, 10, 11, 12, or 15.

1. 1805

2. 1776

3. 1710

4. 168,300

5. 20,779

Worked Solutions

1. **5 only**   This is divisible by 5, because it ends in 5.

    It isn't divisible by any of the even numbers.

    It isn't divisible by 3 or 9, because the sum of the digits is 14.

    It isn't divisible by 11, because the alternate digits have sums of 1 and 13, and that difference is 12.

2. **2, 3, 4, 6, 8, 12**   This is divisible by 2, because it ends in a 6.

    This is divisible by 3, because the sum of the digits is 21.

    This is divisible by 4, because 4 divides the number 76 evenly.

    This is divisible by 6, because it's divisible by 2 and 3 both.

    This is divisible by 8, because 8 divides the number 776 evenly.

    This is divisible by 12, because it's divisible by 3 and 4 both.

3. **2, 3, 5, 6, 9, 10, 15**   This is divisible by 2, 5, and 10 because it ends in 0.

   This is divisible by 3 and 9, because the sum of the digits is 9.

   This is divisible by 6, because it's divisible by 2 and 3 both.

   This is divisible by 15, because it's divisible by 3 and 5 both.

4. **2, 3, 4, 5, 6, 9, 10, 11, 12, 15**   This is divisible by 2, 5, and 10, because it ends in 0.

   This is divisible by 3 and 9, because the sum of the digits is 18.

   This is divisible by 6, because it's divisible by 2 and 3 both.

   This is divisible by 4, because the last two digits form the number 0, and 4 divides 0 evenly (there's no remainder!).

   This is divisible by 12, because it's divisible by 3 and 4 both.

   This is divisible by 15, because it's divisible by 3 and 5 both.

   This is divisible by 11, because the first, third, and fifth digits add up to 9; the second, fourth, and sixth digits add up to 9; and that difference is 0.

5. **11**   This is divisible by 11, because the first, third, and fifth digits add up to 18; the second and fourth digits add up to 7; and the difference between 18 and 7 is 11.

   It isn't divisible by any of the other numbers.

Prime Numbers

A *prime* number is a number that can be divided evenly only by itself and 1. It's important to know when you have a prime number, because you can stop looking for other numbers to divide it evenly. It will not have a factor in it that's shared by another number.

Twenty-five prime numbers are between 1 and 100 and another 21 prime numbers are between 100 and 200. The higher you get, the more rare the prime numbers become, but they never end. No one has been able to figure out a formula for determining where the prime numbers will show up as you go higher and higher, but they're there—that much is known.

Here is a list of the first 32 prime numbers.

Notice that the number 1 isn't considered prime. The number 2 is the smallest prime and the only even prime.

2	3	5	7	11	13	17	19	23	29	31	37	41	43
47	53	59	61	67	71	73	79	83	89	97	101	103	107
109	113	127	131										

To determine whether or not a number is prime, you can divide by as many numbers as you can think of, use the divisibility rules (see the previous section), and see whether there's anything that divides it evenly. A more efficient way than dividing by all these numbers does exist. The rule is to find the biggest perfect square that's smaller than the number. Then check or divide by only the prime numbers that are equal to or smaller than the square root of that biggest perfect square.

Example Problems

These problems show the answers and solutions.

1. Is 157 a prime number?

   answer: Yes, it's prime.

   The biggest perfect square that's smaller than 157 is 144, which is 12^2. The only primes smaller than 12 are 2, 3, 5, 7, and 11. Check to see whether 157 is divisible by any of them.

   157 isn't divisible by 2, because the last digit is 7.

   157 isn't divisible by 3, because the sum of the digits is 13.

   157 isn't divisible by 5, because it doesn't end in 0 or 5.

   157 isn't divisible by 7, because $157 \div 7 = 22$, $R3$.

   157 isn't divisible by 11, because the sums of the alternating digits don't have a difference of 0 or 11.

   Therefore, 157 must be prime. It probably seemed like you had to check a lot of numbers, but only five primes were listed, which is still a lot easier than trying half the numbers smaller than 157! And, this way you can be sure you haven't missed anything.

2. Is 407 prime?

   answer: No, it isn't prime.

   The biggest perfect square that's smaller than 407 is 400, which is 20^2. The only primes smaller than 20 are 2, 3, 5, 7, 11, 13, 17, and 19. Check to see whether 407 is divisible by any of them.

   407 isn't divisible by 2, because the last digit is 7.

   407 isn't divisible by 3, because the sum of the digits is 11.

   407 isn't divisible by 5, because it doesn't end in 0 or 5.

   407 isn't divisible by 7, because $407 \div 7 = 58$, $R1$.

   407 **is** divisible by 11, because the sum of the first and third digits is 11 and the only other digit is 0; their difference is 11, so 407 is divisible by 11.

   Because 407 is divisible by something other than 1 and itself, it's **not** prime.

Work Problems

Use these problems to give yourself additional practice.

Determine whether or not the following numbers are prime.

1. 959

2. 253

3. 463

4. 307

5. 2009

Worked Solutions

1. **Not prime**   The largest perfect square smaller than 959 is 900, which is 30^2. The only numbers that need to be checked to see whether 959 is divisible by them are the primes smaller than 30. They are 2, 3, 5, 7, 11, 13, 17, 19, 23, 29. As it turns out, you don't have to go very far before finding out that 959 is divisible by 7. Therefore, 959 is **not** prime.

2. **Not Prime**   The largest perfect square smaller than 253 is 225, which is 15^2. The only numbers that need to be checked to see whether 253 is divisible by them are the primes smaller than 15. They are 2, 3, 5, 7, 11, and 13. You should have found that 253 is divisible by 11. Therefore, 253 is **not** prime.

3. **Is Prime**   The largest perfect square smaller than 463 is 441, which is 21^2. The only numbers that need to be checked to see whether 463 is divisible by them are the primes smaller than 21. They are 2, 3, 5, 7, 11, 13, 17, and 19. None of these numbers divides 463 evenly. Therefore, 463 must be prime.

4. **Is Prime**   The largest perfect square smaller than 307 is 289, which is 17^2. The only numbers that need to be checked to see whether 307 is divisible by them are the primes equal to or smaller than 17. They are 2, 3, 5, 7, 11, 13, and 17. None of these numbers divides 307 evenly. Therefore, 307 must be prime.

5. **Not Prime**   The largest perfect square smaller than 2009 is 1936, which is 44^2. You need to check many numbers to see whether 2009 is divisible by them. The primes smaller than 44 are 2, 3, 5, 7, 11, 13, 17, 19, 23, 29, 31, 37, 41, and 43. As it turns out, you don't have to go very far before finding out that 2009 is divisible by 7. Therefore, 2009 is **not** prime.

Prime Factorization

A prime factorization of a number is a unique (only one possible) listing of all the prime numbers whose product gives you that number. Every number has a prime factorization, and it's not shared with any other number. The prime factorization of 12 is $2 \cdot 2 \cdot 3 = 2^2 \cdot 3$. Those three factors are the only three prime factors that result in a 12. A prime factorization helps you find common factors between two or more expressions. The prime factorization of a prime number is just that number, because 1 isn't prime. To find the prime factorization of a number, you can start by thinking of any two numbers whose product is that number, and then take each of those two numbers and think of two numbers whose product is each of them, and so on until only prime numbers are being multiplied.

Example Problems

These problems show the answers and solutions.

1. Find the prime factorization of 396.

    answer: $2^2 \cdot 3^2 \cdot 11$

First think of two numbers whose product is 396. How about 6 and 66? $396 = 6 \cdot 66$

Now think of two numbers whose product is 6 and two numbers whose product is 66. The number 6 is $2 \cdot 3$ and the number 66 is $6 \cdot 11$.

So $396 = 6 \cdot 66 = 2 \cdot 3 \cdot 6 \cdot 11$.

The only number in that factorization that isn't a prime is the 6. Changing it to $2 \cdot 3$, the factorization now reads $396 = 2 \cdot 3 \cdot 2 \cdot 3 \cdot 11$. Rewriting the primes in order, $396 = 2 \cdot 2 \cdot 3 \cdot 3 \cdot 11 = 2^2 \cdot 3^2 \cdot 11$.

Although this method is very effective, it does seem a bit disjointed. It will work every time, even if you start with different combinations of numbers. Another way to do this, which is more organized, uses a sort of upside-down division. You start with the first prime that divides the number evenly and do divisions until the number doesn't divide any more. Then go to the next prime and divide by it. Continue until the number at the bottom is a prime. All the primes along the side and the last number are part of the prime factorization.

2. Use the division method to find the prime factorization of 396.

answer: $2 \cdot 2 \cdot 3 \cdot 3 \cdot 11 = 2^2 \cdot 3^2 \cdot 11$

Write 396 in the upside-down division symbol. Write the smallest prime that will divide it outside, to the left. Write the answer below and divide again.

First step, divide 396 by the first prime, 2:
$$2\overline{)396} \\ 198$$

Second step, divide the result, 198, by 2:
$$2\overline{)396} \\ 2\overline{)198} \\ 99$$

Third step, divide the result, 99, by 3:
$$2\overline{)396} \\ 2\overline{)198} \\ 3\overline{)99} \\ 33$$

Fourth step, divide the result, 33, by 3:
$$2\overline{)396} \\ 2\overline{)198} \\ 3\overline{)99} \\ 3\overline{)33} \\ 11$$

The last result is 11, which is prime, so you stop. The numbers in the prime factorization are listed, in order, along the left side. $2 \cdot 2 \cdot 3 \cdot 3 \cdot 11$.

3. Find the prime factorization for 11,250.

    answer: $2 \cdot 3^2 \cdot 5^4$

    First step, divide 11,250 by the first prime, 2:
    $$2\overline{\smash{)}11250}$$
    $$5625$$

    Second step, divide the result, 5625, by 3:
    $$2\overline{\smash{)}11250}$$
    $$3\overline{\smash{)}5625}$$
    $$1875$$

    Third step, divide the result, 1875, by 3:
    $$2\overline{\smash{)}11250}$$
    $$3\overline{\smash{)}5625}$$
    $$3\overline{\smash{)}1875}$$
    $$625$$

    Fourth step, divide the result, 625, by 5:
    $$2\overline{\smash{)}11250}$$
    $$3\overline{\smash{)}5625}$$
    $$3\overline{\smash{)}1875}$$
    $$5\overline{\smash{)}625}$$
    $$125$$

    Fifth step, divide the result, 125, by 5:
    $$2\overline{\smash{)}11250}$$
    $$3\overline{\smash{)}5625}$$
    $$3\overline{\smash{)}1875}$$
    $$5\overline{\smash{)}625}$$
    $$5\overline{\smash{)}125}$$
    $$25$$

    Last step, divide the result, 25, by 5:
    $$2\overline{\smash{)}11250}$$
    $$3\overline{\smash{)}5625}$$
    $$3\overline{\smash{)}1875}$$
    $$5\overline{\smash{)}625}$$
    $$5\overline{\smash{)}125}$$
    $$5\overline{\smash{)}25}$$
    $$5$$

    The prime factorization can be read along the left side and bottom:
    $11,250 = 2 \cdot 3 \cdot 3 \cdot 5 \cdot 5 \cdot 5 \cdot 5 = 2 \cdot 3^2 \cdot 5^4$

Work Problems

Use these problems to give yourself additional practice.

Write the prime factorization of each of the following numbers.

1. 450

2. 255

3. 648

4. 928

5. 1575

Worked Solutions

$$2\,|\,\underline{450}$$

$$3\,|\,\underline{225}$$

1. $2 \cdot 3^2 \cdot 5^2$ $3\,|\,\underline{75}$ So $450 = 2 \cdot 3 \cdot 3 \cdot 5 \cdot 5 = 2 \cdot 3^2 \cdot 5^2$

$$5\,|\,\underline{25}$$

$$5$$

$$3\,|\,\underline{255}$$

2. $3 \cdot 5 \cdot 17$ $5\,|\,\underline{85}$ So $255 = 3 \cdot 5 \cdot 17$

$$17$$

$$2\,|\,\underline{648}$$

$$2\,|\,\underline{324}$$

$$2\,|\,\underline{162}$$

3. $2^3 \cdot 3^4$ $3\,|\,\underline{81}$ So $648 = 2 \cdot 2 \cdot 2 \cdot 3 \cdot 3 \cdot 3 \cdot 3 = 2^3 \cdot 3^4$

$$3\,|\,\underline{27}$$

$$3\,|\,\underline{9}$$

$$3$$

$$2\,|\,\underline{928}$$

$$2\,|\,\underline{464}$$

$$2\,|\,\underline{232}$$

4. $2^5 \cdot 29$ So $928 = 2 \cdot 2 \cdot 2 \cdot 2 \cdot 2 \cdot 29 = 2^5 \cdot 29$

$$2\,|\,\underline{116}$$

$$2\,|\,\underline{58}$$

$$29$$

$$3\underline{|1575}$$
$$3\underline{|525}$$
5. $3^2 \cdot 5^2 \cdot 7$ $\quad 5\underline{|175}$ So $1575 = 3 \cdot 3 \cdot 5 \cdot 5 \cdot 7 = 3^2 \cdot 5^2 \cdot 7$
$$5\underline{|35}$$
$$7$$

Chapter Problems and Solutions

Problems

1. Simplify $\dfrac{16-8}{2} + 5 \cdot 11 - 4^2 - \sqrt{8 \cdot 2}$.

2. Simplify $5\left(3^2 + 2\right) - 3(6+1) + \dfrac{12(5-2)}{8+1}$.

3. Simplify $6(4x + 5x) + 2y + x$.

4. Rewrite the following in a simpler form: $5 \cdot 5mmmmnnpppppp$.

5. Simplify the expression. Assume that x is positive. $\sqrt{16x^2} - |-3x|$

6. Simplify $8x^3 + 3x^2 + 5x - 2x^3 + 4x^2 + 1$.

7. Simplify $4(2a + a) + 3(4a - 2a)$.

8. Complete the statement to make it true: $-47 + (\_\_) = 0$.

9. Complete the statement to make it true: $14 + (6 + 9) = (\_\_ + 6) + 9$.

10. Simplify using the rules for powers: $\left(2ab^2\right)^3 + c^{-6}$.

11. Simplify using the rules for powers: $x^4 x^3 - \dfrac{3y^2 y^7}{y^5}$.

12. Solve $\sqrt{169}$.

13. Solve $\sqrt[3]{125}$.

14. Use the guess-divide-average method to find $\sqrt{6}$ correct to one decimal place.

15. Use the guess-divide-average method to find $\sqrt{180}$ correct to one decimal place.

16. Add the radicals after simplifying them: $\sqrt{8} + \sqrt{72}$.

17. Rationalize the fraction $\dfrac{4}{\sqrt{6}}$.

18. Simplify $\sqrt[4]{a^2} \sqrt[4]{a^3}$.

19. Simplify $\dfrac{27^{5/3}}{27^{1/3}}$.

20. Determine whether or not the number 800,000,001 is divisible by 9.

21. Determine whether or not the number 4,200,000,008 is divisible by 12.

22. Determine whether or not the number 891 is prime.

23. Determine whether or not the number 1001 is prime.

24. Determine the prime factorization of 108.

25. Determine the prime factorization of 1485.

Answers and Solutions

1. **Answer: 39**

$$\frac{16-8}{2}+5\cdot 11-4^2-\sqrt{8\cdot 2}=\frac{8}{2}+55-16-\sqrt{16}$$
$$=4+55-16-4$$
$$=59-16-4$$
$$=43-4=39$$

2. **Answer: 38**

$$5\left(3^2+2\right)-3(6+1)+\frac{12(5-2)}{8+1}=5(9+2)-3(7)+\frac{12(3)}{9}$$
$$=5(11)-21+\frac{36}{9}$$
$$=55-21+4$$
$$=34+4=38$$

3. **Answer: $55x + 2y$**

$$6(4x + 5x) + 2y + x = 6(9x) + 2y + x$$
$$=54x + 2y + x = 55x + 2y$$

4. **Answer: $25\,m^3 n^2 p^6$**

$$5\cdot 5mmmnnpppppp=25m^3n^2p^6$$

5. **Answer: x** $\sqrt{16x^2}-|-3x|=4x-3x=x$

6. **Answer: $6x^3 + 7x^2 + 5x + 1$**

$$8x^3 + 3x^2 + 5x - 2x^3 + 4x^2 + 1 = 8x^3 - 2x^3 + 3x^2 + 4x^2 + 5x + 1$$
$$=6x^3 + 7x^2 + 5x + 1$$

7. **Answer: $18a$**

$$4(2a + a) + 3(4a - 2a) = 4(3a) + 3(2a)$$
$$= 12a + 6a = 18a$$

8. **Answer: 47**   Complete the statement to make it true.

$$-47 + (\underline{}) = 0$$
$$-47 + (47) = 0$$

9. **Answer: 14**   Complete the statement to make it true.

$$14 + (6 + 9) = (\underline{} + 6) + 9$$
$$14 + (6 + 9) = (14 + 6) + 9$$

10. **Answer:** $8\,a^{3}b^{6} + \dfrac{1}{c^{6}}$

$$\left(2ab^{2}\right)^{3} + c^{-6} = 2^{3}a^{3}(b^{2})^{3} + \frac{1}{c^{6}}$$
$$= 8a^{3}b^{6} + \frac{1}{c^{6}}$$

11. **Answer:** $x^{7} - 3y^{4}$

$$x^{4}x^{3} - \frac{3y^{2}\,y^{7}}{y^{5}} = x^{4+3} - \frac{3y^{2+7}}{y^{5}} = x^{7} - \frac{3y^{9}}{y^{5}}$$
$$= x^{7} - 3y^{9-5} = x^{7} - 3y^{4}$$

12. **Answer: 13**

$$\sqrt{169} = \sqrt{13 \cdot 13} = 13$$

13. **Answer: 5**

$$\sqrt[3]{125} = \sqrt[3]{5 \cdot 5 \cdot 5} = 5$$

14. **Answer: 2.4**   Guess 2.5. (Guesses will vary.) Divide 6 by 2.5 and get 2.4. Average 2.5 and 2.4.

$$\frac{2.5 + 2.4}{2} = \frac{4.9}{2} = 2.45.$$

This can round to either 2.5 or 2.4. Check to see which is closer by squaring them. $2.5^{2} = 6.25$ and $2.4^{2} = 5.46$. Because the square of 2.4 is .01 closer, choose the 2.4.

15. **Answer: 13.4**   Guess 13.4 (Guesses will vary.) Divide 180 by 13.4 and get 13.43. Average 13.4 and 13.43 and get 13.4. It was a good guess!

16. **Answer:** $8\sqrt{2}$

$$\sqrt{8} + \sqrt{72} = \sqrt{4}\sqrt{2} + \sqrt{36}\sqrt{2}$$
$$= 2\sqrt{2} + 6\sqrt{2} = 8\sqrt{2}$$

17. **Answer:** $\dfrac{2\sqrt{6}}{3}$

$$\frac{4}{\sqrt{6}} \cdot \frac{\sqrt{6}}{\sqrt{6}} = \frac{4\sqrt{6}}{6} = \frac{2\sqrt{6}}{3}$$

18. **Answer:** $a\sqrt[4]{a}$

$$\sqrt[4]{a^2}\,\sqrt[4]{a^3} = a^{2/4} \cdot a^{3/4} = a^{2/4+3/4} = a^{5/4} = a^{1+1/4} = a^1 a^{1/4} = a\sqrt[4]{a}$$

19. **Answer: 81**

$$\frac{27^{5/3}}{27^{1/3}} = 27^{5/3-1/3} = 27^{4/3} = \left(3^3\right)^{4/3} = 3^{3\cdot\frac{4}{3}} = 3^4 = 81$$

20. **Answer: Yes**　　Yes, it's divisible by 9, because the sum of the digits is 9.

21. **Answer: No**　　No, it isn't divisible by 12, because it isn't divisible by 3. The sum of the digits is 14, and 14 is not divisible by 3.

22. **Answer: No**　　No, it isn't prime, because it's divisible by 3 and 9 and 11. The sum of the digits is 18, which is divisible by both 3 and 9. The alternating digits both add up to 9, so the difference between them is 0.

23. **Answer: No**　　No, it isn't prime, because it's divisible by 7 and 11 and 13. The divisibility by 11 was apparent, because the sums of the alternate digits were both 1.

24. **Answer:** $2^2 \cdot 3^3$

$$108 = 2^2 \cdot 3^3$$

25. **Answer:** $3^3 \cdot 5 \cdot 11$

$$1485 = 3^3 \cdot 5 \cdot 11$$

Supplemental Chapter Problems

Problems

1. Simplify $\dfrac{2(14-3)}{5+6} - 1$.

2. Simplify $\sqrt{12 \cdot 3} + 8 \cdot 2 - 3$.

3. Simplify $a + b + c + 2a + 3b + 4c$.

4. Rewrite the following using exponents: $3 \cdot 3 \cdot 3 \cdot x \cdot x \cdot x \cdot x \cdot x \cdot x \cdot x \cdot y \cdot y$.

5. Simplify (assume x is positive) $\sqrt{25x^2} + |x| - |-5|$.

6. Simplify $3m^3 + 4m - 2m^3 + m$.

7. Simplify $5(x + x) - 2(2x + x)$.

8. Complete the statement to make it true: $4(6 \cdot 3) = (\underline{} \cdot \underline{})3$.

9. Complete the statement to make it true: $\frac{3}{4} \cdot \underline{} = 1$.

10. Simplify $\dfrac{(6x^2 y^3)^2}{(2xy^2)^3}$.

11. Simplify $a^2 a^3 a^{-4} + \dfrac{2b^2}{b}$.

12. Solve $\sqrt{289}$.

13. Solve $\sqrt[4]{625}$.

14. Use guess-divide-average to find $\sqrt{10}$ correct to one decimal place.

15. Use guess-divide-average to find $\sqrt{150}$ correct to one decimal place.

16. Simplify and then subtract $\sqrt{200} - \sqrt{128}$.

17. Rationalize $\dfrac{12}{\sqrt{3}}$.

18. Simplify $x^{1/2} \cdot x^{2/3}$.

19. Simplify $\dfrac{x^{7/6}}{x^{1/3}}$.

20. Is 10,011 divisible by 3?

21. Is 104,422 divisible by 4?

22. Is 123,123 prime?

23. Is 991 prime?

24. Determine the prime factorization of 400.

25. Determine the prime factorization of 165.

Answers

1. 1 (Order of Operations, p. 31)

2. 19 (Order of Operations, p. 31)

3. $3a + 4b + 5c$ (Basic Math Operations, p. 35)

4. $3^3 x^7 y^2$ (Powers (Exponents), p. 36)

5. $6x - 5$ (Absolute Value, p. 37)

6. $m^3 + 5m$ (Combining "Like" Terms, p. 38)

7. $4x$ (Combining "Like" Terms, p. 38)

8. $4 \cdot 6$ (Properties of Algebraic Expressions, p. 41)

9. $\frac{4}{3}$ (Properties of Algebraic Expressions, p. 41)

10. $\frac{9}{2}x$ (Integer Powers (Exponents), p. 44)

11. $a + 2b$ (Integer Powers (Exponents), p. 44)

12. 17 (Square Roots and Cube Roots, p. 48)

13. 5 (Square Roots and Cube Roots, p. 48)

14. 3.2 (Approximating Square Roots, p. 50)

15. 12.2 (Approximating Square Roots, p. 50)

16. $2\sqrt{2}$ (Simplifying Square Roots, p. 52)

17. $4\sqrt{3}$ (Rationalizing Fractions, p. 54)

18. $x^{7/6}$ (Exponents for Roots, p. 56)

19. $x^{5/6}$ (Exponents for Roots, p. 56)

20. Yes (Divisibility Rules, p. 58)

21. No (Divisibility Rules, p. 58)

22. No (Prime Numbers, p. 61)

23. Yes (Prime Numbers, p. 61)

24. $2^4 \cdot 5^2$ (Prime Factorization, p. 63)

25. $3 \cdot 5 \cdot 11$ (Prime Factorization, p. 63)

Chapter 2
Numbers

Numbers, numbers everywhere. What do they look like? How can they be written? How do they compare to one another? What do they do when they're added or multiplied to one another? This chapter deals with positive and negative numbers, fractions, decimals, and percents. The situations that were once just purely numbers are now extended to those with variables mixed into the expressions. You'll also see how the properties of numbers hold when using the variables.

Signed Numbers

All numbers can be considered to be signed numbers, either positive or negative, except for the number 0. When you don't see a + sign or any sign in front of a number, you just assume that it's positive. It would take too long and use up too much ink and space to put a + sign in front of every number that isn't negative. Of course, you have to indicate the − sign when writing about negative numbers. A negative number indicates that the value of the number is less than 0, like degrees below 0 in the winter or a negative balance in an account. The number 0 doesn't have a sign; it's neither positive nor negative. It separates the positives from the negatives and so is neutral. This is best illustrated on a number line.

Number Lines

A number line is usually a horizontal line, with numbers listed above or below it, going from smallest to largest as you read from left to right. For example, a number line with the integers from −5 to 5 would look like the following:

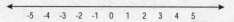

The numbers to the right of 0 are positive, but they're assumed positive, so the + signs aren't needed. To show that a particular number has been selected on a number line, a heavy black dot is used. For instance, to show the numbers −3, 0, 4, and 5, the number line would look like the following:

You can also show that all the values between two integers are selected. For instance, to show that you want all the numbers between −2 and 3, put the dots on the number line and shade between them:

You can also show all the numbers between −2 and 3, including the −2 but not including the 3. This would indicate that you want everything up to the 3—all the fractions and decimals close to it—but you don't want the exact number 3 included. To do this, use a hollow circle over the 3.

The collections of numbers that are shown on a number line can also be described with symbols and mathematical expressions. One way uses greater-than and less-than symbols, called **inequalities**. The other way uses interval notation, which uses parentheses and brackets. Look at the following ways of describing numbers on the number line.

Number Line	*Inequality Notation*	*Interval Notation*
2 — 6	$2 \leq x < 6$	[2,6)
-8 — 0	$-8 \leq x \leq 0$	[−8,0]
-4 — 2	$-4 < x < 2$	(−4,2)
-3 —	$x \geq -3$	[−3,∞)

Inequality notation uses an *x* or some variable to represent all the numbers that are under consideration. Interval notation doesn't use any variable. It's a very efficient notation but causes some confusion when the context isn't clear. The interval (−4,2) could be mistaken for the point (−4,2) in a graphing problem. The arrow pointing to the right in the last example in the table indicates infinity and means that there's no end to the numbers.

Example Problems

These problems show the answers and solutions.

1. Write $-4 \leq x < 2$ in interval notation and graph the numbers on the number line.

    answer: ←•———○→ and [−4,2)
            -4   2

    This inequality notation indicates that the number −4 should be included, but 2 should not be. There'll be a solid dot over the −4 and a hollow dot over the 2 in the number line. Using interval notation, the bracket, [, will be used to show that the −4 is included; a parenthesis,), will show that the 2 is not included.

2. Write the corresponding inequality and interval notation for this number line. ←———○→
                                                         6

    answer: The inequality is $x < 6$, and the interval notation is (−∞,6).

3. Given the interval [18,35], write this as an inequality and draw the number line.

    answer: $18 \leq x \leq 35$ and ←○———•→
                                        -18   35

Work Problems

Use these problems to give yourself additional practice.

1.   Write the interval (−6,18] using inequality notation and draw the number line.

2.   Write the inequality and interval notation for .

3.   Write the inequality and draw the number line for the interval (− ∞,0).

Worked Solutions

1.   −6 < x ≤ 18 and

2.   [−15,−11) and −15 ≤ x < −11

3.   x < 0 and

Addition of Signed Numbers

Signed numbers can be added together, resulting in another signed number (or zero). The rule for adding signed numbers is based on considering them in pairs, two at a time. The first thing to determine is whether their signs are the same or different. If the signs are the same (both positive or both negative), then you ignore the signs—look at their absolute value—and add the two numbers together. The sign that they share is the sign of the answer.

Signs the Same

When the signs are the same, you find the sum, and the sign of the sum is the same as the signs of the numbers.

Example Problems

These problems show the answers and solutions.

1.   Add (−4) + (−8).

     answer: −12

     Add the absolute values of the numbers, the 4 and the 8, to get 12. Because both numbers are negative, the answer will be negative. (−4) + (−8) = −12

2.   Add (+6) + (+2).

     answer: +8

     The + signs are shown here for emphasis. You usually wouldn't show the + signs, because you can assume that a number is positive if there's no sign in front of it. Applying the rule given previously, you add the absolute values of the numbers, the 6 and 2, to get 8. Because both are positive, the answer is positive.

3. Add $(-3) + (-19) + (-11)$.

answer: -33

The rule applies to pairs of numbers, so two of the numbers should be grouped and added and then the result added to the third. $[(-3) + (-19)] + (-11) = [-22] + (-11)$ Grouping the first two numbers and adding, the sum is -22. This is then added to the last number. $(-22) + (-11) = -33$

The same result would have been obtained if the last two numbers had been added first and that result added to the first. The Associative Property still works with these signed numbers. For more on the Associate Property, see, "Associative Property of Addition and Multiplication," in Chapter 1 (page 41).

Signs Different

The other part of the rule for adding signed numbers has to do with when the signs of the two numbers are different. If the signs are different, you consider the absolute values of the numbers and find the difference between them. Yes, this is actually subtraction in an addition rule, but it's what works. After finding the difference between the absolute values, you then determine which original number has the greater absolute value (is farther from zero on the number line). That number is the one that determines the sign of the answer.

When the signs are different, you find the difference (and the sign of the sum is the same as the sign of the number that is farther from zero on the number line). Sorry, there's no nice ditty for this one.

Example Problems

These problems show the answers and solutions.

1. $(-7) + (+3)$

answer: -4

The positive sign on the 3 is there for emphasis. The signs are different, so consider the absolute values of the numbers, 7 and 3. The difference between 7 and 3 is 4. Because -7 has the greater absolute value (it's farther from 0), its sign determines the sign of the answer. The answer is a negative number. $(-7) + (+3) = -4$

2. $(-10) + (+13)$

answer: $+3$

The signs are different. The absolute values are 10 and 13. The difference between 10 and 13 is 3. Because the 13 has the greater absolute value, it determines that the sign of the answer is positive.

3. $(+15) + (-11) + (-2)$

answer: $+2$

The first two numbers will be grouped, added, and then the result added to the last term.

$$[(+15) + (-11)] + (-2) = [+4] + (-2), (+4) + (-2) = +2$$

The numbers could have been grouped differently and even changed in order. Addition is both associative and commutative, so you can take advantage of being able to switch the numbers around to create simpler sums. In the preceding case, it would have been nicer to group the last two numbers together, because their signs are the same, and then to add the result to the first. It's generally an easier task to add (find the sum of) numbers than to find their difference. Grouping the preceding problem differently, $(+15) + [(-11) + (-2)] = (+15) + [-13] = +2$.

4. Add $(-9) + (+9)$.

    answer: 0

    The signs are different, so you find the difference between the absolute values, 9 and 9. The difference is 0. If you try to determine a sign for that result, you're stuck. They're the same distance from 0 on the number line, so it's a tie. No sign is attached to the 0. The number 0 doesn't have a sign; it's neither positive nor negative.

Work Problems

Use these problems to give yourself additional practice.

1. Find the sum of $(-6) + (+5)$.

2. Find the sum of $(-18) + (-2)$.

3. Find the sum of $(+15) + (-13) + (-9)$.

4. Find the sum of $(-5) + (6) + (5) + (-6)$.

5. Find the sum of $(-11) + (29) + (-13)$.

Worked Solutions

1. **−1**   The signs are different. The difference between the absolute values, 6 and 5, is 1. Because the −6 is farther from 0 on the number line, the sign of the answer is negative. $(-6) + (+5) = -1$

2. **−20**   The signs are the same, so find the sum of the absolute values, 18 and 2. The sign of the answer is the same as the numbers, negative. $(-18) + (-2) = -20$

3. **−7**   Group the last two numbers together, because their signs are the same, the sum is negative, and the sum of their absolute values is 22. Then add that answer to the first number. The two numbers then added have different signs, so the difference between their absolute values is determined.

$$(+15) + [(-13) + (-9)] = (+15) + [-22]$$
$$= -7$$

4. **0**   Notice that the + signs have been dropped from the positive numbers. In this case, if you group the two sets of numbers with the same sign together, then there's only one sum of numbers with opposite signs. This will mean using the commutative property of addition, which also allows you to change the order.

$$(-5) + (6) + (5) + (-6) = (-5) + (5) + (6) + (-6)$$
$$= [(-5) + (5)] + [(6) + (-6)]$$
$$= [0] + [0]$$
$$= 0$$

5. **5**   This time change the order to get the two negative numbers together, add them, and then add the result to the positive number.

$$(-11) + (29) + (-13) = (-11) + (-13) + (29)$$
$$= [(-11) + (-13)] + (29)$$
$$= [-24] + 29$$
$$= 5$$

Subtraction of Signed Numbers

Subtraction of signed numbers could be rather complicated if you had to consider the signs of the numbers (whether they're the same or different), the order in which they're subtracted, and their distance from zero on the number line. Eight different rules could be necessary to cover all the situations. So, to avoid all this hassle, the procedure is just to change any subtraction problem into an addition problem and use the addition rules you already know. This probably doesn't sound quite right to you. How can you just change the problem to an addition problem? You know that subtracting $9 - 4$ and adding $9 + 4$ are two different problems with two different answers. Just changing subtraction to addition doesn't get you any closer to the answer to the subtraction problem. A second step in this procedure **will** give you the correct answer to the subtraction problem. The second step is to also change the sign of the second number to its opposite (change the sign).

To change a subtraction problem to addition, change two signs: the sign of the problem from subtraction to addition and then the sign of the second number to its opposite. This creates an equivalent problem. It's a different problem from the original, but it has the same answer as the original. The problem is now one for which there are established rules—rules for adding signed numbers.

Example Problems

These problems show the answers and solutions.

1.   Subtract $9 - 4$.

   answer: 5

   This is a very familiar problem and one that you can do easily, but I'll approach it in terms of the rule for subtracting signed numbers. Change the problem to addition and change the second number, the positive 4, to its opposite, -4. $9 - 4 = 9 + (-4) = 5$

2. Subtract −8 − (−3).

answer: −5

Change the subtraction to addition, and change the −3 to +3. −8 − (−3) = −8 + (+3) = −5

3. Subtract 40 − (−8).

answer: 48

Change subtraction to addition and change −8 to +8. 40 − (−8) = 40 + (+8) = 48

Work Problems
Use these problems to give yourself additional practice.

1. Subtract −15 − (−16).

2. Subtract 3 − (+14).

3. Subtract −5 − (−6).

4. Subtract −6 − (−6).

5. Subtract 5 − 13 − (−6).

Worked Solutions

1. **1** Change the subtraction to addition and change the −16 to +16.

$$-15 - (-16) = -15 + (+16) = 1$$

2. **−11** Change the subtraction to addition and change the +14 to −14.

$$3 - (+14) = 3 + (-14) = -11$$

3. **1** Change the subtraction to addition and change the −6 to +6.

$$-5 - (-6) = -5 + (+6) = +1$$

4. **0** Change the subtraction to addition and change the −6 to +6.

$$-6 - (-6) = -6 + (+6) = 0$$

5. **−2** In this case, because subtraction is not commutative or associative, you can't rearrange the order or group in any convenient manner. The subtraction has to be done in order, moving from left to right. So the subtraction will be performed on the first two terms, and then the result will subtract the last term.

$$5 - 13 - (-6) = [5 - 13] - (-6)$$
$$= [5 + (-13)] - (-6)$$
$$= [-8] - (-6)$$
$$= [-8] + (+6)$$
$$= -2$$

Multiplication and Division of Signed Numbers

Multiplying and dividing signed numbers is really much easier than adding and subtracting them. Instead of having to determine the order in which numbers are written and judging their comparative sizes and changing from one operation to another, the only concern when performing these operations is how many negative signs are in the problem. The multiplication and division go on as usual. The only extra problem is in counting negatives. If you are multiplying or dividing an even number of negative signs (or a mixture of the two), then the final answer is positive. If the problem has an odd number of negative signs, then the answer is negative. Just do the multiplication and/or division as directed—ignoring the negative signs at first—and then go back and count the number of negative signs in the problem to assign the correct sign to the answer.

Example Problems

These problems show the answers and solutions.

1. Multiply $(-4)(6)$.

    answer: -24

    The product of 4 and 6 is 24. There is one negative sign. Because 1 is an odd number, the answer will be negative. $(-4)(6) = -24$.

2. Multiply $(-3)(-4)(-2)(-3)$.

    answer: 72

    The product of $3 \cdot 4 \cdot 2 \cdot 3$ is 72. There are four negative signs. Because 4 is an even number, the answer will be positive. $(-3)(-4)(-2)(-3) = 72$.

3. Perform the operations $\frac{(-5)(-4)}{-10}$.

    answer: -2

    The product in the numerator is 20, because there are two negative signs.

    $$\frac{(-5)(-4)}{-10} = \frac{20}{-10}$$

    The quotient is -2, because the result of dividing 20 by 10 is 2, and there's one negative sign in this problem.

    $$\frac{(-5)(-4)}{-10} = \frac{20}{-10} = -2$$

In a problem like this, where all the operations are multiplication and division, you can perform the operations without regard to sign, and then you can determine the sign of the answer at the very end.

Think of the problem as being $\frac{5 \cdot 4}{10}$, which is $\frac{20}{10}$ or 2. Then count the negative signs—there are three of them—which makes the final answer negative. This all can be done because multiplication and division are on the same level in the Order of Operations (see "Order of Operations" in Chapter 1 for more on this) and because multiplication and division of signed numbers have exactly the same rules, as far as the signs are concerned.

4. Multiply $(-3)(4a)(-a^2)$.

   answer: $12a^3$

   The same rules apply if you're multiplying variables. First, multiply without considering the signs. $(3)(4a)(a^2) = 3 \cdot 4 \cdot a \cdot a^2 = 12a^3$. Now, go back and count the negatives. There are 2 negative signs, so the answer is positive. $(-3)(4a)(-a^2) = 12a^3$

5. Perform the operations $\frac{(6x^2)(-2x^3)(5x)}{(-3x^4)(-x)}$.

   answer: $-20x$

   Ignore the signs for now. Multiply the numerator and denominator separately. Then divide the results.

   $$\frac{(6x^2)(2x^3)(5x)}{(3x^4)(x)} = \frac{6 \cdot 2 \cdot 5 \cdot x^2 \cdot x^3 \cdot x}{3 \cdot x^4 \cdot x} = \frac{60x^6}{3x^5} = 20x$$

   There are three negative signs in the original problem, so the answer is negative.

   $$\frac{(6x^2)(-2x^3)(5x)}{(-3x^4)(-x)} = -20x$$

   When distributing a negative number over two or more terms within a grouping symbol, each sign of every term within the grouping symbol changes. One of the most common errors in algebraic work is to forget to distribute the negative over **each and every** term.

6. Simplify $-y(-2 - 3y + 4y^2)$.

   answer: $2y + 3y^2 - 4y^3$

   Notice how the result of every term's multiplication in the parentheses has changed.

   $$-y(-2 - 3y + 4y^2) = (-y)(-2) + (-y)(-3y) + (-y)(4y^2)$$
   $$= 2y + 3y^2 - 4y^3$$

Work Problems

Use these problems to give yourself additional practice.

1. Multiply $(-2)(-3)(5)$.

2. Perform the operations $\frac{-10(-4)}{-20}$.

3. Perform the operation $(-1)^{22}$.

4. Distribute and simplify $-3(-6+4)$.

5. Distribute and simplify $-a(a^4 - 3a^3 + a^2 - 7)$.

Worked Solutions

1. **30**   The product of $2 \cdot 3 \cdot 5$ is 30. There are two negative signs, so the product is positive.

$$(-2)(-3)(5) = 30$$

2. **-2**   The product of $10 \cdot 4$ is 40. Divide 40 by 20, and you get 2. There are three negative signs, so the final result is negative.

$$\frac{-10(-4)}{-20} = \frac{40}{-20} = -2$$

3. **1**   This is, basically, -1 multiplied 22 times. Repeatedly multiplying 1s always gives you the number 1. Since there are 22 negative signs, and 22 is even, then the result is positive.

$$(-1)^{22} = 1$$

4. **6**   Multiply each term by -3.

$$-3(-6+4) = (-3)(-6) + (-3)(4)$$
$$= 18 - 12 = 6$$

5. $-a^5 + 3a^4 - a^3 + 7a$

Multiply each term by $-a$.

$$-a(a^4 - 3a^3 + a^2 - 7) = (-a)(a^4) + (-a)(-3a^3) + (-a)(a^2) + (-a)(-7)$$
$$= -a^5 + 3a^4 - a^3 + 7a$$

Exponents That Are Signed Numbers

The same properties that apply to exponents that are positive integers or fractions also apply to signed exponents. The rules or properties for exponents are found in the section, "Integer Powers (Exponents)," in Chapter 1. Applying these rules in this section involves keeping in mind both the rules for exponents **and** the rules for signed numbers. There's an added step involved here, too. Results often have to be written without using negative exponents. This does depend on the application, but most of the examples here will require that there be no negative exponents in the answer.

In addition to the rules for exponents already listed in Chapter 1, two more rules can be added here.

First, recall the rule that $a^{-n} = \frac{1}{a^n}$. This will now be extended to the additional rule that $\frac{1}{a^{-n}} = a^n$. This says that if there's a factor with a negative exponent in the denominator, then the exponent changes to positive when it's brought up to the numerator.

Second, a rule for dealing with fractions is that $\left(\dfrac{a}{b}\right)^{-n} = \left(\dfrac{b}{a}\right)^{n}$. This says that if an entire fraction is raised to a negative exponent, that exponent becomes positive if the fraction is flipped.

Example Problems

These problems show the answers and solutions.

1. Simplify, leaving no negative exponents $\dfrac{x^2 x^{-3}}{x^{-4} x^6}$.

   answer: $\dfrac{1}{x^3}$

   The factors in the numerator and denominator will be simplified, separately. Because the bases, x's, are the same, you can multiply by adding the exponents.

   $$\frac{x^2 x^{-3}}{x^{-4} x^6} = \frac{x^{2+(-3)}}{x^{-4+6}} = \frac{x^{-1}}{x^2}$$

   Now, the numerator, x^{-1}, will be replaced by $\dfrac{1}{x^1}$, which moves that factor to the denominator and leaves a 1 in the numerator. $\dfrac{x^{-1}}{x^2} = \dfrac{1}{x^1 \cdot x^2}$

   And, now, the denominator can be simplified by multiplying the factors and adding the exponents.

   $$\frac{1}{x^1 \cdot x^2} = \frac{1}{x^{1+2}} = \frac{1}{x^3}$$

   Another way that the fraction $\dfrac{x^{-1}}{x^2}$ could have been dealt with would have been to do the division of the two values, subtracting exponents, and then rewriting the answer.

   $$\frac{x^{-1}}{x^2} = x^{-1-2} = x^{-3} = \frac{1}{x^3}$$

2. Simplify $\dfrac{4a^3 b^2 c^8}{a^4 b^8 c^{-4}}$.

   answer: $\dfrac{4c^{12}}{a^1 b^6}$

   This problem will be dealt with much like the second method used in the preceding example. It'll be treated as a division problem, subtracting exponents, and rewriting the answer.

   $$\frac{4a^3 b^2 c^8}{a^4 b^8 c^{-4}} = 4a^{3-4} b^{2-8} c^{8-(-4)} = 4a^{-1} b^{-6} c^{12}$$

   To rewrite the answer, each factor with a negative exponent will drop to the denominator of a fraction. The factors with positive exponents will stay in the numerator.

   $$4a^{-1} b^{-6} c^{12} = \frac{4c^{12}}{a^1 b^6}$$

   Note that the 4 is in the numerator. You assume that it's 4 raised to the first power.

3. Simplify $\left(2m^2 n^{-2} z\right)^{-3}$.

   answer: $\dfrac{n^6}{8m^6 z^3}$

   The first step is to raise each factor, independently, to the −third power.

   $$\left(2m^2 n^{-2} z\right)^{-3} = (2)^{-3}\left(m^2\right)^{-3}\left(n^{-2}\right)^{-3}(z)^{-3}$$
   $$= (2)^{-3} m^{-6} n^6 z^{-3}$$

   Next, the result can be rewritten without negative exponents. The factors with negative exponents are moved to the denominator.

   $$2^{-3} m^{-6} n^6 z^{-3} = \frac{n^6}{2^3 m^6 z^3} = \frac{n^6}{8m^6 z^3}$$

4. Simplify $\left(\dfrac{3x^2 y^{-2} z^{-1}}{2x^{-1} y^2 z^6}\right)^{-2}$.

   answer: $\dfrac{4y^8 z^{14}}{9x^6}$

   Your first inclination might be to raise each factor in the numerator and denominator to the −second power. Although this would work, it would mean multiplying by a negative number eight times. Instead, take advantage of the rule that allows you to flip the fraction when there's a negative exponent.

   $$\left(\frac{3x^2 y^{-2} z^{-1}}{2x^{-1} y^2 z^6}\right)^{-2} = \left(\frac{2x^{-1} y^2 z^6}{3x^2 y^{-2} z^{-1}}\right)^{2}$$

   Notice that the exponent outside the parenthesis is now positive. Before raising each factor to the second power, there's some simplifying that can be done inside the parenthesis. Factors with negative exponents will switch places. If a factor with a negative exponent is in the numerator, move it to the denominator to change the exponent to a positive. And vice versa with those in the denominator.

   $$\left(\frac{2x^{-1} y^2 z^6}{3x^2 y^{-2} z^{-1}}\right)^{2} = \left(\frac{2y^2 y^2 z^6 z^1}{3x^2 x^1}\right)^{2} = \left(\frac{2y^4 z^7}{3x^3}\right)^{2}$$

   Now each factor can be raised to the second power.

   $$\left(\frac{2y^4 z^7}{3x^3}\right)^{2} = \frac{2^2 \left(y^4\right)^2 \left(z^7\right)^2}{3^2 \left(x^3\right)^2} = \frac{4y^8 z^{14}}{9x^6}$$

Work Problems

Use these problems to give yourself additional practice.

Simplify each, rewriting the answers so that there are no negative exponents.

1. $\dfrac{a^2 b^{-2} c^{-4} d}{a^{-1} bc^{-3} d^2}$

2. $\left(\dfrac{3x^2 y^4}{2x^3 y}\right)^3$

3. $\left(2x^{-3} y^2 z^3\right)^{-1}$

4. $\left(\dfrac{4a^2 y^{-2} z}{3a^{-3} yz^{-4}}\right)^{-3}$

5. $\left[\left(\dfrac{3^3 xyz}{6x^2 yz^{-2}}\right)^{-2}\right]^{-1}$

Worked Solutions

1. $\dfrac{a^3}{b^3 cd}$ First rewrite the fraction by dividing. When dividing factors with the same base, subtract the exponents.

$$\frac{a^2 b^{-2} c^{-4} d}{a^{-1} bc^{-3} d^2} = a^{2-(-1)} b^{-2-1} c^{-4-(-3)} d^{1-2}$$
$$= a^3 b^{-3} c^{-1} d^{-1}$$

Now rewrite the result as a fraction, so there will be no negative exponents.

$$a^3 b^{-3} c^{-1} d^{-1} = \frac{a^3}{b^3 cd}$$

2. $\dfrac{27y^9}{8x^3}$ First do the division within the parenthesis and then rewrite as a fraction. Lastly, raise each factor to the third power.

$$\left(\frac{3x^2 y^4}{2x^3 y}\right)^3 = \left(\frac{3}{2} x^{2-3} y^{4-1}\right)^3 = \left(\frac{3}{2} x^{-1} y^3\right)^3$$
$$= \left(\frac{3y^3}{2x}\right)^3 = \frac{3^3 (y^3)^3}{2^3 x^3} = \frac{27y^9}{8x^3}$$

3. $\dfrac{x^3}{2y^2 z^3}$ Raise each factor to the −first power and then rewrite without negative exponents.

$$\left(2x^{-3} y^2 z^3\right)^{-1} = 2^{-1} (x^{-3})^{-1} (y^2)^{-1} (z^3)^{-1}$$
$$= 2^{-1} x^3 y^{-2} z^{-3}$$
$$= \frac{x^3}{2y^2 z^3}$$

4. $\dfrac{27y^9}{64a^{15}z^{15}}$ First flip the fraction to change the -3 exponent to $+3$. Then divide and rewrite.

$$\left(\frac{4a^2\,y^{-2}\,z}{3a^{-3}\,yz^{-4}}\right)^{-3}=\left(\frac{3a^{-3}\,yz^{-4}}{4a^2\,y^{-2}\,z}\right)^{3}$$

$$=\left(\frac{3}{4}\,a^{-3-2}\,y^{1-(-2)}\,z^{-4-1}\right)^{3}=\left(\frac{3}{4}\,a^{-5}\,y^{3}\,z^{-5}\right)^{3}$$

$$=\left(\frac{3y^3}{4a^5z^5}\right)^{3}=\frac{3^3(y^3)^3}{4^3(a^5)^3(z^5)^3}$$

$$=\frac{27y^9}{64a^{15}z^{15}}$$

5. $\dfrac{81z^6}{4x^2}$ In this case, the two outside powers can be multiplied together, resulting in a positive power. Then the division and rewriting can be done before raising everything to the second power.

$$\left[\left(\frac{3^3\,xyz}{6x^2\,yz^{-2}}\right)^{-2}\right]^{-1}=\left(\frac{3^3\,xyz}{6x^2\,yz^{-2}}\right)^{2}$$

$$\left(\frac{3^3\,xyz}{6x^2\,yz^{-2}}\right)^{2}=\left(\frac{3^3}{6}\,x^{1-2}\,y^{1-1}\,z^{1-(-2)}\right)^{2}=\left(\frac{27}{6}\,x^{-1}\,y^{0}\,z^{3}\right)^{2}=\left(\frac{27^9}{6_2}\,x^{-1}\,z^{3}\right)^{2}$$

$$\left(\frac{9z^3}{2x}\right)^{2}=\frac{81z^6}{4x^2}$$

Fractions

Many algebraic expressions or formulas involve fractions. Knowing how to perform all basic operations on fractions—addition, subtraction, multiplication, and division—is therefore a necessary part of calculating algebraic expressions.

Lowest Terms and Equivalent Fractions

Fractional answers are written in their lowest terms, which means that there shouldn't be any common factors between the numerator and denominator. Having smaller numbers, when possible, makes comparisons and other operations easier. On the other hand, in order to add or subtract fractions there has to be a common denominator, so you need to multiply or increase the number of common factors in a fraction. The key to both of these procedures is the number 1. To reduce a fraction to lowest terms, divide by 1. It won't actually be the number 1, it'll be something equivalent to it like: $\dfrac{3}{3}$, $\dfrac{2x}{2x}$, $\dfrac{yz^2}{yz^2}$, or whatever the fraction's numerator and denominator have in common. To rewrite a fraction as an equivalent fraction, so it can be added or subtracted, you multiply by 1, again, in the form of $\dfrac{4}{4}$, etc.

Example Problems
These problems show the answers and solutions.

1. Write the fraction in lowest terms (reduce it): $\dfrac{20x^2 yz^3}{24xy}$.

   answer: $\dfrac{5xz^3}{6}$

   The greatest common divisor of the numerator and denominator is $4xy$. This divides both numerator and denominator evenly. So you make a fraction with $4xy$ in both the numerator and denominator. That way it's equal to 1. Then you divide the fraction by 1.

   $$\frac{20x^2 yz^3}{24xy} \div \frac{4xy}{4xy} = \frac{5xz^3}{6}$$

2. Write the fraction in lowest terms: $\dfrac{36a^2 bc^4 - 20ab^2 c^2}{8ac^3 + 32b^2 c^2}$.

   answer: $\dfrac{9a^2 bc^2 - 5ab^2}{2ac + 8b^2}$

   This fraction has four terms in it: two in the numerator and two in the denominator. Each term has to be considered when looking for the greatest common divisor. In this case, the factor $4c^2$ divides each term evenly. Therefore, the fraction to divide by is $\dfrac{4c^2}{4c^2}$.

   $$\frac{36a^2 bc^4 - 20ab^2 c^2}{8ac^3 + 32b^2 c^2} \div \frac{4c^2}{4c^2} = \frac{9a^2 bc^2 - 5ab^2}{2ac + 8b^2}$$

3. Add the fractions and reduce the answers to the lowest terms: $\dfrac{5}{6} + \dfrac{1}{10}$.

   answer: $\dfrac{14}{15}$

   The least common denominator for the two fractions is 30. Each fraction will be multiplied by 1 (in the form of $\frac{5}{5}$ and $\frac{3}{3}$) before adding them together.

   $$\frac{5}{6} + \frac{1}{10} = \frac{5}{6} \cdot \frac{5}{5} + \frac{1}{10} \cdot \frac{3}{3} = \frac{25}{30} + \frac{3}{30} = \frac{28}{30}$$

   Now the answer can be reduced by dividing by 1 (in the form of $\frac{2}{2}$). $\dfrac{28}{30} \div \dfrac{2}{2} = \dfrac{14}{15}$

4. Add the fractions and reduce to lowest terms: $\dfrac{2}{3a} + \dfrac{4}{a^2}$.

   answer: $\dfrac{2a + 12}{3a^2}$

   The least common denominator is $3a^2$. This contains all of the factors of both denominators. Each denominator divides it evenly. To write the first fraction with this denominator, you multiply by $\frac{a}{a}$. To write the second fraction with that same denominator, you multiply by $\frac{3}{3}$.

   $$\frac{2}{3a} + \frac{4}{a^2} = \frac{2}{3a} \cdot \frac{a}{a} + \frac{4}{a^2} \cdot \frac{3}{3} = \frac{2a}{3a^2} + \frac{12}{3a^2} = \frac{2a + 12}{3a^2}$$

This answer has three terms, two in the numerator and one in the denominator. There's no common factor shared by all three at the same time, so this can't be reduced. It's in lowest terms.

5. Add the fractions and reduce to lowest terms: $\dfrac{1+x}{4x^2 y} - \dfrac{2-y}{6xy^2}$.

answer: $\dfrac{3y + 5xy - 4x}{12x^2 y^2}$

The least common denominator has to contain all of the factors of all of the denominators. In this case, the number 12 will be in the least common denominator, because it contains all of the factors of both 4 and 6. Also, the least common denominator will contain x^2 and y^2. The denominator is $12x^2 y^2$.

$$\frac{1+x}{4x^2 y} - \frac{2-y}{6xy^2} = \frac{1+x}{4x^2 y} \cdot \frac{3y}{3y} - \frac{2-y}{6xy^2} \cdot \frac{2x}{2x}$$

The distributive property will be used when multiplying both of the original numerators by the fractions needed to produce the common denominator.

$$\frac{1+x}{4x^2 y} \cdot \frac{3y}{3y} - \frac{2-y}{6xy^2} \cdot \frac{2x}{2x} = \frac{1(3y) + x(3y)}{12x^2 y^2} - \frac{2(2x) - y(2x)}{12x^2 y^2}$$

$$= \frac{3y + 3xy}{12x^2 y^2} - \frac{4x - 2xy}{12x^2 y^2}$$

$$= \frac{3y + 3xy - 4x + 2xy}{12x^2 y^2}$$

Notice that the negative sign was distributed over the two terms in the second fraction.

$$\frac{3y + 3xy - 4x + 2xy}{12x^2 y^2} = \frac{3y + 5xy - 4x}{12x^2 y^2}$$

Because no factor is common to all four terms in the fraction, this is in lowest terms.

Work Problems

Use these problems to give yourself additional practice.

1. Reduce the fraction to lowest terms: $\dfrac{45x^2}{60xy}$.

2. Reduce to lowest terms: $\dfrac{12x^4 y^2 - 27xy^6}{30x^5 y + 18y^4}$.

3. Find an equivalent fraction that has a denominator of $15x^2 y^3$. $\dfrac{4x - 3y}{5xy^3}$,

4. Add the fractions: $\dfrac{8}{3x^2 y} + \dfrac{y}{6x}$.

5. Subtract the fractions: $\dfrac{6-b}{4ab^2} - \dfrac{7-a}{9a^2 b}$.

Worked Solutions

1. $\dfrac{3x}{4y}$ Both the numerator and denominator are divisible by 15x. $\dfrac{45x^2}{60xy} \div \dfrac{15x}{15x} = \dfrac{3x}{4y}$

2. $\dfrac{4x^4 y - 9xy^5}{10x^5 + 6y^3}$ Reduce to lowest terms by dividing both numerator and denominator by 3y.

$$\frac{12x^4 y^2 - 27xy^6}{30x^5 y + 18y^4} \div \frac{3y}{3y} = \frac{4x^4 y - 9xy^5}{10x^5 + 6y^3}$$

3. $\dfrac{12x^2 - 9xy}{15x^2 y^3}$ Multiply numerator and denominator by 3x.

$$\frac{4x - 3y}{5xy^3} \cdot \frac{3x}{3x} = \frac{(4x - 3y) \cdot 3x}{15x^2 y^3} = \frac{12x^2 - 9xy}{15x^2 y^3}$$

4. $\dfrac{16 + xy^2}{6x^2 y}$ The common denominator for the two fractions is $6x^2 y$. Rewrite each fraction with that denominator before adding.

$$\frac{8}{3x^2 y} + \frac{y}{6x} = \frac{8}{3x^2 y} \cdot \frac{2}{2} + \frac{y}{6x} \cdot \frac{xy}{xy} = \frac{16}{6x^2 y} + \frac{xy^2}{6x^2 y}$$

$$= \frac{16 + xy^2}{6x^2 y}$$

5. $\dfrac{54a - 5ab - 28b}{36a^2 b^2}$ First write the fractions as equivalent fractions with the common denominator $36a^2 b^2$ and then subtract.

$$\frac{6 - b}{4ab^2} - \frac{7 - a}{9a^2 b} = \frac{6 - b}{4ab^2} \cdot \frac{9a}{9a} - \frac{7 - a}{9a^2 b} \cdot \frac{4b}{4b}$$

$$= \frac{54a - 9ab}{36a^2 b^2} - \frac{28b - 4ab}{36a^2 b^2}$$

$$= \frac{54a - 9ab - 28b + 4ab}{36a^2 b^2}$$

$$= \frac{54a - 5ab - 28b}{36a^2 b^2}$$

Multiplying and Dividing Fractions

When multiplying and dividing fractions, you don't need a common denominator like you do when adding or subtracting fractions. The main challenge here is to have a result that's in lowest terms. This can be done in one or both of two ways: reduce the fractions **before** multiplying or dividing or reduce the result **after** multiplying or dividing. If you can reduce the fractions before performing the operations, the processes will be in terms of smaller and more manageable factors.

Basically, when multiplying fractions, multiply the two numerators together and multiply the two denominators together. When dividing, change the problem to a multiplication problem, because multiplication is so much easier! To change a division to a multiplication problem, multiply the first (left) fraction by the **reciprocal** of the second (right) fraction. The reciprocal of a fraction is the fraction obtained when the numerator and denominator trade places.

Example Problems

These problems show the answers and solutions.

1. Multiply the fractions: $\frac{6}{25} \cdot \frac{20}{21}$.

   answer: $\frac{8}{35}$

   $\frac{6}{25} \cdot \frac{20}{21} = \frac{120}{525}$ Oops! Stop. These numbers are much too large. Reduce the fractions first by finding common factors from **either** numerator and denominator. $\frac{6}{25} \cdot \frac{20}{21} = \frac{\overset{2}{6}}{\underset{5}{25}} \cdot \frac{\overset{4}{20}}{\underset{7}{21}} = \frac{8}{35}$

2. Divide the fractions: $\frac{40}{33} \div \frac{50}{18}$.

   answer: $\frac{24}{55}$

   First change this to a multiplication problem in which the first fraction multiplies the reciprocal of the second $\frac{40}{33} \div \frac{50}{18} = \frac{40}{33} \cdot \frac{18}{50}$. Now complete the reducing of the problem and the multiplication.

$$\frac{40}{33} \cdot \frac{18}{50} = \frac{\overset{4}{40}}{\underset{11}{33}} \cdot \frac{\overset{6}{18}}{\underset{5}{50}} = \frac{24}{55}$$

   The most wonderful part of reducing the fraction before multiplying is that it's completely reduced after it's multiplied.

3. Multiply: $\frac{12x^2 y}{49} \cdot \frac{21x}{32y^3}$.

   answer: $\frac{9x^3}{56y^2}$

   First reduce the fractions. Then multiply and simplify.

$$\frac{\overset{3}{12}x^2 \overset{}{\cancel{y}}}{\underset{7}{49}} \cdot \frac{\overset{3}{21}x}{\underset{8}{32}\overset{}{\cancel{y^3}}_{y^2}} = \frac{3x^2 \cdot 3x}{7 \cdot 8y^2} = \frac{9x^3}{56y^2}$$

4. Divide: $\frac{8a^5(b+1)^2}{15} \div \frac{16a(b+1)^4}{27}$.

   answer: $\frac{9a^4}{10(b+1)^2}$

   First change the problem to multiplication.

$$\frac{8a^5(b+1)^2}{15} \div \frac{16a(b+1)^4}{27} = \frac{8a^5(b+1)^2}{15} \cdot \frac{27}{16a(b+1)^4}$$

   Then reduce the fractions and multiply.

$$\frac{8a^5(b+1)^2}{15} \cdot \frac{27}{16a(b+1)^4} = \frac{\overset{1}{8}\overset{a^4}{a^5}(b+1)^2}{\underset{5}{15}} \cdot \frac{\overset{9}{27}}{\underset{2}{16}\underset{1}{a}(b+1)^4_{(b+1)^2}}$$

$$= \frac{a^4 \cdot 9}{5 \cdot 2(b+1)^2} = \frac{9a^4}{10(b+1)^2}$$

Work Problems

Use these problems to give yourself additional practice.

1. $\dfrac{18}{55} \cdot \dfrac{22}{27}$

2. $\dfrac{90}{49} \div \dfrac{81}{28}$

3. $\dfrac{16a}{35} \cdot \dfrac{77}{24ax}$

4. $\dfrac{18x^2 y^3}{11ab} \div \dfrac{45xy^4}{22a^2}$

5. $\dfrac{3(x+y)^3}{4(a+b)} \cdot \dfrac{28(a+b)^4}{27(x+y)^5}$

Worked Solutions

1. $\dfrac{4}{15}$ $\dfrac{18}{55} \cdot \dfrac{22}{27} = \dfrac{{}^2\cancel{18}}{{}_5\cancel{55}} \cdot \dfrac{\cancel{22}^2}{\cancel{27}_3} = \dfrac{4}{15}$

2. $\dfrac{40}{63}$ $\dfrac{90}{49} \div \dfrac{81}{28} = \dfrac{{}^{10}\cancel{90}}{{}_7\cancel{49}} \cdot \dfrac{\cancel{28}^4}{\cancel{81}_9} = \dfrac{40}{63}$

3. $\dfrac{22}{15x}$ $\dfrac{16a}{35} \cdot \dfrac{77}{24ax} = \dfrac{{}^2\cancel{16a}}{{}_5\cancel{35}} \cdot \dfrac{\cancel{77}^{11}}{{}_3\cancel{24ax}} = \dfrac{22}{15x}$

4. $\dfrac{4ax}{5by}$

$$\dfrac{18x^2 y^3}{11ab} \div \dfrac{45xy^4}{22a^2} = \dfrac{18x^2 y^3}{11ab} \cdot \dfrac{22a^2}{45xy^4} = \dfrac{{}^2\cancel{18x}^{2x}\,\cancel{y}^3}{{}_1\cancel{11ab}} \cdot \dfrac{{}^2\cancel{22a}^{2a^1}}{{}_5\cancel{45xy}^{4y^1}}$$

$$= \dfrac{2x \cdot 2a^1}{b \cdot 5y} = \dfrac{4ax}{5by}$$

5. $\dfrac{7(a+b)^3}{9(x+y)^2}$

$$\dfrac{3(x+y)^3}{4(a+b)} \cdot \dfrac{28(a+b)^4}{27(x+y)^5} = \dfrac{\cancel{3}(x+y)^3}{\cancel{4}(a+b)} \cdot \dfrac{{}^7\cancel{28}(a+b)^{4(a+b)^3}}{{}_9\cancel{27}(x+y)^5_{(x+y)^2}}$$

$$= \dfrac{7(a+b)^3}{9(x+y)^2}$$

Mixed Numbers and Complex Fractions

To multiply and divide fractions, you need all mixed numbers, those with an integer and fraction such as $4\frac{2}{3}$, written as single, improper fractions with no integers in front. To multiply $4\frac{1}{8} \cdot 6\frac{2}{3}$, first change the mixed numbers to $\dfrac{33}{8} \cdot \dfrac{20}{3}$ and then multiply as usual. Your final answer should be written in the original form of a mixed number. So, $\dfrac{33}{8} \cdot \dfrac{20}{3} = \dfrac{{}^{11}\cancel{33}}{{}_2\cancel{8}} \cdot \dfrac{\cancel{20}^5}{\cancel{3}} = \dfrac{55}{2} = 27\frac{1}{2}$.

A complex fraction is a fraction within a fraction. There's a fraction in either the numerator or the denominator or both. Examples of complex fractions are: $\dfrac{\frac{2}{3}}{4}$ or $\dfrac{6}{\frac{1}{2}}$ or $\dfrac{\frac{5}{6}}{\frac{3}{8}}$. Complex fractions can be simplified by multiplying the numerator of the fraction by the reciprocal of the denominator.

Example Problems

These problems show the answers and solutions.

1. Simplify $\dfrac{\frac{2}{3}}{4}$.

   answer: $\dfrac{1}{6}$

   Multiply the numerator by the reciprocal of the denominator.

   $$\frac{\frac{2}{3}}{4} = \frac{2}{3} \cdot \frac{1}{4} = \frac{\cancel{2}}{3} \cdot \frac{1}{\cancel{4}_2} = \frac{1}{6}$$

2. Simplify $\dfrac{6}{\frac{1}{2}}$.

   answer: 12

   Multiply the numerator by the reciprocal of the denominator.

   $$\frac{6}{\frac{1}{2}} = \frac{6}{1} \cdot \frac{2}{1} = \frac{12}{1} = 12$$

3. Simplify $\dfrac{\frac{5}{6}}{\frac{3}{8}}$.

   answer: $2\dfrac{2}{9}$

   Multiply the numerator by the reciprocal of the denominator.

   $$\frac{\frac{5}{6}}{\frac{3}{8}} = \frac{5}{6} \cdot \frac{8}{3} = \frac{5}{{}_3\cancel{6}} \cdot \frac{\cancel{8}^{4}}{3} = \frac{20}{9} = 2\frac{2}{9}$$

4. Divide $5\dfrac{2}{5} \div 3\dfrac{9}{10}$.

   answer: $1\dfrac{5}{13}$

   Change the mixed numbers to improper fractions. Then change the division to multiplication.

   $$5\frac{2}{5} \div 3\frac{9}{10} = \frac{27}{5} \div \frac{39}{10} = \frac{27}{5} \cdot \frac{10}{39}$$

   Now reduce and multiply.

   $$\frac{27}{5} \cdot \frac{10}{39} = \frac{\cancel{27}^{9}}{\cancel{5}} \cdot \frac{\cancel{10}^{2}}{\cancel{39}_{13}} = \frac{18}{13} = 1\frac{5}{13}$$

5. Simplify $\dfrac{4\frac{1}{12}}{6\frac{1}{8}}$.

answer: $\dfrac{2}{3}$

Change the mixed numbers to improper fractions; then multiply the numerator by the reciprocal of the denominator.

$$\frac{4\frac{1}{12}}{6\frac{1}{8}} = \frac{\frac{49}{12}}{\frac{49}{8}} = \frac{49}{12} \cdot \frac{8}{49} = \frac{\cancel{49}}{\cancel{12}_{3}} \cdot \frac{\cancel{8}^{2}}{\cancel{49}} = \frac{2}{3}$$

6. Simplify $\dfrac{\frac{ab}{cd}}{\frac{a^2}{d^3}}$.

answer: $\dfrac{bd^2}{ac}$

Multiply the numerator by the reciprocal of the denominator.

$$\frac{\frac{ab}{cd}}{\frac{a^2}{d^3}} = \frac{ab}{cd} \cdot \frac{d^3}{a^2} = \frac{\cancel{a}b}{c\cancel{d}} \cdot \frac{\cancel{d}^{\,d^2}}{\cancel{a}^{2}_{\,a}} = \frac{bd^2}{ac}$$

Work Problems

Use these problems to give yourself additional practice.

1. Multiply $7\frac{1}{7} \cdot 4\frac{9}{10}$.

2. Divide $2\frac{3}{11} \div 1\frac{2}{33}$.

3. Simplify $\dfrac{\frac{2}{3}}{8}$.

4. Simplify $\dfrac{2x}{\frac{x^2}{y}}$.

5. Simplify $\dfrac{\frac{2(a+1)}{15b^3}}{\frac{(a+1)^7}{10b}}$.

Worked Solutions

1. **35** First change the mixed numbers to improper fractions. Then multiply.

$$7\frac{1}{7} \cdot 4\frac{9}{10} = \frac{50}{7} \cdot \frac{49}{10} = \frac{\cancel{50}^{5}}{\cancel{7}_{1}} \cdot \frac{\cancel{49}^{7}}{\cancel{10}_{1}} = \frac{35}{1} = 35$$

2. $2\frac{1}{7}$ First change the mixed numbers to improper fractions. Then multiply the first fraction by the reciprocal of the second fraction.

$$2\frac{3}{11} \div 1\frac{2}{33} = \frac{25}{11} \div \frac{35}{33} = \frac{25}{11} \cdot \frac{33}{35} = \frac{\overset{5}{\cancel{25}}}{\underset{1}{\cancel{11}}} \cdot \frac{\overset{3}{\cancel{33}}}{\underset{7}{\cancel{35}}} = \frac{15}{7} = 2\frac{1}{7}$$

3. $\frac{1}{12}$ Multiply the numerator by the reciprocal of the denominator.

$$\frac{\frac{2}{3}}{8} = \frac{2}{3} \cdot \frac{1}{8} = \frac{\overset{1}{\cancel{2}}}{3} \cdot \frac{1}{\underset{4}{\cancel{8}}} = \frac{1}{12}$$

4. $\frac{2y}{x}$ Multiply the numerator by the reciprocal of the denominator.

$$\frac{\frac{2x}{x^2}}{y} = \frac{2x}{1} \cdot \frac{y}{x^2} = \frac{2\cancel{x}}{1} \cdot \frac{y}{\underset{x}{\cancel{x^2}}} = \frac{2y}{x}$$

5. $\dfrac{4}{3(a+1)^6 b^2}$ Multiply the numerator by the reciprocal of the denominator.

$$\frac{\frac{2(a+1)}{15b^3}}{\frac{(a+1)^7}{10b}} = \frac{2(a+1)}{15b^3} \cdot \frac{10b}{(a+1)^7} = \frac{2(a+1)}{\underset{3\,15b^3}{\cancel{15b^3}}_{b^2}} \cdot \frac{\overset{2}{\cancel{10b}}}{\underset{(a+1)^6}{\cancel{(a+1)^7}}} = \frac{4}{3(a+1)^6 b^2}$$

Decimals

Numbers can be written as fractions and decimals—even when they're whole numbers or integers. Depending on the situation, a number may be more useable when it's a fraction or, in another situation, as a decimal. Fractional forms are considered to be exact values. When fractions are changed to decimals, you may be dealing with an approximate value—if the decimal is rounded to a certain number of decimal places. Some decimals that represent fractions **terminate**—they end eventually. Some decimals that are equivalent to fractions **repeat**—they never end, but their pattern of digits repeats forever.

Here are some common fractions and their decimal equivalents. Decimals that repeat are shown with an ellipsis, ..., after them, and there are enough digits to show what's repeating.

$\frac{1}{2} = .5$	$\frac{3}{2} = 1.5$		
$\frac{1}{3} = .333...$	$\frac{2}{3} = .666...$	$\frac{4}{3} = 1.333...$	$\frac{5}{3} = 1.666...$
$\frac{1}{4} = .25$	$\frac{3}{4} = .75$	$\frac{5}{4} = 1.25$	$\frac{7}{4} = 1.75$
$\frac{1}{5} = .2$	$\frac{2}{5} = .4$	$\frac{3}{5} = .6$	$\frac{4}{5} = .8$
$\frac{1}{6} = .1666...$	$\frac{5}{6} = .8333...$	$\frac{7}{6} = 1.1666...$	$\frac{11}{6} = 1.8333...$

Fractions to Decimals

To change a fraction into its equivalent decimal, divide the denominator into the numerator until the division process either terminates or you begin to see a pattern repeating itself. A fraction will have a terminating decimal only if the factors of the denominator are 2 or 5 or any powers of 2 or 5 or any multiples of powers of 2 and 5. If there are any other factors in the denominator, the decimal equivalent will repeat.

Example Problems

1. Change the fraction $\frac{7}{16}$ into a decimal.

    answer: .4375

    $16\overline{)7.0000}$ $\quad$ This decimal terminates, and this is an exact value. You might want to round it, however, to two or three decimal places. Rounding .4375 to two places gives you .44, and rounding it to three places gives you .438.

2. Find the decimal value of $\frac{3}{5}$ and round it to three decimal places.

    answer: The decimal equivalent is .6. This terminates and is .600 correct to three decimal places.

3. Find the decimal values of $\frac{6}{7}$ and $\frac{9}{11}$ and round them to three decimal places.

    answer: .857 and .818

    The decimal equivalent of $\frac{6}{7}$ is .857142857142.... This repeats, because the digits 8, 5, 7, 1, 4, and 2 keep repeating. Rounded to three places, it's .857. The decimal equivalent of $\frac{9}{11}$ is .818181... with the digits 8 and 1 repeating forever. Rounded to three places, it's .818.

Decimals to Fractions

To change a **terminating** decimal back into a fraction, you can place the digits in the decimal in the numerator of a fraction. Then put a power of 10 that has as many zeros as there are digits in the decimal into the denominator. Then reduce the fraction. To change a **repeating** decimal back into a fraction, this can be done rather simply if all of the digits in the decimal are in the repeating part. If that's the case, then the repeating digits are put into the numerator of a fraction and as many 9s as there are digits that repeat go in the denominator. Reduce the fraction.

Example Problems

These problems show the answers and solutions.

1. Change .325 into a fraction.

    answer: $\frac{13}{40}$

    Place the digits 325 over 1000. Then reduce the fraction: $\frac{325}{1000} = \frac{13}{40}$.

2. Change .0008 into a fraction.

    answer: $\frac{1}{1250}$

Place the digit 8 (the zeros in front aren't necessary) over 10000. The zeros aren't needed in the numerator, but they're counted for the denominator. Then reduce the fraction.

$$\frac{8}{10000} = \frac{1}{1250}$$

3. Change the repeating decimal .454545... into a fraction.

   answer: $\frac{5}{11}$

   Place the repeating part, 45, over 99. $\frac{45}{99} = \frac{5}{11}$

4. Change the repeating decimal .108108108... into a fraction.

   answer: $\frac{4}{37}$

   Place the repeating part, 108, over 999. $\frac{108}{999} = \frac{12}{111} = \frac{4}{37}$

If the repeating decimal has some digits that don't repeat, such as in .1666... or .15999..., then another technique has to be used that includes solving equations. See "Solving Linear Equations" on page 107 for the method used to do this.

Work Problems

Use these problems to give yourself additional practice.

1. Change the fraction $\frac{4}{25}$ to a decimal and round to three decimal places.

2. Change the fraction $\frac{7}{36}$ to a decimal and round to three decimal places.

3. Change the fraction $\frac{15}{32}$ to a decimal and round to three decimal places.

4. Change the decimal .6875 to a fraction in lowest terms.

5. Change the decimal .242424... to a fraction in lowest terms.

Worked Solutions

1. **.160**   Divide 25 into 4, $25\overline{)4.00}$ with .16 above. The decimal .16 is .160 to three decimal places.

2. **.194**   Divide 36 into 7, $36\overline{)7.00}$ with .19444... above. The decimal .19444... rounded to three places is .194.

3. **.469**   Divide 32 into 15, $32\overline{)15.00000}$ with .46875 above. The decimal .46875 rounded to three places is .469.

4. $\frac{11}{16}$ Place the digits 6875 over 10000 and reduce the fraction. $\frac{6875}{10000} = \frac{11}{16}$

5. $\frac{8}{33}$ Place the digits 24 over 99 and reduce the fraction. $\frac{24}{99} = \frac{8}{33}$

Percents

Fractions, decimals, and percents are all closely tied together. A fraction has an equivalent value as a decimal and as a percent. Fractional forms are useful at times; decimal forms and percents are useful at other times. It all depends on the application and the preferred format. You're exposed to percents in all sorts of printed and spoken language every day. There's the percent chance of rain, the percent of people voting in an election, the percent change in price, and so on. Percents are much easier to grasp than most fractions, because they're the comparison of a number and 100. Most of us have a fairly good sense of what 40 out of a hundred means, because we can picture the relationship between those two amounts. If you're comparing amounts, it's easier to compare $5\frac{1}{2}$% to $22\frac{1}{2}$% than it is to compare $\frac{11}{200}$ to $\frac{9}{40}$ (these are the fraction equivalents of those two percents).

There's a step between fractions and percents, and that's the decimal value. To change a fraction to a decimal, see the previous section. To change a decimal to a percent, move the decimal point two places to the right. To change a percent back into a decimal, move the decimal point two places to the left.

Decimals to Percents

.45	.45.	45%
.02	.02.	2%
.315	.31.5	31.5%

Percents to Decimals

16%	.16.	.16
4.5%	.04.5	.045
$13\frac{1}{2}$%	.13.5	.135

When percents appear in applications or mathematics problems, they need to be changed back into their decimal form so that they can be used in the various operations.

Example Problems

These problems show the answers and solutions.

1. You read that 14% of the students bring their own lunch to school. If there are 550 students in the school, how many bring their lunches?

   answer: 77 students

   14% of 550 = .14(550) = 77 students

2. You hear that, on average, it's cloudy 21 days during the month of January. What percent is that?

   answer: 67.7%

First, there are 31 days in January. $\frac{21}{31} \approx .677$ (dividing, it's .677 rounded to three places), which is 67.7%.

3. The local shoe store is offering all shoes at 15% off the ticketed price. How much is a $60 pair of shoes after the discount?

answer: $51

If the price is 15% **off**, then the remaining cost, after the 15% is subtracted, is 100%−15% = 85%. To get the sale price of the $60 shoes, find 85% of $60.   0.85(60) = $51

4. The cost of a television at the local store is $480 plus a sales tax of $7\frac{1}{4}$%. What is the total cost of the television?

answer: $514.80

The $7\frac{1}{4}$% tax can be written as 7.25%. Then that is changed to .0725 before multiplying. To get the total amount, the cost of the television plus the tax, multiply $480 times 1.0725. The 1.0725 = 1 + .0725 from which the 1 represents the original cost of the television, and the .0725 represents the additional tax. 480(1.0725) = 514.80. The total cost is $514.80, so, if the television is $480, subtract that from the $514.80 to get the tax of $34.80.

Work Problems

Use these problems to give yourself additional practice.

1. The population of Centerville, USA, is 63% second-generation Centervillians. If there are 4,300 people in that city, how many are second-generation residents?

2. Fourteen out of seventeen people in one club have paid their dues. Ten out of thirteen people in another club have paid their dues. Which club has the better dues paying percentage?

3. An $80 dress is on sale at 30% off. What does the dress cost after the discount?

4. What is the total price of a car at $24,000 plus $5\frac{1}{2}$% sales tax?

5. One company offers a salary of $100 per week plus 8% commission on total sales. Another company offers $50 per week plus 10% commission. Which would be better if your total sales per week average $2000?

Worked Solutions

1. **2,709 people**   Multiply 63(4300) = 2,709 are second-generation residents.

2. **The first club is better.** $\frac{14}{17} \approx .8235$ and $\frac{10}{13} \approx .7692$

   The first club has about 82% paying their dues, and the second club has about 77% paying their dues. It's much easier to compare these when using percentages.

3. **$56**   Multiply .70(80) = 56. The dress will cost $56.

4. **$25,320**   The car will cost $24,000 plus .055 times the $24,000.

$$(1.055)\,24,000 = 25,320$$

5. **The first company.**   The first company will pay $100 + .08(2000) = 100 + 160 = $260.

The second company will pay $50 + .10(2000) = 50 + 200 = $250.

It will pay better to work for the first company. There will be a point, however, at which it will be better to work for the second company—when your average total sales get to a higher point.

Scientific Notation

Scientific notation is so named because it is used extensively by the scientific community to more efficiently write very large and very small numbers. Some numbers, like distances to the stars, could take up more than one whole line on a page. Scientific notation takes care of that problem and also allows for easier computations involving very large and very small numbers.

The form for a number written in scientific notation is $a \times 10^p$ where a is some number between 1 and 10 (usually with numbers after the decimal point), and p is an integer. The value of p is determined when the decimal point in the original number is moved to the left or right to create the value a. The sign of p is positive for large numbers and negative for small numbers.

To change the number 147,000,000,000 into scientific notation, move the decimal place from the very right-hand end until it's between the 1 and the 4. The number 1.47 is between 1 and 10, so it's the a. The decimal was moved 11 places to the left, so the value of p is 11.

$$147,000,000,000 = 1.47 \times 10^{11}$$

Here are some other numbers written in scientific notation.

$$2,000,000,000,000,000 = 2 \times 10^{15}$$
$$13,000,000 = 1.3 \times 10^{7}$$
$$6,379,123,456 = 6.379123456 \times 10^{9}$$

That last number didn't really help as far as saving space in writing the number. Numbers that large are usually rounded off before writing them in scientific notation. For instance, rounding it to the nearer millionth, 6,379,123,456 becomes 6,379,000,000. Then, in scientific notation, $6,379,000,000 = 6.379 \times 10^{9}$. This is much more manageable.

Scientific notation makes it much easier to compare numbers. For instance, which is larger?

417,200,000,000,000,000,000,000 or 53,000,000,000,000,000,000,000,000 ?

In scientific notation, these two numbers are: 4.172×10^{23} and 5.3×10^{25}.

The second number is much larger, because of the exponent. The exponent on the second number is 2 greater than the first, which means it's 10^2 or 100 times as great.

Very small numbers can be written in scientific notation by using negative exponents. Because the decimal point has to be moved to the right instead of to the left, the negative sign is used to indicate this. For example,

$$.00000000032 = 3.2 \times 10^{-10}$$
$$.0000000413 = 4.13 \times 10^{-8}$$

The decimal point is moved until the value of *a* is between 1 and 10.

Multiplying and dividing numbers in scientific notation involves performing the operation on the two separate portions of the numbers and then simplifying the result.

Example Problems

These problems show the answers and solutions.

1. Multiply 5.25×10^8 and 2.6×10^4.

    answer: 1.365×10^{13}

    $$(5.25 \times 10^8)(2.6 \times 10^4) = (5.25 \times 2.6)(10^8 \times 10^4)$$
    $$= (13.65)(10^{12})$$

    The 13.65 isn't between 1 and 10, so it'll have to be rewritten in scientific notation and combined with the power of 10.

    $$(13.65)(10^{12}) = (1.365 \times 10^1)(10^{12})$$
    $$= 1.365 \times 10^{13}$$

2. Multiply $(1.23 \times 10^6)(6 \times 10^{-14})$.

    answer: 7.38×10^{-8}

    $$(1.23 \times 10^6)(6 \times 10^{-14}) = (1.23 \times 6)(10^6 \times 10^{-14})$$
    $$= (7.38)(10^{-8})$$
    $$= 7.38 \times 10^{-8}$$

3. Divide 4.392×10^{10} by 3.6×10^5.

    answer: 1.22×10^5

    $$\frac{4.392 \times 10^{10}}{3.6 \times 10^5} = \frac{4.392}{3.6} \times \frac{10^{10}}{10^5}$$
    $$= 1.22 \times 10^5$$

4. Divide 1.72×10^{-5} by 4.8×10^{-9}.

    answer: 3.58×10^3

    $$\frac{1.72 \times 10^{-5}}{4.8 \times 10^{-9}} = \frac{1.72}{4.8} \times \frac{10^{-5}}{10^{-9}}$$
    $$= .358333... \times 10^{-5-(-9)}$$

The decimal part continues forever, so round it to three decimal places.

$$.358333... \times 10^{-5-(-9)} \approx .358 \times 10^4$$

The first number isn't between 1 and 10, so change it to scientific notation and combine it with the power of 10 that follows.

$$.358 \times 10^4 = (3.58 \times 10^{-1}) \times 10^4$$
$$= 3.58 \times 10^3$$

Work Problems

Use these problems to give yourself additional practice.

1. Change to scientific notation: 830,000,000,000,000,000.

2. Change to scientific notation: .0000000000001723.

3. Multiply and write the answer in scientific notation. $(3.412 \times 10^{-18})(6.417 \times 10^{-3})$

4. Divide and write the answer in scientific notation. $(1.221 \times 10^3) \div (3.3 \times 10^{-4})$

5. Which is larger? 82,000,000,000,000,000,000,000 or 81,000,000,000,000,000,000,000,000

Worked Solutions

1. $830,000,000,000,000,000 = 8.3 \times 10^{17}$

2. $.0000000000001723 = 1.723 \times 10^{-13}$

3. $2.1894804 \times 10^{-20}$

$$(3.412 \times 10^{-18})(6.417 \times 10^{-3}) = (3.412 \times 6.417)(10^{-18} \times 10^{-3})$$
$$= (21.894804) \times (10^{-21})$$
$$= (2.1894804 \times 10^1) \times (10^{-21})$$
$$= 2.1894804 \times 10^{-20}$$

4. 3.7×10^6

$$(1.221 \times 10^3) \div (3.3 \times 10^{-4}) = \frac{1.221}{3.3} \times \frac{10^3}{10^{-4}}$$
$$= .37 \times 10^7$$
$$= 3.7 \times 10^{-1} \times 10^7$$
$$= 3.7 \times 10^6$$

5. **The second is larger.** $82,000,000,000,000,000,000,000 = 8.2 \times 10^{22}$
$81,000,000,000,000,000,000,000,000 = 8.1 \times 10^{25}$ So, the second number is much larger.

Chapter Problems and Solutions

Problems

1. Write $x \geq 2$ in interval notation.

2. Write $-2 < x \leq 1$ in interval notation.

3. Graph the values on the interval $[-3,5]$.

4. Graph the inequality $2 < x < 9$.

5. Add $-8 + 5$.

6. Add $-13 + (-11)$.

7. Subtract $4 - (-18)$.

8. Subtract $-6 - 3$.

9. Multiply $-8(-3)$.

10. Multiply $4(-2)$.

11. Divide $\dfrac{-16}{2}$.

12. Simplify $\dfrac{-6(-8)}{-4}$.

13. Distribute $-3(6a - 2b + 1)$.

14. Distribute $-y(y^3 - y + 3)$.

15. Simplify $\dfrac{8x^6 y^{-3}}{4x^{-4} y^3}$.

16. Simplify $\left(\dfrac{2x^2 y^{-3}}{3xy^{-4}}\right)^{-2}$.

17. Write in lowest terms $\dfrac{40a^3 b^2 - 6ab^3}{12ab^4 - 8a^4 b^2}$.

18. Subtract and write in lowest terms $\dfrac{5}{4y^3} - \dfrac{3}{16y}$.

19. Multiply and write in lowest terms $\dfrac{16x^2}{25y^3} \cdot \dfrac{35y^6}{20x^4}$.

20. Simplify $\dfrac{\dfrac{6mn^2}{5xy}}{\dfrac{8m^2 n^2}{25xy}}$.

21. Change $\dfrac{4}{5}$ to a decimal and a percent.

22. Change $.425$ to a fraction.

23. Change .3636... to a fraction.

24. Change 45% to a fraction.

25. Change $\frac{1}{20}$ to a percent.

26. Write 46,000,000,000,000 in scientific notation.

27. Change 3.524×10^{-18} back to a decimal number.

28. Simplify $(4.23 \times 10^9)(2.4 \times 10^8)$.

Answers and Solutions

1. **Answer: [2, ∞)**   The bracket with the 2 shows that the 2 is included in the answer.

2. **Answer: (−2, 1]**   The parenthesis with the −2 shows that −2 is not included in the answer.

3. **Answer:**   The solid dots indicate that both −3 and 5 are included.

4. **Answer:**   The hollow dots indicate that the 2 and 9 are not included.

5. **Answer: −3**   The signs are different, so you find the difference between the numbers.

6. **Answer: −24**   The signs are the same, so you find the sum.

7. **Answer: 22**   Change the subtraction to addition; change the second number to a positive; and then add.

8. **Answer: −9**   Change the subtraction to addition and change the second number to a negative.

9. **Answer: 24**   There are two negative signs, so the product is positive.

10. **Answer: −8**   There is one negative sign, so the answer is negative.

11. **Answer: −8**   There is one negative sign, so the answer is negative.

12. **Answer: −12**   There are three negative signs, so the answer is negative.

13. **Answer: −18*a* + 6*b* −3**   Each sign in the product changes from those in the parentheses.

14. **Answer: $-y^4 + y^2 - 3y$**   Each sign in the product changes from those in the parentheses.

15. **Answer: $\dfrac{2x^{10}}{y^6}$**   Move the factors with negative exponents; then simplify.

$$\frac{8x^6 y^{-3}}{4x^{-4} y^3} = \frac{8x^6 x^4}{4y^3 y^3} = \frac{2x^{10}}{y^6}$$

16. **Answer: $\dfrac{9}{4x^2 y^2}$**

$$\left(\frac{2x^2 y^{-3}}{3xy^{-4}}\right)^{-2} = \left(\frac{2x^2 y^4}{3xy^3}\right)^{-2} = \left(\frac{2x^2 y^4}{3xy^3}\right)^{-2} = \left(\frac{2xy}{3}\right)^{-2} = \left(\frac{3}{2xy}\right)^2 = \frac{9}{4x^2 y^2}$$

17. **Answer:** $\dfrac{20a^2 - 3b}{6b^2 - 4a^3}$ Divide each term by the greatest common factor, $2ab^2$.

$$\frac{40a^3 b^2 - 6ab^3}{12ab^4 - 8a^4 b^2} = \frac{40^{20} a^{\cancel{3}2} b^{\cancel{2}} - \cancel{6}^3 \cancel{a}b^{\cancel{3}1}}{\cancel{12}^6 \cancel{a}b^{\cancel{4}2} - \cancel{8}^4 a^{\cancel{4}3} b^{\cancel{2}}} = \frac{20a^2 - 3b}{6b^2 - 4a^3}$$

18. **Answer:** $\dfrac{20 - 3y^2}{16y^3}$ Find a common denominator before subtracting.

$$\frac{5}{4y^3} - \frac{3}{16y} = \frac{5}{4y^3} \cdot \frac{4}{4} - \frac{3}{16y} \cdot \frac{y^2}{y^2} = \frac{20}{16y^3} - \frac{3y^2}{16y^3} = \frac{20 - 3y^2}{16y^3}$$

19. **Answer:** $\dfrac{28y^3}{25x^2}$ Reduce first and then multiply. $\dfrac{\cancel{16}^4 x^{\cancel{2}}}{\cancel{25}^5 y^{\cancel{3}}} \cdot \dfrac{\cancel{35}^7 y^{\cancel{6}3}}{\cancel{20}^5 x^{\cancel{4}2}} = \dfrac{28y^3}{25x^2}$

20. **Answer:** $\dfrac{15}{4m}$ Multiply the numerator by the reciprocal of the denominator.

$$\frac{\frac{6mn^2}{5xy}}{\frac{8m^2 n^2}{25xy}} = \frac{\cancel{6}^3 m\cancel{n}^{\cancel{2}}}{\cancel{5}xy} \cdot \frac{\cancel{25}^5 \cancel{xy}}{\cancel{8}^4 m^{\cancel{2}1} \cancel{n}^{\cancel{2}}} = \frac{15}{4m}$$

21. **Answer: .8 and 80%**   Divide 5 into 4.

22. **Answer:** $\dfrac{17}{40}$ Put 425 over 1000 and reduce the fraction, $\dfrac{425}{1000} = \dfrac{17}{40}$.

23. **Answer:** $\dfrac{4}{11}$ Put 36 over 99 and reduce the fraction, $\dfrac{36}{99} = \dfrac{4}{11}$.

24. **Answer:** $\dfrac{9}{20}$ First, 45% is equivalent to the decimal .45. Then put 45 over 100 and reduce the fraction.

25. **Answer: 5%**   First, divide the 1 by 20 to get the decimal .05. Then move the decimal point two places to the right to get the percent.

26. **Answer:** 4.6×10^{13}

27. **Answer: .000000000000000003524**

28. **Answer:** 1.0152×10^{18} First, rearrange the factors for multiplication, $(4.23 \cdot 2.4)(10^9 \times 10^8) = 10.152 \times 10^{17}$. This isn't in scientific notation, because the first number isn't between 0 and 10. Move the decimal place and recombine. $10.152 \times 10^{17} = 1.0152 \times 10^1 \times 10^{17} = 1.0152 \times 10^{18}$.

Supplemental Chapter Problems

Problems

1. Write the interval $[-1,1)$ in inequality notation.

2. Graph the inequality $x < 8$.

3. Write the inequality $-12 \le x < 1$ in interval notation.

4. Add $(-4) + 9$.

5. Add $6 + (-15)$.

6. Subtract $-2 - (-8)$.

7. Subtract $-5 - 10$.

8. Multiply $-5(6)$.

9. Multiply $-11(-1)$.

10. Divide $\frac{-24}{-3}$.

11. Simplify $\frac{5(-12)}{-6}$.

12. Distribute $-(4x - 3y + 8)$.

13. Distribute $-a(2a^2 - 3a + 1)$.

14. Simplify $\frac{36m^{-3}n^{-2}p^2}{12m^{-4}n^2p^{-2}}$.

15. Simplify $\left(\frac{18y^{-2}z}{12y^{-2}z^{-3}}\right)^{-1}$.

16. Add and write in lowest terms $\frac{1}{4x^2} + \frac{5}{6x}$.

17. Multiply and reduce $\frac{a^2(b+c)^3}{9b^2(a-c)^3} \cdot \frac{15b(a-c)^4}{2a^2(b+c)^2}$.

18. Divide $\frac{48xyz}{35a^2b^2c^2} \div \frac{16x^2y^3z^4}{25ab^3c^2}$.

19. Simplify $\frac{\frac{9}{14a}}{\frac{18}{35a^4}}$.

20. Change $\frac{13}{25}$ to a decimal and a percent.

21. Change $.625$ to a fraction.

22. Change $.405405...$ to a fraction.

23. Change 6% to a fraction.

24. $\frac{3}{20}$ to a percent.

25. Write in scientific notation $.0000000001635$.

26. Change back to a decimal form 9.2×10^{11}.

27. Simplify $(5.2 \times 10^{-6}) \div (4 \times 10^{-10})$.

Answers

1. $-1 \leq x < 1$ (Number Lines, p. 73)

2.   (Number Lines, p. 73)

3. $[-12,1)$ (Number Lines, p. 73)

4. 5   (Addition of Signed Numbers, p. 75)

5. -9 (Addition of Signed Numbers, p. 75)

6. 6   (Subtraction of Signed Numbers, p. 78)

7. -15 (Subtraction of Signed Numbers, p. 78)

8. -30 (Multiplication and Division of Signed Numbers, p. 80)

9. 11   (Multiplication and Division of Signed Numbers, p. 80)

10. 8   (Multiplication and Division of Signed Numbers, p. 80)

11. 10   (Multiplication and Division of Signed Numbers, p. 80)

12. $-4x + 3y - 8$ (Multiplication and Division of Signed Numbers, p. 80)

13. $-2a^3 + 3a^2 - a$ (Multiplication and Division of Signed Numbers, p. 80)

14. $\dfrac{3mp^4}{n^4}$ (Exponents That Are Signed Numbers, p. 82)

15. $\dfrac{2}{3z^4}$ (Exponents That Are Signed Numbers, p. 82)

16. $\dfrac{3 + 10x}{12x^2}$ (Lowest Terms and Equivalent Fractions, p. 86)

17. $\dfrac{5(b+c)(a-c)}{6b}$ (Multiplying and Dividing Fractions, p. 89)

18. $\dfrac{15b}{7axy^2 z^3}$ (Multiplying and Dividing Fractions, p. 89)

19. $\dfrac{5a^3}{4}$ (Mixed Numbers and Complex Fractions, p. 91)

20. .52 and 52%   (Decimals, p. 94)

21. $\dfrac{5}{8}$ (Decimals, p. 94)

22. $\dfrac{15}{37}$ (Decimals, p. 94)

23. $\dfrac{3}{50}$ (Percents, p. 97)

24. 15%   (Percents, p. 97)

25. 1.635×10^{-10} (Scientific Notation, p. 99)

26. 920,000,000,000   (Scientific Notation, p. 99)

27. 1.3×10^4 (Scientific Notation, p. 99)

Chapter 3

Linear Equations and Algebraic Fractions

Much of algebra involves learning how to manipulate symbols and operations—changing format and rewriting expressions for more convenience. The material in this chapter is aimed at applying the rules of algebra to situations where a solution or an answer to a practical problem is needed. These linear equations are the first of many types of equations covered in this book. Many times, more complicated equations can be changed to this simple linear form for solving.

Solving Linear Equations

Solving an equation means to determine what the variable represents so that the statement will be true. To solve a linear equation, the answer is determined when the variable is by itself on one side of the equal sign and the numerical answer is on the other side. This is what you're always working toward as you perform operations to each side of the equation. A linear equation is one in which the variable is to the first power and there's only one solution. A property of equations that allows you to solve them is that if you add the same thing to each side, subtract the same thing from each side, multiply each side by the same number (except 0), or divide each side by the same number (again, except 0), then you don't change the "truth" or the solution of the equation. It remains balanced, and this altered equation has the same answer as the original. Here are the properties written symbolically:

If $a = b$, then $a + c = b + c$.

If $a = b$, then $a - c = b - c$.

If $a = b$, then $a \cdot c = b \cdot c$

If $a = b$, then $a \div c = b \div c$

Here's how adding the same thing to each side can help solve the linear equation $4 = x - 1$.

$$4 = x - 1$$

Add 1 to each side: $\underline{+1 \qquad +1}$

$$5 = x$$

The solution is that the variable x is equal to 5.

Here's how subtracting the same thing from each side can help solve the linear equation $8 = y + 3$.

$$8 = y + 3$$

Subtract 3 from each side: $\underline{-3 \qquad -3}$

$$5 = y$$

The solution is that the variable y is equal to 5.

Here's how multiplying the same thing by each side can help solve the linear equation $\frac{z}{2} = 7$.

Multiply each side by 2: $2 \cdot \frac{z}{2} = 7 \cdot 2$.

$$\cancel{2} \cdot \frac{z}{\cancel{2}} = 7 \cdot 2$$
$$z = 14$$

Here's how dividing each side by the same thing can help to solve the linear equation $3w = 6$.

Divide each side by 3: $\dfrac{\cancel{3}w}{\cancel{3}} = \dfrac{6}{3}$
$$w = 2$$

In every case, an inverse was used to get the variable by itself on one side of the equation. When something was adding or subtracting the variable, the opposite was done. Additive inverses are the opposite of one another, and their sum is 0. When something was multiplying or dividing the variable, the opposite was done. Multiplicative inverses have a product of 1 when they're multiplied together. For more information on inverses, see "Properties of Algebraic Expressions," in Chapter 1.

Example Problems

These problems show the answers and solutions.

1. Solve for x in $x - 13 = 11$.

    answer: 24

    The goal is to get the x by itself on one side. Because 13 is being subtracted from x, do the opposite and add 13 to each side of the equation.

    $$\begin{array}{r} x - 13 = 11 \\ \underline{+13 + 13} \\ x = 24 \end{array}$$

2. Solve for z in $-9z = 45$.

    answer: -5

    The goal is to get z by itself on one side. Because -9 is multiplying z, do the opposite and divide by -9.

    $$\frac{-9z}{-9} = \frac{45}{-9}$$
    $$\frac{-\cancel{9}z}{-\cancel{9}} = \frac{45}{-9}$$
    $$z = -5$$

3. Solve for y in $\frac{y}{6} = 3$.

    answer: 18

Get y by itself by multiplying each side by 6. That's the opposite of the dividing by 6 that's being done right now.

$$6 \cdot \frac{y}{6} = 3 \cdot 6$$

$$\cancel{6} \cdot \frac{y}{\cancel{6}} = 3 \cdot 6$$

$$y = 18$$

4. Solve for x in $9 + x = -11$.

    answer: -20

    Get x by itself by adding -9 to each side. That's the opposite of the 9 being added.

$$\begin{array}{r} 9 + x = -11 \\ \underline{-9 \qquad -9} \\ x = -20 \end{array}$$

5. Solve for y in $2y - 4 = y$.

    answer: 4

    In this equation, there are two terms with a y in them. Subtract $2y$ from each side to get the y terms to the right.

$$\begin{array}{r} 2y - 4 = y \\ \underline{-2y \qquad -2y} \\ -4 = -y \end{array}$$

This equation has the y term by itself, but it's negative y, not y. Think of the negative sign as being -1. Then, by dividing each side by -1, the value of y can be determined.

$$\frac{-4}{-1} = \frac{-y}{-1}$$

$$\frac{-4}{-1} = \frac{-\cancel{1}y}{-\cancel{1}}$$

$$4 = y$$

6. Solve for x in $5x - 3x = 20$.

    answer: 10

    In this case, it will require simplifying the terms on the left before each side can be divided by the same thing to solve for x.

$$5x - 3x = 20$$

$$2x = 20$$

$$\frac{\cancel{2}x}{\cancel{2}} = \frac{20}{2}$$

$$x = 10$$

Work Problems

Use these problems to give yourself additional practice.

In each of the following problems, solve for the value of the variable.

1. $y + 4 = 13$

2. $4x = 40$

3. $x - 5 = -6$

4. $\dfrac{z}{-8} = 2$

5. $4x + 5x = -36$

Worked Solutions

1. **9**   Add −4 to each side (or subtract 4 from each side).

$$
\begin{array}{r}
y + 4 = 13 \\
\underline{-4 \quad -4} \\
y \quad\quad = 9
\end{array}
$$

2. **10**   Divide each side by 4.

$$
\frac{\cancel{4}x}{\cancel{4}} = \frac{40}{4}
$$
$$
x = 10
$$

3. **−1**   Add 5 to each side.

$$
\begin{array}{r}
x - 5 = -6 \\
\underline{\quad 5 \quad\quad 5} \\
x \quad\quad = -1
\end{array}
$$

4. **−16**   Multiply each side by −8.

$$
-\cancel{8} \cdot \frac{z}{-\cancel{8}} = 2 \cdot (-8)
$$
$$
z = -16
$$

5. **−4**   First combine the terms on the left. Then divide each side by the coefficient of x.

$$
9x = -36
$$
$$
\frac{\cancel{9}x}{\cancel{9}} = \frac{-36}{9}
$$
$$
x = -4
$$

Solving Linear Equations with More Than One Operation

When more than one operation is being performed on the variable in a linear equation, there will have to be an opposite operation for each of these to solve for that variable. To solve it most efficiently and correctly, there's an important order in which to perform these operations.

First, simplify each side of the equation independently, distributing and combining like terms.

Second, get all terms with variables in them on one side of the equation, and the terms without variables in them on the other side, by adding or subtracting terms.

Third, multiply or divide to solve for the variable.

Example Problems

These problems show the answers and solutions.

1. Solve for x in $13x - 4 = 5x + 12$.

   answer: 2

   Move the variables to the left and the numbers to the right by adding opposites or subtracting. Subtract $5x$ from each side and add 4 to each side.

   $$
   \begin{array}{rcr}
   13x - 4 &=& 5x + 12 \\
   -5x & & -5x \\
   \hline
   8x - 4 &=& 12 \\
   +4 & & +4 \\
   \hline
   8x &=& 16
   \end{array}
   $$

   Now divide each side by 8.

   $$\frac{\cancel{8}x}{\cancel{8}} = \frac{16}{8}$$

   $$x = 2$$

2. Solve for y in $5y + 2(y - 1) = 3y + 10$.

   answer: 3

   First distribute the 2 on the left side and combine like terms.

   $$
   \begin{aligned}
   5y + 2(y - 1) &= 3y + 10 \\
   5y + 2y - 2 &= 3y + 10 \\
   7y - 2 &= 3y + 10
   \end{aligned}
   $$

   Next, subtract $3y$ from each side and add 2 to each side.

$$7y - 2 = 3y + 10$$
$$\underline{-3y \quad\quad -3y}$$
$$4y - 2 = \quad\quad 10$$
$$\underline{\quad\quad +2 \quad\quad +2}$$
$$4y \quad = \quad\quad 12$$

Now, divide each side by 4.

$$\frac{\cancel{4}y}{\cancel{4}} = \frac{12}{4}$$
$$y = 3$$

3. Solve for z in $3z + 2z - 1 = 8z - 19$.

   answer: 6

   First, combine the like terms on the left. Then subtract $8z$ from each side.

   $$3z + 2z - 1 = 8z - 19$$
   $$5z - 1 = 8z - 19$$
   $$\underline{-8z \quad\quad -8z}$$
   $$-3z - 1 = \quad\quad -19$$

   By subtracting the $8z$ from each side, the variable term is on the left, as is frequently preferred. Some would rather not have the negative coefficient, though, and would have subtracted $5z$ from each side, instead. Either way works fine—it's just personal preference.

   Now add 1 to each side and then divide by the coefficient of z.

   $$-3z - 1 = -19$$
   $$\underline{\quad +1 \quad +1}$$
   $$-3z \quad = -18$$
   $$\frac{\cancel{-3}z}{\cancel{-3}} = \frac{-18}{-3}$$
   $$z = 6$$

4. Solve for w in $\frac{13w - 6}{4} = 5w + 2$.

   answer: −2

   The fraction can be dealt with in several ways, but the preferred way is to just get rid of it by multiplying each side by 4.

   $$4 \cdot \frac{13w - 6}{4} = (5w + 2) \cdot 4$$
   $$\cancel{4} \cdot \frac{13w - 6}{\cancel{4}} = 20w + 8$$
   $$13w - 6 = 20w + 8$$

Now subtract 20w from each side and add 6 to each side.

$$
\begin{aligned}
13w - 6 &= 20w + 8 \\
-20w &\quad -20w \\
\hline
-7w - 6 &= +8 \\
+6 &\quad +6 \\
\hline
-7w &= +14
\end{aligned}
$$

Lastly, divide each side by −7.

$$
\frac{-7w}{-7} = \frac{14}{-7}
$$
$$
w = -2
$$

The "Decimals," section of Chapter 2 on page 94, provides some examples of how to change from repeating decimals back to fractions. When all of the digits repeat, the procedure is simple. If some digits at the beginning of the decimal don't repeat, then it's a little more complicated. It takes solving linear equations to be able to do these decimal to fraction changes, so the technique is discussed here.

5. Change the repeating decimal .19444... into its fraction.

answer: $\frac{7}{36}$

To do this, first write an equation using the decimal and setting it equal to N.

$$
N = .19444...
$$

Next, multiply each side of the equation by whatever power of 10 will move the decimal point so that the nonrepeating parts of the decimal will be on the left side and only the repeating digits will be on the right. Because in this problem two digits don't repeat, multiplying by 100 will move the decimal point where you want it.

$$
N = .19444...
$$
$$
100N = 19.444...
$$

Now, each side of that new equation is multiplied by 10 to create a second equation with only the repeating digits on the right.

$$
1000N = 194.444...
$$

Place the two equations on top of one another—with the larger coefficient of N on the top—and subtract.

$$
\begin{aligned}
1000N &= 194.444... \\
-100N &= -19.444... \\
\hline
900N &= 175
\end{aligned}
$$

All of the repeating 4s subtract one another and disappear.

Now divide each side of the equation and reduce the fraction.

$$\frac{\cancel{900}N}{\cancel{900}} = \frac{175}{900}$$

$$N = \frac{175}{900}$$

$$N = \frac{7}{36}$$

Work Problems

Use these problems to give yourself additional practice.

1. Solve for m. $8m - 3 = 5m - 6$

2. Solve for t. $3(t - 4) + 2t = 4t - 7$

3. Solve for x. $5 - x = 7(x + 4) + 1$

4. Solve for z. $\frac{3z + 2}{5} = 10 - z$

5. Find the fractional value for the repeating decimal 0.1777....

Worked Solutions

1. **−1**   Subtract $5m$ from each side and add 3 to each side.

$$
\begin{array}{rcl}
8m - 3 &=& 5m - 6 \\
-5m && -5m \\
\hline
3m - 3 &=& -6 \\
+3 && +3 \\
\hline
3m &=& -3
\end{array}
$$

Now divide each side by 3.

$$\frac{\cancel{3}m}{\cancel{3}} = \frac{-3}{3}$$

$$m = -1$$

2. **5**   First, distribute the 3 and combine like terms on the left.

$$3(t - 4) + 2t = 4t - 7$$
$$3t - 12 + 2t = 4t - 7$$
$$5t - 12 = 4t - 7$$

Subtract $4t$ from each side and add 12 to each side.

$$5t - 12 = 4t - 7$$
$$\underline{-4t \qquad -4t}$$
$$t - 12 = \quad -7$$
$$\underline{\qquad +12 \quad +12}$$
$$t \quad = \quad 5$$

3. **–3** Distribute the 7 and combine like terms on the right.

$$5 - x = 7(x + 4) + 1$$
$$5 - x = 7x + 28 + 1$$
$$5 - x = 7x + 29$$

Add x to each side and subtract 29 from each side.

$$5 - x = 7x + 29$$
$$\underline{+x \quad +x}$$
$$5 \quad = 8x + 29$$
$$\underline{-29 \qquad -29}$$
$$-24 = 8x$$

Now, divide each side by 8.

$$\frac{-24}{8} = \frac{\cancel{8}x}{\cancel{8}}$$
$$-3 \; = x$$

4. **6** First, multiply each side by 5 to get rid of the fraction.

$$\cancel{5} \cdot \frac{3z + 2}{\cancel{5}} = (10 - z) \cdot 5$$
$$3z + 2 = 50 - 5z$$

Now, add $5z$ to each side and subtract 2 from each side.

$$3z + 2 = 50 - 5z$$
$$\underline{5z \qquad\qquad 5z}$$
$$8z + 2 = 50$$
$$\underline{-2 \qquad -2}$$
$$8z \quad = 48$$

Now, divide each side by 8.

$$\frac{\cancel{8}z}{\cancel{8}} = \frac{48}{8}$$
$$z = 6$$

5. $\frac{8}{45}$ Let $N = .1777\ldots$

Multiply each side by 10 to get the repeating part alone on the right.

$$10N = 1.777\ldots$$

Multiply each side by 10, again, to get another equation.

$$100N = 17.777\ldots$$

Subtract the second equation from the third.

$$100N = 17.777\ldots$$
$$-10N = -1.777\ldots$$
$$90N = 16$$

Divide each side by 90.

$$\frac{90N}{90} = \frac{16}{90}$$

$$N = \frac{16}{90} = \frac{8}{45}$$

Solving Linear Formulas

Formulas are the backbone of science and business and any other mathematical application. In geometry, there's $P = 2l + 2w$, $A = \frac{1}{2}h(b_1 + b_2)$, $C = 2\pi r$; in finance there's $P = R - C$; and when you're traveling, there's $d = rt$ or $°C = \frac{5}{9}(°F - 32)$.

Frequently, it's more efficient to solve a formula for a particular variable—other than the one it's already solved for—to save on repeating the same operations. For instance, if you want to use 400 feet of fencing to build a rectangular yard and want to figure out what the dimensions should be, you start with the formula for the perimeter of a rectangle, $P = 2l + 2w$. The P is the perimeter, or the distance around the outside. You know that will be the 400 feet. You want to be able to choose a length and then determine what the width will have to be. To do this, take the original perimeter formula and solve for w.

Get the variable w by itself on the right side by subtracting $2l$ from each side.

$$P = 2l + 2w$$
$$-2l - 2l$$
$$P - 2l = 2w$$

The P and the ls don't combine—they're two different variables, so the left side can't be simplified. Now divide each side by 2.

$$\frac{P - 2l}{2} = \frac{2w}{2}$$

$$\frac{P - 2l}{2} = w$$

Substitute 400 for the P and try some fencing dimensions.

$$w = \frac{400 - 2l}{2}$$

If you want a length of 40 feet, what will the width be?

$$w = \frac{400 - 2(40)}{2} = \frac{400 - 80}{2} = \frac{320}{2} = 160$$

That would be a long, narrow yard. Now try a length of 100 feet.

$$w = \frac{400 - 2(100)}{2} = \frac{400 - 200}{2} = \frac{200}{2} = 100$$

That would be a square yard—the length and width are the same.

Example Problems

These problems show the answers and solutions.

1. Given the formula for the area of a rectangle, $A = lw$, solve for l.

   answer: $l = \frac{A}{w}$

   Divide each side by w.

   $$\frac{A}{w} = \frac{l\cancel{w}}{\cancel{w}}$$

   $$l = \frac{A}{w}$$

2. Given the formula for the area of a trapezoid, $A = \frac{1}{2}h(b_1 + b_2)$, solve for b_1.

   answer: $b_1 = \frac{2A - b_2 h}{h}$

   A trapezoid is a four-sided figure with the two bases (b_1 *and* b_2) parallel to one another. The distance between those two bases is the height, h.

   To solve for either of the bases, in the parenthesis, first multiply each side by 2 to get rid of the fraction, and then divide each side by h, leaving the parentheses alone on the right.

   $$2 \cdot A = \cancel{2} \cdot \frac{1}{\cancel{2}} h(b_1 + b_2)$$

   $$2A = h(b_1 + b_2)$$

   $$\frac{2A}{h} = \frac{\cancel{h}(b_1 + b_2)}{\cancel{h}}$$

   $$\frac{2A}{h} = b_1 + b_2$$

   Now subtract b_2 from each side.

$$\frac{2A}{h} - b_2 = b_1 + b_2 - b_2$$

$$b_1 = \frac{2A}{h} - b_2$$

You can write the right side as one fraction if the common denominator, h, is used in the second term.

$$b_1 = \frac{2A}{h} - \frac{b_2 h}{h} = \frac{2A - b_2 h}{h}$$

3. Solve for h in the formula for the surface area of a cylinder (think of the sides, top, and bottom of a soup can), $A = 2\pi r(r + h)$.

    answer: $h = \dfrac{A - 2\pi r^2}{2\pi r}$

    This could be done in a similar fashion to the previous example, dividing by the $2\pi r$ and then subtracting the r from each side. Instead, here's another approach—distributing the $2\pi r$ first.

$$A = 2\pi r(r + h)$$
$$A = 2\pi r(r) + 2\pi r(h)$$
$$A = 2\pi r^2 + 2\pi rh$$

Now, the term on the right without the h in it is subtracted from each side.

$$A = 2\pi r^2 + 2\pi rh$$
$$\underline{-2\pi r^2 \quad -2\pi r^2}$$
$$A - 2\pi r^2 = 2\pi rh$$

Lastly, divide each side by $2\pi r$ to solve for h.

$$\frac{A - 2\pi r^2}{2\pi r} = \frac{2\pi rh}{2\pi r}$$

$$\frac{A - 2\pi r^2}{2\pi r} = h$$

Work Problems

Use these problems to give yourself additional practice.

1. Given the formula for profit equaling Revenue minus Cost, solve for Cost, C. $P = R - C$

2. Given the formula for the area of a trapezoid, solve for the height, h. $A = \frac{1}{2}h(b_1 + b_2)$

3. Given the formula for the distance traveled based on the time and rate of speed, solve for the time traveled. $d = rt$

4. Given the formula for determining degrees Celsius when given degrees Fahrenheit, solve for the degrees Fahrenheit, F. $°C = \frac{5}{9}(°F - 32)$

5. Given the formula for the total amount in an account, the Principal that was initially invested, and the interest added in, solve for the amount of time that the Principal was invested. $A = P + Prt$

Worked Solutions

1. **$C = R - P$** To solve for C, start with $P = R - C$ and subtract R from each side. You'll have a negative C, but that can be remedied by multiplying each side by -1.

$$P = R - C$$
$$\underline{-R \qquad -R}$$
$$P - R = -C$$
$$-1(P - R) = -1(-C)$$
$$-P + R = C \text{ or } C = R - P$$

2. **$h = \frac{2A}{(b_1 + b_2)}$** Multiply each side by 2 to get rid of the fraction and then divide each side by the values in the parenthesis.

$$2 \cdot A = 2 \cdot \frac{1}{2} h(b_1 + b_2)$$
$$2A = h(b_1 + b_2)$$
$$\frac{2A}{(b_1 + b_2)} = \frac{h(b_1 + b_2)}{(b_1 + b_2)}$$
$$\frac{2A}{(b_1 + b_2)} = h$$

3. **$t = \frac{d}{r}$** Start with $d = rt$. Divide each side by the rate, r.

$$\frac{d}{r} = \frac{rt}{r}$$
$$\frac{d}{r} = t$$

4. **$°F = \frac{9}{5} \cdot °C + 32$** Start with $°C = \frac{5}{9}(°F - 32)$.

First multiply each side by the multiplicative inverse of $\frac{5}{9}$. Then add 32 to each side.

$$\frac{9}{5} \cdot °C = \frac{9}{5} \cdot \frac{5}{9}(°F - 32)$$
$$\frac{9}{5} \cdot °C = °F - 32$$
$$\underline{+32 \qquad +32}$$
$$\frac{9}{5} \cdot °C + 32 = °F$$

5. $t = \dfrac{A - P}{Pr}$ Start with $A = P + Prt$.

First, subtract P from each side. Then, divide each side by the coefficient of t, Pr.

$$A = P + Prt$$

$$\underline{-P \quad -P}$$

$$A - P = Prt$$

$$\frac{A - P}{Pr} = \frac{\cancel{P}rt}{\cancel{P}\cancel{r}}$$

$$\frac{A - P}{Pr} = t$$

Ratios and Proportions

A ratio is an expression or a way in which you can compare numbers to one another. If you're talking about "the ratio of 3 to 4," then you're comparing 3 and 4. A convenient way of doing this is to write it as a fraction, $\frac{3}{4}$. Then you can get more information about how 3 compares to 4 by changing it to a decimal and a percent. The fraction $\frac{3}{4}$ is equal to .75 as a decimal and 75% as a percent. Three is 75% of 4. If you reverse the order and want to talk about the ratio of 4 to 3, the fraction becomes $\frac{4}{3}$, which is 1.333... (round that to 1.33) and is equivalent to 133%. So 4 is 133% of 3. For more on fractions, decimals, and percents, see "Decimals," in Chapter 2. Another popular way to write the ratio 3 to 4 is to use a colon, 3:4. In this section, however, ratios will be written as fractions, because they're the easiest to deal with in algebra.

When two ratios are written in an equation as being equal to one another, you have a *proportion*. Proportions have several handy properties that make them popular for use in solving algebraic applications and their equations.

Some properties of the proportion $\frac{a}{b} = \frac{c}{d}$:

$\frac{a}{b} = \frac{c}{d}$ is equivalent to $a \cdot d = b \cdot c$. This means that their cross products are equal.

$\frac{a}{b} = \frac{c}{d}$ is equivalent to $\frac{b}{a} = \frac{d}{c}$. This means that you can flip a proportion for more ease in solving a particular problem.

$\frac{a}{b} = \frac{c}{d}$ can be reduced in four ways. Look for common factors in a and b, c and d, a and c, or b and d. You can reduce fractions vertically or horizontally.

Consider the proportion $\frac{12}{15} = \frac{20}{25}$.

The cross products are equal: $12 \cdot 25 = 15 \cdot 20$, $300 = 300$.

The flipped proportion is equivalent: $\frac{15}{12} = \frac{25}{20}$, $15 \cdot 20 = 12 \cdot 25$, $300 = 300$.

Reduce fractions vertically: $\frac{12}{15} = \frac{20}{25}$, $\frac{\overset{4}{\cancel{12}}}{\underset{5}{\cancel{15}}} = \frac{20}{25}$, $4 \cdot 25 = 5 \cdot 20$, $100 = 100$

$\frac{12}{15} = \frac{20}{25}$, $\frac{12}{15} = \frac{\overset{4}{\cancel{20}}}{\underset{5}{\cancel{25}}}$ $12 \cdot 5 = 15 \cdot 4$, $60 = 60$

Reduce fractions horizontally: $\frac{12}{15} = \frac{20}{25}$, $\frac{\cancel{12}^3}{15} = \frac{\cancel{20}^5}{25}$, $3 \cdot 25 = 15 \cdot 5$, $75 = 75$

$\frac{12}{15} = \frac{20}{25}$, $\frac{12}{\cancel{15}_3} = \frac{20}{\cancel{25}_5}$ $12 \cdot 5 = 3 \cdot 20$, $60 = 60$

The following examples show how using the properties of proportions, shown previously, helps in solving problems.

Example Problems

These problems show the answers and solutions.

1. Solve for x in the proportion, $\frac{x}{35} = \frac{16}{28}$.

    answer: 20

    Reduce the proportion and then cross multiply and solve the equation for x.

    $$\frac{x}{35} = \frac{\cancel{16}^4}{\cancel{28}_7}$$

    $$7x = 35 \cdot 4 = 140$$

    $$7x = 140$$

    $$\frac{7x}{7} = \frac{140}{7}$$

    $$x = 20$$

2. Solve for z in the proportion $\frac{27}{z} = \frac{54}{26}$.

    answer: 13

    $$\frac{27}{z} = \frac{54}{26}$$

    $$\frac{\cancel{27}^1}{z} = \frac{\cancel{54}^2}{26}$$

    By flipping this new proportion, the problem suddenly becomes much easier.

    $$\frac{z}{1} = \frac{26}{2}$$

    $$z = \frac{26}{2}$$

    $$z = 13$$

3. Two out of three of the people who signed up for the course are biology majors. If 285 people signed up for the course, how many are biology majors?

    answer: 190 people

Write the ratio 2 out of 3 as $\frac{2}{3}$ where $\frac{2}{3}$: $\dfrac{\text{biology majors}}{\text{people signed up for the course}}$, and write a

corresponding ratio $\frac{x}{285}$: $\dfrac{\text{number of biology majors}}{\text{people signed up for the course}}$.

Now set them equal and solve the proportion for x.

$$\frac{2}{3} = \frac{x}{285}$$

$$\frac{2}{\cancel{3}_1} = \frac{x}{\cancel{285}_{95}}$$

$$2 \cdot 95 = 1 \cdot x$$

$$x = 190$$

4. A television commercial claims that 9 out of every 10 dentists surveyed recommended Bliss toothpaste. If 450 of the dentists surveyed recommended Bliss, then how many did **not** recommend it?

   answer: 50 dentists

   The ratio of those who recommended Bliss to dentists surveyed is $\frac{9}{10}$. This could be set equal to $\frac{450}{x}$, because the numbers recommending are on the top, and the total number is on the bottom. So x represents the *total number* surveyed, not the number who didn't recommend it. Solve for that total number and determine how many didn't recommend it after that.

$$\frac{\cancel{9}^1}{10} = \frac{\cancel{450}^{50}}{x}$$

$$1 \cdot x = 10 \cdot 50$$

$$x = 500$$

   So, if 450 of the total recommended Bliss, then $500 - 450 = 50$ who did **not** recommend it.

5. Solve for y in $\frac{30}{40} = \frac{27}{y+9}$.

   answer: 27

   First, reduce either vertically or horizontally. (Actually, both work, and you can do them in either order.)

$$\frac{\cancel{30}^3}{\cancel{40}_4} = \frac{27}{y+9}$$

$$\frac{\cancel{3}^1}{4} = \frac{\cancel{27}^9}{y+9}$$

$$\frac{1}{4} = \frac{9}{y+9}$$

Now cross multiply and solve the equation.

$$1(y+9) = 4 \cdot 9$$
$$y + 9 = 36$$
$$\underline{-9 \quad -9}$$
$$y \quad = 27$$

Work Problems

Use these problems to give yourself additional practice.

1. Solve the proportion for x. $\frac{18}{x} = \frac{63}{35}$

2. Solve the proportion for y. $\frac{y}{30} = \frac{13}{100}$

3. Solve the proportion for n. $\frac{n+3}{20} = \frac{n-3}{4}$

4. A cookie recipe calls for 3 eggs and $4\frac{1}{2}$ cups of flour. You're in charge of the big bake sale, and you'll make cookies using the 360 cups of flour you have on hand. How many eggs will you need?

5. In a local school, the records show that 5 out of every 8 students have a computer at home. If there are 1856 students in the school, how many **don't** have a computer at home?

Worked Solutions

1. **10**   Start with $\frac{18}{x} = \frac{63}{35}$. Reduce both horizontally and vertically before cross multiplying and solving for x.

$$\frac{^2\cancel{18}}{x} = \frac{\cancel{63}^7}{35}$$

$$\frac{2}{x} = \frac{\cancel{7}^1}{\cancel{35}_5}$$

$$\frac{2}{x} = \frac{1}{5}$$

$$2 \cdot 5 = x \cdot 1$$

$$x = 10$$

2. **3.9**   Start with the proportion $\frac{y}{30} = \frac{13}{100}$. First, reduce horizontally before cross multiplying and solving for y.

$$\frac{y}{\cancel{30}_3} = \frac{13}{\cancel{100}_{10}}$$

$$\frac{y}{3} = \frac{13}{10}$$

$$y \cdot 10 = 3 \cdot 13$$

$$10y = 39$$

$$\frac{10y}{10} = \frac{39}{10}$$

$$y = \frac{39}{10} = 3.9$$

3. **4.5** Start with $\frac{n+3}{20} = \frac{n-3}{4}$. Reduce horizontally. Then cross multiply.

$$\frac{n+3}{\cancel{20}_5} = \frac{n-3}{\cancel{4}_1}$$

$$(n+3) \cdot 1 = 5(n-3)$$

$$n+3 = 5n - 15$$

$$\underline{-5n \qquad -5n}$$

$$-4n + 3 = -15$$

$$\underline{\quad -3 \quad -3}$$

$$-4n \quad = -18$$

$$\frac{-4n}{-4} = \frac{-18}{-4}$$

$$n = \frac{18}{4} = \frac{9}{2} = 4.5$$

4. **240 eggs**   Write a proportion with flour in the numerators and eggs in the denominators. Let one ratio be for the original recipe and the other ratio be for the bake sale.

$$\frac{\text{recipe flour}}{\text{recipe eggs}} = \frac{\text{bake sale flour}}{\text{bake sale eggs}}$$

$$\frac{4.5}{3} = \frac{360}{x}$$

$$4.5x = 3 \cdot 360$$

$$4.5x = 1080$$

$$\frac{4.5x}{4.5} = \frac{1080}{4.5}$$

$$x = 240$$

5. **696 students**   First, consider the ratio 5 to 8. This is the ratio of students who **do** have a computer to the total number of students. This means that the ratio of students who **don't** have a computer is 3 to 8. If $\frac{5}{8}$ of the students have a computer, then $\frac{3}{8}$ don't. Write a proportion with the students who don't have a computer in the numerators and the total number of students in the denominators.

$$\frac{\text{ratio who don't have a computer}}{\text{ratio of all students}} = \frac{\text{number who don't have a computer}}{\text{total number of students}}$$

$$\frac{3}{8} = \frac{n}{1856}$$

$$\frac{3}{\cancel{8}} = \frac{n}{\cancel{1856}_{232}}$$

$$\frac{3}{1} = \frac{n}{232}$$

$$3 \cdot 232 = 1 \cdot n$$

$$n = 696$$

Operations with Algebraic Fractions

The two major groupings of operations on fractions are the multiplication and division and then the addition and subtraction. The multiplication and division problems are sometimes considered to be nicer to deal with because you don't need to find a common denominator, but they can have their own challenges when you get to the size of the answers.

Multiplying and Dividing Algebraic Fractions

When multiplying two or more fractions, the numerators are all multiplied together, as are the denominators. That's simple enough, and it's carried over as the basis of multiplying fractions that contain variables, as well. One technique used to make multiplying fractions easier is to reduce the problem before multiplying—if that's possible. When multiplying fractions, as long as a common factor is found in one numerator and one denominator, they don't even have to be in the same fraction. For example, multiplying the numbers $\frac{10}{21} \cdot \frac{14}{39} \cdot \frac{42}{35}$ would contain huge numbers that would have to be reduced in the end. But, by reducing first, the multiplication problem becomes more manageable. The 10 and 35 have a common factor of 5. The 14 and 21 have a common factor of 7. And, the 42 and 39 have a common factor of 3. You can combine numbers in other ways, and the equation still may be reducible after all this, but it makes the problem so much easier to do it in this way:

$$\frac{\overset{2}{\cancel{10}}}{\underset{3}{\cancel{21}}} \cdot \frac{\overset{2}{\cancel{14}}}{\cancel{39}_{13}} \cdot \frac{\overset{14}{\cancel{42}}}{\cancel{35}_{7}} = \frac{2}{3} \cdot \frac{2}{13} \cdot \frac{\overset{2}{\cancel{14}}}{\cancel{7}_{1}} = \frac{2}{3} \cdot \frac{2}{13} \cdot \frac{2}{1} = \frac{8}{39}$$

Example Problems

These problems show the answers and solutions.

1. Multiply $\frac{4x^2}{3y^3} \cdot \frac{15xy}{22}$.

    answer: $\frac{10x^3}{11y^2}$

The 4 and 22 have a common factor of 2; the 3 and 15 have a common factor of 3; and there's a common factor of y.

$$\frac{\overset{2}{\cancel{4}}x^2}{\underset{y}{\cancel{3}}\,\cancel{y^3}_{y^2}} \cdot \frac{\overset{5}{\cancel{15}}x\cancel{y}}{\cancel{22}_{11}} = \frac{2x^2}{y^2} \cdot \frac{5x}{11}$$

Now, multiply the numerators and denominators.

$$\frac{2x^2}{y^2} \cdot \frac{5x}{11} = \frac{10x^3}{11y^2}$$

2. Multiply $\dfrac{10a(a+3)^2}{21b(b-2)^3} \cdot \dfrac{24b^3(b-2)^2}{25(a+3)}$.

    answer: $\dfrac{16ab^2(a+3)}{35(b-2)}$

First, reduce the common factors.

$$\frac{\overset{2}{\cancel{10}}a(a+3)^2}{\underset{7}{\cancel{21}}b(b-2)^3} \cdot \frac{\overset{8}{\cancel{24}}b^3(b-2)^2}{\underset{5}{\cancel{25}}(a+3)} = \frac{2a(a+3)^2}{7b(b-2)^3} \cdot \frac{8b^3(b-2)^2}{5(a+3)}$$

When reducing the binomials, like $(a + 3)$, just mark off the exponents rather than the whole thing, unless the entire factor is divided out.

$$\frac{2a(a+3)^2}{7b(b-2)^3} \cdot \frac{8b^3(b-2)^2}{5(a+3)} = \frac{2a(a+3)^{\cancel{2}1}}{7b(b-2)^{\cancel{3}1}} \cdot \frac{8b^3\cancel{(b-2)^2}}{5\cancel{(a+3)}}$$

$$= \frac{2a(a+3)}{7b(b-2)} \cdot \frac{8b^3}{5}$$

$$= \frac{16ab^3(a+3)}{35b(b-2)}$$

$$= \frac{16ab^2(a+3)}{35(b-2)}$$

3. Divide $\dfrac{12(x+3)(y-2)^2}{25xyz} \div \dfrac{18(x+3)^2}{5x^2yz^3}$.

    answer: $\dfrac{2xz^2(y-2)^2}{15(x+3)}$

When dividing fractions, the problem is changed to multiplication by flipping the second fraction. Any reducing of common factors cannot be done until the problem is changed to multiplication.

$$\frac{12(x+3)(y-2)^2}{25xyz} \div \frac{18(x+3)^2}{5x^2yz^3} = \frac{12(x+3)(y-2)^2}{25xyz} \cdot \frac{5x^2yz^3}{18(x+3)^2}$$

Now the common factors can be divided out.

$$\frac{^2\cancel{12}(x+3)(y-2)^2}{{}_5\cancel{25}xyz} \cdot \frac{^1\cancel{5}x^2yz^3}{{}_3\cancel{18}(x+3)^2} = \frac{2(x+3)(y-2)^2}{5xyz} \cdot \frac{x^2yz^3}{3(x+3)^2}$$

$$= \frac{2\cancel{(x+3)}(y-2)^2}{5\cancel{x}\cancel{y}\cancel{z}} \cdot \frac{x^{\cancel{2}1}\cancel{y}z^{\cancel{3}2}}{3(x+3)^{\cancel{2}1}}$$

$$= \frac{2(y-2)^2}{5} \cdot \frac{xz^2}{3(x+3)}$$

$$= \frac{2xz^2(y-2)^2}{15(x+3)}$$

Work Problems

Use these problems to give yourself additional practice.

1. Multiply $\dfrac{14a^2b}{15xy^2} \cdot \dfrac{40x^2y^8}{21a^4b^4}$.

2. Multiply $\dfrac{16(x+3)}{27(y-2)} \cdot \dfrac{9(y-2)^3}{8(x+3)}$.

3. Multiply $\dfrac{42xy(y-2)^2}{11a^3} \cdot \dfrac{55a^2b^3}{21x(y-1)^4}$.

4. Divide $\dfrac{(a-1)(b+2)c}{(x-1)(y+2)z} \div \dfrac{(a-1)^2c^3}{(x-1)(y+2)^3}$.

5. Divide $\dfrac{9a^2b^3(c+8)}{20ab^4(c-8)^3} \div \dfrac{6a^3b^2(c+8)^3}{5(c-8)^3}$.

Worked Solutions

1. $\dfrac{16xy^6}{9a^2b^3}$ First, reduce the fractions and then multiply.

    $$\frac{^2\cancel{14}a^2\cancel{b}}{{}_3\cancel{15}\cancel{x}y^2} \cdot \frac{^8\cancel{40}x^{\cancel{2}1}y^{\cancel{8}6}}{{}_3\cancel{21}a^{\cancel{4}2}b^{\cancel{4}3}} = \frac{2}{3} \cdot \frac{8xy^6}{3a^2b^3} = \frac{16xy^6}{9a^2b^3}$$

2. $\dfrac{2(y-2)^2}{3}$ First, reduce the fractions and then multiply.

    $$\frac{^2\cancel{16}\cancel{(x+3)}}{{}_3\cancel{27}\cancel{(y-2)}} \cdot \frac{^1\cancel{9}(y-2)^{\cancel{3}2}}{\cancel{8}\cancel{(x+3)}} = \frac{2}{3} \cdot \frac{(y-2)^2}{1} = \frac{2(y-2)^2}{3}$$

3. $\dfrac{10b^3y(y-2)^2}{a(y-1)^4}$ First, reduce the fractions and then multiply.

    $$\frac{^2\cancel{42}\cancel{x}y(y-2)^2}{\cancel{11}a^{\cancel{3}1}} \cdot \frac{^5\cancel{55}a^{\cancel{2}}b^3}{\cancel{21}\cancel{x}(y-1)^4} = \frac{2y(y-2)^2}{a} \cdot \frac{5b^3}{(y-1)^4} = \frac{10b^3y(y-2)^2b^3}{a(y-1)^4}$$

4. $\dfrac{(b+2)(y+2)^2}{(a-1)c^2z}$　First, flip the second fraction and change it to multiplication. Then reduce and multiply.

$$\frac{(a-1)(b+2)c}{(x-1)(y+2)z} \div \frac{(a-1)^2c^3}{(x-1)(y+2)^3} = \frac{(a-1)(b+2)c}{(x-1)(y+2)z} \cdot \frac{(x-1)(y+2)^3}{(a-1)^2c^3}$$

$$\frac{\cancel{(a-1)}(b+2)\cancel{c}}{\cancel{(x-1)}\cancel{(y+2)}z} \cdot \frac{\cancel{(x-1)}(y+2)^{\cancel{3}2}}{(a-1)^{\cancel{2}1}c^{\cancel{3}2}} = \frac{(b+2)}{z} \cdot \frac{(y+2)^2}{(a-1)c^2} = \frac{(b+2)(y+2)^2}{(a-1)c^2z}$$

5. $\dfrac{3}{8a^2b^3(c+8)^2}$　First, change the problem to multiplication. Then reduce and multiply.

$$\frac{9a^2b^3(c+8)}{20ab^4(c-8)^3} \div \frac{6a^3b^2(c+8)^3}{5(c-8)^3} = \frac{9a^2b^3(c+8)}{20ab^4(c-8)^3} \cdot \frac{5(c-8)^3}{6a^3b^2(c+8)^3}$$

$$\frac{^3\cancel{9}\cancel{a^2}b^{\cancel{3}1}\cancel{(c+8)}}{_4\cancel{20}ab^4\cancel{(c-8)^3}} \cdot \frac{\cancel{5}\cancel{(c-8)^3}}{_2\cancel{6}a^{\cancel{3}1}(c+8)^{\cancel{3}2}} = \frac{3\cancel{b}}{4ab^{\cancel{4}3}} \cdot \frac{1}{2a(c+8)^2} = \frac{3}{8a^2b^3(c+8)^2}$$

Sometimes, you don't catch all the factors that need to be divided—especially when it's a long, complicated problem. Just do a second round of reducing, if necessary.

Adding and Subtracting Algebraic Fractions

To add or subtract fractions, you need common denominators. Every fraction involved has to have exactly the same denominator—the same factors and the same number of each factor. The two fractions here have the same factors, but they don't have the same **number** of each factor.

$$\frac{3}{x(x+1)^3(x-3)^4} \text{ and } \frac{2}{x^3(x+1)^2(x-3)^2}$$

A common denominator—needed to add these two fractions together—has to contain the greater number of each factor. So a common denominator for these would contain three factors of x, three factors of $(x+1)$, and four factors of $(x-3)$. The fractions have to be multiplied by 1 so they'll still be equivalent to the same value, but they'll have these common denominators. The 1 that they're multiplied by consists of a fraction made up of the missing factors from the common denominator. For the two fractions given previously:

$$\frac{3}{x(x+1)^3(x-3)^4} \cdot \frac{x^2}{x^2} \text{ The 1 multiplying the fraction is made up of the missing factors.}$$

$$\frac{2}{x^3(x+1)^2(x-3)^2} \cdot \frac{(x+1)(x-3)^2}{(x+1)(x-3)^2} \text{ Now this fraction has all of the factors needed.}$$

$$\frac{3}{x(x+1)^3(x-3)^4} \cdot \frac{x^2}{x^2} + \frac{2}{x^3(x+1)^2(x-3)^2} \cdot \frac{(x+1)(x-3)^2}{(x+1)(x-3)^2} =$$

$$\frac{3x^2}{x^3(x+1)^3(x-3)^4} + \frac{2(x+1)(x-3)^2}{x^3(x+1)^3(x-3)^4}$$

These fractions can be added together, because they have the same denominator.

Example Problems

These problems show the answers and solutions.

1. Add $\dfrac{5}{x^2 y} + \dfrac{8}{3xy^3}$.

    answer: $\dfrac{15y^2 + 8x}{3x^2 y^3}$

    The common denominator has to contain all of the factors of the denominator of each fraction—with the greater amount of a factor that either has. The common denominator in this case is $3x^2 y^3$. Rewriting the two fractions in the problem as equivalent fractions with that denominator:

    $$\frac{5}{x^2 y} \cdot \frac{3y^2}{3y^2} = \frac{15y^2}{3x^2 y^3}$$

    $$\frac{8}{3xy^3} \cdot \frac{x}{x} = \frac{8x}{3x^2 y^3}$$

    Now, the fractions can be added together.

    $$\frac{15y^2}{3x^2 y^3} + \frac{8x}{3x^2 y^3} = \frac{15y^2 + 8x}{3x^2 y^3}$$

2. Subtract $\dfrac{2a}{3b} - \dfrac{4b}{5a}$.

    answer: $\dfrac{10a^2 - 12b^2}{15ab}$

    The common denominator has to contain all of the factors of the denominator of each fraction. These denominators have nothing in common, so each fraction will have to be multiplied by the denominator of the other.

    $$\frac{2a}{3b} \cdot \frac{5a}{5a} = \frac{10a^2}{15ab}$$

    $$\frac{4b}{5a} \cdot \frac{3b}{3b} = \frac{12b^2}{15ab}$$

    Now, they can be subtracted.

    $$\frac{10a^2}{15ab} - \frac{12b^2}{15ab} = \frac{10a^2 - 12b^2}{15ab}$$

3. Add $\dfrac{5}{x+2} + \dfrac{3}{x^2} + \dfrac{1}{x(x-2)}$.

    answer: $\dfrac{5x^3 - 6x^2 + 2x - 12}{x^2(x+2)(x-2)}$

    The common denominator has to contain all of the factors of the denominator of each fraction—with the greatest amount of each factor. The common denominator in this case is $x^2(x+2)(x-2)$. Rewrite each of the fractions with the common denominator:

    $$\frac{5}{x+2} \cdot \frac{x^2(x-2)}{x^2(x-2)} = \frac{5x^2(x-2)}{x^2(x+2)(x-2)}$$

$$\frac{3}{x^2} \cdot \frac{(x+2)(x-2)}{(x+2)(x-2)} = \frac{3(x+2)(x-2)}{x^2(x+2)(x-2)}$$

$$\frac{1}{x(x-2)} \cdot \frac{x(x+2)}{x(x+2)} = \frac{x(x+2)}{x^2(x+2)(x-2)}$$

Now, they can be added together.

$$\frac{5x^2(x-2)}{x^2(x+2)(x-2)} + \frac{3(x+2)(x-2)}{x^2(x+2)(x-2)} + \frac{x(x+2)}{x^2(x+2)(x-2)} =$$

$$\frac{5x^2(x-2) + 3(x+2)(x-2) + x(x+2)}{x^2(x+2)(x-2)}$$

Distribute the values in the three terms in the numerator.

$$\frac{5x^2(x-2) + 3(x+2)(x-2) + x(x+2)}{x^2(x+2)(x-2)} =$$

$$\frac{5x^3 - 10x^2 + 3x^2 - 12 + x^2 + 2x}{x^2(x+2)(x-2)}$$

Combining the like terms in the numerator, the answer is $\dfrac{5x^3 - 6x^2 + 2x - 12}{x^2(x+2)(x-2)}$.

Work Problems

Use these problems to give yourself additional practice.

1. Find the least common denominator for the fractions $\dfrac{a+1}{a^2 x^2}$ and $\dfrac{1}{ax^3}$.

2. Add $\dfrac{3}{xy^2} + \dfrac{5}{y^4}$.

3. Add $\dfrac{4}{m(m+1)^2} + \dfrac{3(m+1)}{m^2(m+2)}$.

4. Subtract $\dfrac{5}{3cd} - \dfrac{2}{c^2 d^3}$.

5. Subtract $\dfrac{1}{9mn} - \dfrac{5}{12m^3}$.

Worked Solutions

1. $a^2 x^3$ The common denominator has to contain all of the factors of the denominator of each fraction—with the greater amount of a factor that either has. The common denominator in this case is $a^2 x^3$.

2. $\dfrac{3y^2 + 5x}{xy^4}$ Find the common denominator and then write the fractions as equivalent fractions.

$$\frac{3}{xy^2} + \frac{5}{y^4} = \frac{3}{xy^2} \cdot \frac{y^2}{y^2} + \frac{5}{y^4} \cdot \frac{x}{x}$$

$$= \frac{3y^2}{xy^4} + \frac{5x}{xy^4}$$

$$= \frac{3y^2 + 5x}{xy^4}$$

3. $\dfrac{4m(m+2) + 3(m+1)^3}{m^2(m+1)^2(m+2)}$ Find a common denominator and add them.

$$\frac{4}{m(m+1)^2} + \frac{3(m+1)}{m^2(m+2)} = \frac{4}{m(m+1)^2} \cdot \frac{m(m+2)}{m(m+2)} + \frac{3(m+1)}{m^2(m+2)} \cdot \frac{(m+1)^2}{(m+1)^2}$$

$$= \frac{4m(m+2)}{m^2(m+1)^2(m+2)} + \frac{3(m+1)^3}{m^2(m+1)^2(m+2)}$$

$$= \frac{4m(m+2) + 3(m+1)^3}{m^2(m+1)^2(m+2)}$$

This fraction can be further simplified using techniques found in the section, "Multiplying Polynomials," in Chapter 4.

4. $\dfrac{5cd^2 - 6}{3c^2 d^3}$ Find the common denominator, rewrite the fractions, and subtract.

$$\frac{5}{3cd} - \frac{2}{c^2 d^3} = \frac{5}{3cd} \cdot \frac{cd^2}{cd^2} - \frac{2}{c^2 d^3} \cdot \frac{3}{3}$$

$$= \frac{5cd^2}{3c^2 d^3} - \frac{6}{3c^2 d^3}$$

$$= \frac{5cd^2 - 6}{3c^2 d^3}$$

5. $\dfrac{4m^2 - 15n}{36m^3 n}$ Find the common denominator and subtract.

$$\frac{1}{9mn} - \frac{5}{12m^3} = \frac{1}{9mn} \cdot \frac{4m^2}{4m^2} - \frac{5}{12m^3} \cdot \frac{3n}{3n}$$

$$= \frac{4m^2}{36m^3 n} - \frac{15n}{36m^3 n}$$

$$= \frac{4m^2 - 15n}{36m^3 n}$$

Equations with Fractions

Many algebraic equations start out with fractions in them. This isn't necessarily a problem, but most people seem to be more comfortable with equations that don't heavily involve fractions. Sometimes an equation can be cleared of fractions just by multiplying each term in the equation by a well-chosen number. That number is usually the common denominator of all the fractions in the problem. A little more care has to be taken when there's a variable in one or more of the denominators. This situation is dealt with later in "Quadratic Equations" in Chapter 6.

Example Problems

These problems show the answers and solutions.

1. Solve the equation $\dfrac{4}{3}x + \dfrac{1}{2} = \dfrac{5}{6}x$.

   answer: $x = -1$

   To clear this equation of fractions, multiply each term by the common denominator, 6.

$$\frac{^2\cancel{6}}{1}\left(\frac{4}{\cancel{3}}x\right)+\frac{^3\cancel{6}}{1}\left(\frac{1}{\cancel{2}}\right)=\frac{\cancel{6}}{1}\left(\frac{5}{\cancel{6}}x\right)$$

$$8x \quad +3 \quad = \quad 5x$$

Now solve the equation by subtracting 8x from each side.

$$8x + 3 = 5x$$
$$\underline{-8x \qquad -8x}$$
$$3 = -3x$$
$$\frac{3}{-3} = \frac{-3x}{-3}$$
$$-1 = x$$

2. Solve the equation $\frac{2y-1}{5} - y = \frac{4-3y}{4}$.

    answer: $y = 8$

    The common denominator in this case is 20. Clear the fractions by multiplying through by 20.

$$\frac{^4\cancel{20}}{1}\left(\frac{2y-1}{\cancel{5}}\right)-20y=\frac{^5\cancel{20}}{1}\left(\frac{4-3y}{\cancel{4}}\right)$$

$$4(2y-1)-20y=5(4-3y)$$

Now distribute the two terms and solve for y.

$$8y - 4 - 20y = 20 - 15y$$
$$-12y - 4 = 20 - 15y$$
$$\underline{+15y \qquad = \qquad +15y}$$
$$3y - 4 = 20$$
$$\underline{+4 = +4}$$
$$3y \quad = 24$$
$$\frac{3y}{3} = \frac{24}{3}$$
$$y = 8$$

3. Solve for z. $\frac{4z}{7} + \frac{z-6}{14} = \frac{z}{2} + \frac{z+10}{12}$

    answer: $z = 21\frac{1}{5}$

    The common denominator in this case is 84. That's if you use all four denominators. Another approach would be to find a common denominator for each side, separately, and then treat the problem as a proportion, using all of the rules for proportions. For more on rules for proportions, see "Ratios and Proportions" earlier in this chapter.

On the left side, the common denominator is 14; on the right side, the common denominator is 12.

$$\frac{4z}{7} + \frac{z-6}{14} = \frac{2}{2} \cdot \frac{4z}{7} + \frac{z-6}{14} = \frac{8z}{14} + \frac{z-6}{14} = \frac{8z+z-6}{14} = \frac{9z-6}{14}$$

$$\frac{z}{2} + \frac{z+10}{12} = \frac{6}{6} \cdot \frac{z}{2} + \frac{z+10}{12} = \frac{6z}{12} + \frac{z+10}{12} = \frac{6z+z+10}{12} = \frac{7z+10}{12}$$

Set those two fractions equal to one another:

$$\frac{9z-6}{14} = \frac{7z+10}{12}$$

First, reduce horizontally and then cross multiply:

$$\frac{9z-6}{\cancel{14}_7} = \frac{7z+10}{\cancel{12}_6}$$

$$(9z-6) \cdot 6 = 7 \cdot (7z+10)$$

Now distribute and solve for z:

$$(9z-6) \cdot 6 = 7 \cdot (7z+10)$$

$$54z - 36 = 49z + 70$$

$$\underline{-49z \qquad\quad -49z}$$

$$5z - 36 = \qquad 70$$

$$\underline{+36 \qquad\quad +36}$$

$$5z \qquad = 106$$

$$\frac{5z}{5} = \frac{106}{5}$$

$$z = \frac{106}{5} = 21\frac{1}{5}$$

Work Problems

Use these problems to give yourself additional practice.

1. Solve by first multiplying each side of the equation by the same number: $\frac{x+3}{10} = \frac{4}{5}$.

2. Solve for y. $\frac{y}{3} + \frac{3}{4} = \frac{7}{8}$

3. Solve for z. $\frac{z}{6} + \frac{1}{3} = \frac{z}{2}$

4. Solve for n. $\frac{3n}{20} + \frac{1}{10} = \frac{n}{10} + \frac{3}{20}$

5. Solve for t. $\frac{7t}{30} - \frac{1}{15} = \frac{3t}{25} + \frac{4}{25}$

Worked Solutions

1. **x = 5**   Starting with $\frac{x+3}{10} = \frac{4}{5}$, multiply each term by 10.

$$\frac{\cancel{10}}{1} \cdot \frac{x+3}{\cancel{10}} = \frac{\overset{2}{\cancel{10}}}{1} \cdot \frac{4}{\cancel{5}}$$

$$x + 3 = 8$$

$$x = 5$$

2. **y = $\frac{3}{8}$**   Starting with $\frac{y}{3} + \frac{3}{4} = \frac{7}{8}$, multiply each term by 24.

$$\frac{\overset{8}{\cancel{24}}}{1} \cdot \frac{y}{\cancel{3}} + \frac{\overset{6}{\cancel{24}}}{1} \cdot \frac{3}{\cancel{4}} = \frac{\overset{3}{\cancel{24}}}{1} \cdot \frac{7}{\cancel{8}}$$

$$8y + \quad 18 = 21$$

$$8y = 3$$

$$y = \frac{3}{8}$$

3. **z = 1**   Starting with $\frac{z}{6} + \frac{1}{3} = \frac{z}{2}$, multiply each term by 6.

$$\frac{\cancel{6}}{1} \cdot \frac{z}{\cancel{6}} + \frac{\overset{2}{\cancel{6}}}{1} \cdot \frac{1}{\cancel{3}} = \frac{\overset{3}{\cancel{6}}}{1} \cdot \frac{z}{\cancel{2}}$$

$$z + 2 = 3z$$

$$2 = 2z$$

$$1 = z$$

4. **n = 1**   Starting with $\frac{3n}{20} + \frac{1}{10} = \frac{n}{10} + \frac{3}{20}$, multiply each term by 20.

$$\frac{\cancel{20}}{1} \cdot \frac{3n}{\cancel{20}} + \frac{\overset{2}{\cancel{20}}}{1} \cdot \frac{1}{\cancel{10}} = \frac{\overset{2}{\cancel{20}}}{1} \cdot \frac{n}{\cancel{10}} + \frac{\cancel{20}}{1} \cdot \frac{3}{\cancel{20}}$$

$$3n + 2 = 2n + 3$$

$$n = 1$$

5. **t = 2**   Starting with $\frac{7t}{30} - \frac{1}{15} = \frac{3t}{25} + \frac{4}{25}$, write the left side all as one term after finding the common denominator. Add the two terms on the right. Then solve the proportion.

$$\frac{7t}{30} - \frac{2}{2} \cdot \frac{1}{15} = \frac{3t}{25} + \frac{4}{25}$$

$$\frac{7t}{30} - \frac{2}{30} = \frac{3t+4}{25}$$

$$\frac{7t-2}{30_6} = \frac{3t+4}{25_5}$$

$$(7t - 2) \cdot 5 = 6(3t + 4)$$

$$35t - 10 = 18t + 24$$

$$17t = 34$$

$$t = 2$$

Chapter Problems and Solutions

Problems

In problems 1 through 9, solve for the value of the variable.

1. $x + 3 = -8$

2. $y - 5 = 14$

3. $6z = -18$

4. $\frac{w}{5} = -2$

5. $5x - 2 = 3x + 6$

6. $3(y - 2) = 4y + 1$

7. $\frac{x+3}{4} = 5$

8. $3(y - 2) = 6(y + 2)$

9. $8t - 2 - 5t = 2t + 1$

10. Find the fractional value for the repeating decimal .41666....

11. Solve for h in $A = \frac{1}{2} bh$.

12. Solve for r in $A = P(1 + r)$.

13. Solve for x in $\frac{x}{16} = \frac{35}{40}$.

14. Solve for z in $\frac{56}{z} = \frac{49}{14}$.

15. Solve for n in $\frac{n+6}{9} = \frac{40}{15}$.

16. The apple pie recipe calls for $\frac{3}{4}$ cup of sugar per 4 cups of sliced apples. If you're making pies out of 350 cups of slice apples, how much sugar will you need?

17. Multiply $\frac{10a^3(b+1)^2}{49(c+3)} \cdot \frac{21(c+3)^4}{25a}$.

18. Divide $\frac{2x^2(x+1)(x-1)}{15(x+2)(x-2)} \div \frac{4x^2(x-1)^3}{55(x+2)^2}$.

19. Add $\frac{4}{2x} + \frac{3}{x^2}$.

20. Subtract $\frac{8}{3x(x-1)} - \frac{x}{x^2(x-1)}$.

21. Solve for x in $\frac{x}{4} + \frac{x}{2} = \frac{2x+1}{3}$.

22. Solve for z in $\frac{4z}{5} - \frac{z}{4} = \frac{3z}{10} + \frac{2z}{8}$.

Answers and Solutions

1. **Answer: -11**   Subtract 3 from each side.

$$x + 3 = -8$$
$$\underline{-3 \quad -3}$$
$$x \quad = -11$$

2. **Answer: 19**   Add 5 to each side of the equation.

$$y - 5 = 14$$
$$\underline{+5 \quad +5}$$
$$y \quad = 19$$

3. **Answer: -3**   Divide each side by 6.

$$\frac{6z}{6} = \frac{-18}{6}$$
$$z = -3$$

4. **Answer: -10**   Multiply each side by 5.

$$5 \cdot \frac{w}{5} = -2 \cdot 5$$
$$w = -10$$

5. **Answer: 4**   Add 2 to each side and subtract $3x$ from each side.

$$5x - 2 = 3x + 6$$
$$\underline{+2 \qquad +2}$$
$$5x \quad = 3x + 8$$
$$\underline{-3x \qquad -3x}$$
$$2x \quad = 8$$

Now divide each side by 2.

$$\frac{2x}{2} = \frac{8}{2}$$
$$x = 4$$

6. **Answer: -7**   Distribute the 3 on the left side.

$$3(y - 2) = 4y + 1$$
$$3y - 6 = 4y + 1$$

Add 6 to each side and subtract $4y$ from each side.

$$3y - 6 = 4y + 1$$
$$\underline{-4y + 6 \quad -4y + 6}$$
$$-y \quad = \quad 7$$

Multiply each side by −1.

$$-1(-y) = -1(7)$$
$$y = -7$$

7. **Answer: 17**   Multiply each side by 4. Then subtract 3 from each side.

$$4 \cdot \frac{x+3}{4} = 5 \cdot 4$$
$$x + 3 = 20$$
$$\underline{\quad -3 \quad -3}$$
$$x = 17$$

8. **Answer: −6**   Divide each side by 3. Then distribute the 2 on the right side.

$$\frac{3(y-2)}{3} = \frac{6(y+2)}{3}$$
$$y - 2 = 2(y+2)$$
$$y - 2 = 2y + 4$$

Subtract 4 from each side and subtract y from each side.

$$y - 2 = 2y + 4$$
$$\underline{-y - 4 \quad -y - 4}$$
$$-6 = y$$

9. **Answer: 3**   Combine the t terms on the left.

$$8t - 2 - 5t = 2t + 1$$
$$3t - 2 = 2t + 1$$

Add 2 to each side and subtract $2t$ from each side.

$$3t - 2 = 2t + 1$$
$$\underline{-2t + 2 \quad -2t + 2}$$
$$t \quad = \quad 3$$

10. **Answer: $\frac{5}{12}$**   Let $N = .41666...$

Then $100N = 41.666...$

And $1000N = 416.666...$

Subtract $100N$ from $1000N$.

$$1000N = 416.666...$$
$$-100N = -41.666...$$
$$900N = 375$$

Divide each side by 900 and reduce the fraction.

$$N = \frac{375}{900} = \frac{5}{12}$$

11. **Answer:** $h = \dfrac{2A}{b}$ Multiply each side by 2.

$$2 \cdot A = 2 \cdot \frac{1}{2} bh$$
$$2 \cdot A = bh$$

Divide each side by b.

$$\frac{2A}{b} = \frac{bh}{b}$$
$$\frac{2A}{b} = h$$

12. **Answer:** $r = \dfrac{A - P}{P}$ Distribute the P on the right.

$$A = P(1 + r)$$
$$A = P + Pr$$

Subtract P from each side; then divide each side by P.

$$A = P + Pr$$
$$\underline{-P - P}$$
$$\frac{A - P}{P} = \frac{Pr}{P}$$
$$\frac{A - P}{P} = r$$

13. **Answer: 14**   Reduce vertically in the right fraction and then horizontally in the denominators.

$$\frac{x}{16} = \frac{35^{7}}{40_{8}}$$
$$\frac{x}{{}_2\!16} = \frac{7}{8_1}$$
$$\frac{x}{2} = \frac{7}{1}$$

Cross multiply.

$$x \cdot 1 = 7 \cdot 2$$
$$x = 14$$

14.   **Answer: 16**   Reduce horizontally in the numerators and vertically on the right side.

$$\frac{{}^{8}\cancel{56}}{z} = \frac{\cancel{49}^{7}}{14}$$

$$\frac{8}{z} = \frac{\cancel{7}^{1}}{\cancel{14}_{2}}$$

$$\frac{8}{z} = \frac{1}{2}$$

Cross multiply.

$$8 \cdot 2 = 1 \cdot z$$
$$16 = z$$

15.   **Answer: 18**   Reduce vertically on the right and horizontally in the denominators.

$$\frac{n+6}{9} = \frac{\cancel{40}^{8}}{\cancel{15}_{3}}$$

$$\frac{n+6}{\cancel{9}_{3}} = \frac{8}{\cancel{3}_{1}}$$

$$\frac{n+6}{3} = \frac{8}{1}$$

Cross multiply; then subtract 6 from each side.

$$(n+6) \cdot 1 = 8 \cdot 3$$
$$n+6 = 24$$
$$\underline{-6 \quad -6}$$
$$n = 18$$

16.   **Answer: $65\frac{5}{8}$**

Set up the proportion with sugar in the numerators and apples in the denominators.

$$\frac{\frac{3}{4}}{4} = \frac{x}{350}$$

Reduce horizontally in the denominators.

$$\frac{\frac{3}{4}}{\cancel{4}_{2}} = \frac{x}{\cancel{350}_{175}}$$

$$\frac{\frac{3}{4}}{2} = \frac{x}{175}$$

Cross multiply; then multiply each side by $\frac{1}{2}$.

$$\frac{3}{4} \cdot 175 = x \cdot 2$$

$$\frac{525}{4} = 2x$$

$$\frac{1}{2} \cdot \frac{525}{4} = \frac{2x}{2}$$

$$\frac{525}{8} = x \ \ or \ \ x = 65\frac{5}{8}$$

17. **Answer:** $\dfrac{6a^2(b+1)^2(c+3)^3}{35}$ Reduce the fractions before multiplying.

$$\frac{{}_2\cancel{10}a^{\cancel{3}2}(b+1)^{\cancel{3}2}}{\cancel{49}^7(\cancel{c+3})} \cdot \frac{\cancel{21}^3(c+3)^{\cancel{4}3}}{{}_5\cancel{25}\cancel{a}} = \frac{6a^2(b+1)^2(c+3)^3}{35}$$

18. **Answer:** $\dfrac{11(x+1)(x+2)}{6(x-2)(x-1)^2}$ Change the problem to multiplication. Then reduce the fractions and multiply.

$$\frac{2x^2(x+1)(x-1)}{15(x+2)(x-2)} \div \frac{4x^2(x-1)^3}{55(x+2)^2} = \frac{2x^2(x+1)(x-1)}{15(x+2)(x-2)} \cdot \frac{55(x+2)^2}{4x^2(x-1)^3}$$

$$= \frac{2x^2(x+1)(\cancel{x-1})}{{}_3\cancel{15}(\cancel{x+2})(x-2)} \cdot \frac{{}^{11}\cancel{55}(x+2)^{\cancel{2}1}}{{}_2\cancel{4}x^2(x-1)^{\cancel{3}2}}$$

$$= \frac{11(x+1)(x+2)}{6(x-2)(x-1)^2}$$

19. **Answer:** $\dfrac{2x+3}{x^2}$ Find a common denominator; then add the fractions.

$$\frac{4}{2x} + \frac{3}{x^2} = \frac{4}{2x} \cdot \frac{x}{x} + \frac{3}{x^2} \cdot \frac{2}{2} = \frac{4x}{2x^2} + \frac{6}{2x^2} = \frac{4x+6}{2x^2}$$

This can be reduced, because each term is divisible by 2. Dividing each term by 2, it's $\dfrac{2x+3}{x^2}$.

20. **Answer:** $\dfrac{5x}{3x^2(x-1)}$ Find a common denominator and subtract.

$$\frac{8}{3x(x-1)} - \frac{x}{x^2(x-1)} = \frac{8}{3x(x-1)} \cdot \frac{x}{x} - \frac{x}{x^2(x-1)} \cdot \frac{3}{3} = \frac{8x}{3x^2(x-1)} - \frac{3x}{3x^2(x-1)}$$

$$= \frac{8x-3x}{3x^2(x-1)} = \frac{5x}{3x^2(x-1)}$$

21. **Answer: 4** Multiply through by 12 to get rid of the fractions.

$$12 \cdot \frac{x}{4} + 12 \cdot \frac{x}{2} = 12 \cdot \frac{2x+1}{3}$$

$$3x + 6x = 4(2x+1)$$

Combine the two terms on the left and distribute the 4 on the right.

$$3x + 6x = 4(2x + 1)$$
$$9x = 8x + 4$$

Subtract 8x from each side.

$$9x = 8x + 4$$
$$\underline{-8x \quad -8x}$$
$$x = 4$$

22. **Answer: Any number**   Subtract the two terms on the left and add the two terms on the right.

$$\frac{4z}{5} - \frac{z}{4} = \frac{3z}{10} + \frac{2z}{8}$$

$$\frac{4z}{5} \cdot \frac{4}{4} - \frac{z}{4} \cdot \frac{5}{5} = \frac{3z}{10} \cdot \frac{4}{4} + \frac{2z}{8} \cdot \frac{5}{5}$$

$$\frac{16z - 5z}{20} = \frac{12z + 10z}{40}$$

$$\frac{11z}{20} = \frac{22z}{40}$$

Reduce horizontally in the numerators and denominators.

$$\frac{\cancel{11}z}{\cancel{20}} = \frac{\cancel{22}z}{\cancel{40}}$$

$$\frac{z}{1} = \frac{2z}{2}$$

$$z = z$$

This statement is always true, so any number will work.

Supplemental Chapter Problems

Problems

In problems 1 through 9, solve for the value of the variable.

1. $p - 14 = 11$

2. $t + 6 = -9$

3. $\dfrac{v}{-3} = -3$

4. $\dfrac{2}{3}w = 8$

5. $4 - 5z = 1 - 6z$

6. $5(x + 2) = x + 6$

7. $3(x + 5) = 3x + 13 - x$

8. $\dfrac{2x - 3}{8} = 12$

9. $4 - x = x - 4$

10. Find the fractional value for the repeating decimal .458333....

11. Solve for r in $A = Prt$.

12. Solve for C in $P = A + B + 2C$.

13. Solve for y in $\dfrac{y}{44} = \dfrac{21}{77}$.

14. Solve for z in $\dfrac{63}{35} = \dfrac{36}{z}$.

15. Solve for n in $\dfrac{20}{n-2} = \dfrac{12}{21}$.

16. In a survey, it was found that 4 out of every 10 people watched the Olympics on TV for at least 2 hours. If 180 people did **not** watch them, then how many **did** watch the Olympics?

17. Multiply $\dfrac{22x^2 y^3 z}{35(a-6)^2} \cdot \dfrac{40(a-6)^2}{33xy^4 z^2}$.

18. Divide $\dfrac{8rt^2}{21(m+3)(m-2)^2} \div \dfrac{12r^2 t}{28(m+3)^2(m-2)^3}$.

19. Add $\dfrac{5}{m^3} + \dfrac{2}{4m}$.

20. Subtract $\dfrac{4}{(a+b)(a-b)} - \dfrac{2}{3(a-b)}$.

21. Solve for y in $\dfrac{y}{6} + \dfrac{1}{2} = \dfrac{y+1}{4}$.

22. Solve for z in $\dfrac{z-1}{100} + \dfrac{3z+1}{20} = \dfrac{z}{50} + \dfrac{z+1}{10}$.

Answers

1. 25   (Solving Linear Equations, p. 107)

2. −15   (Solving Linear Equations, p. 107)

3. 9   (Solving Linear Equations, p. 107)

4. 12   (Solving Linear Equations, p. 107)

5. −3   (Solving Linear Equations with More Than One Operation, p. 111)

6. −1   (Solving Linear Equations with More Than One Operation, p. 111)

7. −2   (Solving Linear Equations with More Than One Operation, p. 111)

8. $49\frac{1}{2}$ (Solving Linear Equations with More Than One Operation, p. 111)

9. 4   (Solving Linear Equations with More Than One Operation, p. 111)

10. $\frac{11}{24}$ (Solving Linear Equations with More Than One Operation, p. 111)

11. $r = \dfrac{A}{Pt}$ (Solving Linear Formulas, p. 116)

12. $C = \dfrac{P - A - B}{2}$ (Solving Linear Formulas, p. 116)

13.   12   (Ratios and Proportions, p. 120)

14.   20   (Ratios and Proportions, p. 120)

15.   37   (Ratios and Proportions, p. 120)

16.   120   (Ratios and Proportions, p. 120)

17. $\dfrac{16x}{21yz}$ (Multiplying and Dividing Algebraic Fractions, p. 125)

18. $\dfrac{8t(m+3)(m-2)}{9r}$ (Multiplying and Dividing Algebraic Fractions, p. 125)

19. $\dfrac{10+m^2}{2m^3}$ (Adding and Subtracting Algebraic Fractions, p. 128)

20. $\dfrac{12-2(a+b)}{3(a+b)(a-b)}$ (Adding and Subtracting Algebraic Fractions, p. 128)

21.   3   (Equations with Fractions, p. 131)

22. $\dfrac{3}{2}$ (or $1\frac{1}{2}$)   (Equations with Fractions, p. 131)

Chapter 4
Polynomials and Factoring

Multiplication and division of algebraic expressions follow the basic rules regarding exponents and variables that are alike or different. Different algebraic tasks require different formats of the same expression. Sometimes you multiply and expand the expression. Other times you factor or divide it. The goal is to change the form but keep the expressions equivalent. This chapter has some special rules for dealing with these forms.

Multiplying Monomials

A monomial is an expression that has only one term. The following are monomials: 4, $4a$, $4a^2$, $4a^2 b$, $4a^2 bcde$, and $4a^2 b(c+d)^3$.

A term is something that has one or more factors multiplied together, as do all of these examples. Monomials, binomials, and trinomials are all special types of polynomials. Poly- means "many" and nom- means "name," so polynomial is the generic word for these expressions that have "many" or any number of terms. Bi- is for two; tri- is for three; and polynomials with more terms than that don't generally have special names—they're just called polynomials.

In Chapter 1, in the section, "Combining 'Like' Terms," examples show that only terms that are alike as far as their variables and the powers of those variables can be added together. You can add the monomials $4xy$ and $6xy$ to get $10xy$, but you cannot add $4xy^2$ to $6x^2 y$, because they are not alike. This situation can be compared to combining fractions. You cannot add or subtract unless you have common denominators—denominators that are exactly alike. Adding and subtracting things are generally more complicated in algebra. Not so with multiplying or dividing. You can multiply monomials by adding exponents of the same variable and then writing the numbers and variables as multipliers or factors of one another.

Example Problems

These problems show the answers and solutions.

1. Multiply $5xy \cdot 4x^2 y$.

   answer: $20x^3 y^2$

$$5xy \cdot 4x^2 y = 5 \cdot 4 \cdot x \cdot x^2 \cdot y \cdot y = 20x^3 y^2$$

A couple of properties are used here: the commutative property to rearrange the factors and the associative property to group the factors for combining. An exponent law also was employed. For more on the properties, see the section "Properties of Algebraic Expressions," in Chapter 1. Exponents are also discussed in Chapter 1.

2. Multiply $-3a^2 b^3 d(-5ab^3 c^4)$.

answer: $15a^3 b^6 c^2 d$

$$-3a^2 b^3 d(-5ab^3 c^2) = (-3)(-5)a^2 \cdot a \cdot b^3 \cdot b^3 \cdot c^2 \cdot d = 15a^3 b^6 c^2 d$$

The distributive property works for multiplying monomials over other expressions of two or more terms. "Properties of Algebraic Expressions" in Chapter 1 shows you how to distribute a number or variable over terms in parentheses. A monomial, which is made up of any number of factors, can be distributed over other terms. The number parts of each product in the distribution are multiplied; the variables that are alike have their exponents added; and terms that are exactly alike can be combined.

3. Distribute $3x$ over the binomial $6x + 2$.

answer: $18x^2 + 6x$

Multiply each term in the binomial by $3x$.

$$3x(6x + 2) = 3x(6x) + 3x(2)$$
$$= 3 \cdot 6 \cdot x \cdot x + 3 \cdot 2 \cdot x$$
$$= 18x^2 + 6x$$

4. Distribute $-9a^2 b$ over the trinomial $2ab - 3a^3 - 1$.

answer: $-18a^3 b^2 + 27a^5 b + 9a^2 b$

When multiplying each term by $-9a^2 b$, don't forget the negative sign.

$$-9a^2 b(2ab - 3a^3 - 1) = -9a^2 b(2ab) - 9a^2 b(-3a^3) - 9a^2 b(-1)$$
$$= -9 \cdot 2 \cdot a^2 \cdot a \cdot b \cdot b - 9(-3) \cdot a^2 \cdot a^3 \cdot b - 9(-1) \cdot a^2 b$$
$$= -18a^3 b^2 + 27a^5 b + 9a^2 b$$

5. Distribute $\dfrac{2r}{tz^2}$ over the binomial $4r^2 t^2 z^2 - 9rtz^3$.

answer: $8r^3 t - 18r^2 z$

In this case, it might be easier to rewrite $\dfrac{2r}{tz^2}$ using negative exponents before distributing. That way, variables can be multiplied together more easily, and the answer can be changed back to a fraction if necessary.

$$\frac{2r}{tz^2} = 2rt^{-1} z^{-2}, \text{ so}$$

$$\frac{2r}{tz^2}(4r^2 t^2 z^2 - 9rtz^3) = 2rt^{-1} z^{-2}(4r^2 t^2 z^2 - 9rtz^3)$$

$$= 2rt^{-1} z^{-2}(4r^2 t^2 z^2) + 2rt^{-1} z^{-2}(-9rtz^3)$$

$$= 2 \cdot 4 \cdot r \cdot r^2 \cdot t^{-1} \cdot t^2 \cdot z^{-2} \cdot z^2 - 18r \cdot r \cdot t^{-1} \cdot t \cdot z^{-2} \cdot z^3$$

$$= 8r^3 t^1 z^0 - 18r^2 t^0 z^1$$

$$= 8r^3 t - 18r^2 z$$

Work Problems

Use these problems to give yourself additional practice.

1. Distribute $3xy(2x^2 y - 5xy^3)$.

2. Distribute $-2a^2 b^2(3ab^2 + 4a^3 b - 3)$.

3. Distribute $4x^{-1} y^{-1}(3xy^2 + 2x^2 y - xy)$.

4. Distribute $\dfrac{5a^2}{c}(2a^3 c^2 + 3ac)$.

5. Distribute $\dfrac{1}{m^2 n^2 p}\left(3mp - 4m^2 n^3 - 5m^3 n^2 p^4\right)$.

Worked Solutions

1. $\mathbf{6x^3 y^2 - 15x^2 y^4}$ Multiply each term by $3xy$.

$$3xy(2x^2 y - 5xy^3) = 3xy(2x^2 y) + 3xy(-5xy^3)$$
$$= 3 \cdot 2 \cdot x \cdot x^2 \cdot y \cdot y + 3(-5) \cdot x \cdot x \cdot y \cdot y^3$$
$$= 6x^3 y^2 - 15x^2 y^4$$

2. $\mathbf{-6a^3 b^4 - 8a^5 b^3 + 6a^2 b^2}$ Multiply each term by $-2a^2 b^2$.

$$-2a^2 b^2(3ab^2 + 4a^3 b - 3) = -2a^2 b^2(3ab^2) - 2a^2 b^2(4a^3 b) - 2a^2 b^2(-3)$$
$$= -2 \cdot 3 \cdot a^2 \cdot a \cdot b^2 \cdot b^2 - 2(4) \cdot a^2 \cdot a^3 \cdot b^2 \cdot b - 2(-3)a^2 b^2$$
$$= -6a^3 b^4 - 8a^5 b^3 + 6a^2 b^2$$

3. $\mathbf{12y + 8x - 4}$ The negative exponent is written as a fraction in the final answer.

$$4x^{-1} y^{-1}(3xy^2 + 2x^2 y - xy) = 4x^{-1} y^{-1}(3xy^2) + 4x^{-1} y^{-1}(2x^2 y) + 4x^{-1} y^{-1}(-xy)$$
$$= 4 \cdot 3 \cdot x^{-1} \cdot x \cdot y^{-1} \cdot y^2 + 4(2)x^{-1} \cdot x^2 \cdot y^{-1} \cdot y + 4(-1)x^{-1} \cdot x \cdot y^{-1} \cdot y$$
$$= 12x^0 y^1 + 8x^1 y^0 - 4x^0 y^0$$
$$= 12y + 8x - 4$$

4. $\mathbf{10a^5 c + 15a^3}$ Rewrite the fraction as a term with a negative exponent.

$$\frac{5a^2}{c}(2a^3 c^2 + 3ac) = 5a^2 c^{-1}(2a^3 c^2 + 3ac)$$
$$= 5(2)a^2 \cdot a^3 \cdot c^{-1} \cdot c^2 + 5(3)a^2 \cdot a \cdot c^{-1} c$$
$$= 10a^5 c^1 + 15a^3 c^0$$
$$= 10a^5 c + 15a^3$$

5. $\dfrac{3}{mn^2} - \dfrac{4n}{p} - 5mp^3$　Rewrite the fraction using negative exponents. Then change the final answer back to a fraction.

$$\frac{1}{m^2 n^2 p}(3mp - 4m^2 n^3 - 5m^3 n^2 p^4) = m^{-2} n^{-2} p^{-1}(3mp - 4m^2 n^3 - 5m^3 n^2 p^4)$$

$$= m^{-2} n^{-2} p^{-1}(3mp) + m^{-2} n^{-2} p^{-1}(-4m^2 n^3) + m^{-2} n^{-2} p^{-1}(-5m^3 n^2 p^4)$$

$$= 3m^{-2} mn^{-2} p^{-1} p - 4m^{-2} m^2 n^{-2} n^3 p^{-1} - 5m^{-2} m^3 n^{-2} n^2 p^{-1} p^4$$

$$= 3m^{-1} n^{-2} p^0 - 4m^0 n^1 p^{-1} - 5m^1 n^0 p^3$$

$$= \frac{3}{mn^2} - \frac{4n}{p} - 5mp^3$$

Multiplying Polynomials

Distributing a monomial through other polynomials is discussed in the previous section. A monomial can multiply through two, three, four, or any number of terms. Multiplying a polynomial with a binomial (two terms) or trinomial (three terms) or higher takes carefully chosen methods and organization. For instance, to multiply two binomials together such as $(x + 2)(a + 3)$, you can use the distributive property. Distribute the $(x + 2)$ over the other two terms.

$$(x + 2)(a + 3) = (x + 2)(a) + (x + 2)(3)$$
$$= x(a) + 2(a) + x(3) + 2(3)$$
$$= ax + 2a + 3x + 6$$

None of the terms are alike, so this can't be simplified further.

The "FOIL" Method

Because multiplying two binomials together is a very common operation in algebra, a rather simple acronym has been associated with this process. It's called FOIL, and its application is a preferred method for multiplying most types of binomials.

The letters in FOIL stand for First, Outer, Inner, and Last. These words describe the positions of the terms in the two binomials. Each term actually will have two different names, because each term is used twice in the process. In the multiplication problems given previously, $(x + 2)(a + 3)$:

The x and the a are the First terms of each binomial.

The x and the 3 are the Outer terms in the two binomials.

The 2 and the a are the Inner terms in the two binomials.

The 2 and the 3 are the Last terms of each binomial.

These pairings tell you what to multiply:

F: multiply $x \cdot a$
O: multiply $x \cdot 3$
I: multiply $2 \cdot a$
L: multiply $2 \cdot 3$

Add these together, $x \cdot a + x \cdot 3 + 2 \cdot a + 2 \cdot 3 = ax + 3x + 2a + 6$. It's the same result, in a slightly different order, as the problem done with distribution.

When using this FOIL method, you'll notice that, when the two binomials are alike—that is, they have the same types of terms—the Outer and Inner terms combine, and the result is a trinomial. If they aren't alike, as shown by this last example, then none of the terms in the solution will combine, and you'll have a four-term polynomial.

Example Problems

These problems show the answers and solutions.

1. Use FOIL to multiply $(a - 4)(a + 2)$.

   answer: $a^2 - 2a - 8$

   Identify the products in FOIL.

   F: $a \cdot a$ O: $a \cdot 2$ I: $-4 \cdot a$ L: $-4 \cdot 2$

   $$(a - 4)(a + 2) = a^2 + 2a - 4a - 8$$
   $$= a^2 - 2a - 8$$

2. Use FOIL to multiply $(2m + 5)(3m - 7)$.

   answer: $6m^2 + m - 35$

   F: $2m \cdot 3m$ O: $2m(-7)$ I: $5 \cdot 3m$ L: $5(-7)$

   $$(2m + 5)(3m - 7) = 6m^2 - 14m + 15m - 35$$
   $$= 6m^2 + m - 35$$

3. Use FOIL to multiply $(4x - 3)(y + 1)$.

   answer: $4xy + 4x - 3y - 3$

   F: $4x \cdot y$ O: $4x \cdot 1$ I: $-3 \cdot y$ L: $-3 \cdot 1$

   $$(4x - 3)(y + 1) = 4xy + 4x - 3y - 3$$

4. Use FOIL to multiply $(a^2 - 9)(a^2 - 11)$.

   answer: $a^4 - 20a^2 + 99$

   F: $a^2 \cdot a^2$ O: $a^2(-11)$ I: $-9 \cdot a^2$ L: $-9(-11)$

   $$(a^2 - 9)(a^2 - 11) = a^4 - 20a^2 + 99$$

Other Products

It's nice to have this pattern or method to multiply two binomials together. Unfortunately, no other really handy tricks exist for multiplying other polynomials. Basically, you just distribute the smaller of the two polynomials over the other polynomial.

Example Problems

These problems show the answers and solutions.

1. Multiply the binomial times the trinomial $(x - 3)(x^2 + 4x - 1)$.

    answer: $x^3 + x^2 - 13x + 3$

    The binomial will be distributed over the trinomial.

    $$(x - 3)(x^2 + 4x - 1) = (x - 3) \cdot x^2 + (x - 3) \cdot 4x + (x - 3)(-1)$$
    $$= x \cdot x^2 - 3 \cdot x^2 + x \cdot 4x - 3 \cdot 4x + x(-1) - 3(-1)$$
    $$= x^3 - 3x^2 + 4x^2 - 12x - x + 3$$
    $$= x^3 + x^2 - 13x + 3$$

2. Multiply the trinomial times the trinomial $(c^2 + 2c - 1)(3c^3 + c^2 + 4)$.

    answer: $3c^5 + 7c^4 - c^3 + 3c^2 + 8c - 4$

    This time, neither polynomial is smaller than the other. Just distribute the first trinomial over the second.

    $$(c^2 + 2c - 1)(3c^3 + c^2 + 4) = (c^2 + 2c - 1) \cdot 3c^3 + (c^2 + 2c - 1) \cdot c^2 + (c^2 + 2c - 1) \cdot 4$$
    $$= c^2 \cdot 3c^3 + 2c \cdot 3c^3 - 1 \cdot 3c^3 + c^2 \cdot c^2 + 2c \cdot c^2 - 1 \cdot c^2 + c^2 \cdot 4 + 2c \cdot 4 - 1 \cdot 4$$
    $$= 3c^5 + 6c^4 - 3c^3 + c^4 + 2c^3 - c^2 + 4c^2 + 8c - 4$$

    There are lots of like terms to be combined to finish this problem.

    $$= 3c^5 + 7c^4 - c^3 + 3c^2 + 8c - 4$$

Work Problems

Use these problems to give yourself additional practice.

1. Multiply using FOIL $(x + 5)(x - 3)$.

2. Multiply using FOIL $(5y - 7)(6y - 1)$.

3. Multiply using FOIL $(3z^2 + 7)(2z^2 + 3)$.

4. Multiply using FOIL $(3a - 2)(a^2 + 2)$.

5. Multiply $(2n^2 + 3)(4n^4 - 6n^2 + 9)$.

Worked Solutions

1. $x^2 + 2x - 15$ Using FOIL, find the four different products and simplify the results.

$$(x + 5)(x - 3) = x \cdot x + x(-3) + 5 \cdot x + 5(-3)$$
$$= x^2 - 3x + 5x - 15$$
$$= x^2 + 2x - 15$$

2. $30y^2 - 47y + 7$ Using FOIL, find the four products and simplify the results.

$$(5y - 7)(6y - 1) = 5y \cdot 6y + 5y(-1) - 7 \cdot 6y - 7(-1)$$
$$= 30y^2 - 5y - 42y + 7$$
$$= 30y^2 - 47y + 7$$

3. $6z^4 + 23z^2 + 21$ The Outer and Inner terms will still combine in this problem.

$$(3z^2 + 7)(2z^2 + 3) = 3z^2 \cdot 2z^2 + 3z^2 \cdot 3 + 7 \cdot 2z^2 + 21$$
$$= 6z^4 + 9z^2 + 14z^2 + 21$$
$$= 6z^4 + 23z^2 + 21$$

4. $3a^3 - 2a^2 + 6a - 4$ In this case, the binomials aren't the same, so the Outer and Inner products won't combine.

$$(3a - 2)(a^2 + 2) = 3a \cdot a^2 + 3a \cdot 2 - 2 \cdot a^2 - 2 \cdot 2$$
$$= 3a^3 + 6a - 2a^2 - 4$$
$$= 3a^3 - 2a^2 + 6a - 4$$

This last step is just cosmetic. It's better form to write the answer in decreasing powers of the variable.

5. $8n^6 + 27$ This is a special product. It's the result of multiplying the sum of two cube roots times their squares and product. You'll see more on this in the next section. You can sometimes save yourself time if you recognize it for being that special arrangement and use the pattern instead of multiplying it out as shown.

$$(2n^2 + 3)(4n^4 - 6n^2 + 9) = (2n^2 + 3) \cdot 4n^4 + (2n^2 + 3)(-6n^2) + (2n^2 + 3) \cdot 9$$
$$= 2n^2 \cdot 4n^4 + 3 \cdot 4n^4 + 2n^2(-6n^2) + 3(-6n^2) + 2n^2 \cdot 9 + 3 \cdot 9$$
$$= 8n^6 + 12n^4 - 12n^4 - 18n^2 + 18n^2 + 27$$
$$= 8n^6 + 27$$

Special Products

Any polynomials can be multiplied together. Many different types of multiplications are classified by the number of terms in the multiplier. Some products, though, are made easier to perform because of patterns that exist in them. These patterns largely are due to the special types of polynomials that are being multiplied together. Whenever you can recognize a special situation and can take advantage of a pattern, you'll save time and be less likely to make an error. Here are the special products:

1. $(a+b)(a-b) = a^2 - b^2$
2. $(a+b)^2 = a^2 + 2ab + b^2$
3. $(a+b)^3 = a^3 + 3a^2 b + 3ab^2 + b^3$ and $(a-b)^3 = a^3 - 3a^2 b + 3ab^2 - b^3$
4. $(a+b)(a^2 - ab + b^2) = a^3 + b^3$ and $(a-b)(a^2 + ab + b^2) = a^3 - b^3$

Special Product #1

$$(a+b)(a-b) = a^2 - b^2$$

This represents the product of two binomials that have the same two terms, but the terms are added in one binomial and subtracted in the other. It's better known as, "The sum and the difference of the same two numbers." Notice that, if FOIL is used here, the Outer product, $-ab$, and the Inner product, ab, are the opposites of one another. This means that they add up to 0, leaving just the first term squared minus the last term squared.

Example Problems

These problems show the answers and solutions.

1. $(x + 4)(x - 4)$

    answer: $x^2 - 16$

    Just square the first and last terms. $(x + 4)(x - 4) = x^2 - 4^2 = x^2 - 16$

2. $(t^3 + 3)(t^3 - 3)$

    answer: $t^6 - 9$

    $$(t^3 + 3)(t^3 - 3) = (t^3)^2 - 3^2 = t^6 - 9$$

3. $(abc + 11y)(abc - 11y)$

    answer: $a^2 b^2 c^2 - 121y^2$

    $$(abc + 11y)(abc - 11y) = (abc)^2 - (11y)^2 = a^2 b^2 c^2 - 121y^2$$

Special Product #2

$$(a+b)^2 = a^2 + 2ab + b^2$$

This product is of a binomial times itself. It's known as, "The perfect square trinomial." The pattern is that the first and last terms in the trinomial are the squares of the two terms in the binomial. The middle term of the trinomial is twice the product of the two original terms. Because the first and last terms are squares, they'll always be positive. The sign of the middle term will depend upon whatever the operation is in the binomial.

Use FOIL on the square of a binomial:

$$(a+b)^2 = (a+b)(a+b) = a \cdot a + a \cdot b + b \cdot a + b \cdot b = a^2 + ab + ab + b^2 = a^2 + 2ab + b^2$$

Using the special product on $(a+b)^2$, first write down the squares of the two terms:

$$a^2 \qquad b^2$$

Then double the product of the two terms: $2ab$. That's the middle term.

$$a^2 + 2ab + b^2$$

Example Problems

These problems show the answers and solutions.

1. $(z+3)^2$

   answer: $z^2 + 6z + 9$

   The squares of z and 3 are z^2 and 9. The product of the two terms is $3z$. Double that to get $6z$.

   $$(z+3)^2 = z^2 + 2 \cdot z \cdot 3 + 3^3 = z^2 + 6z + 9$$

2. $(6-m)^2$

   answer: $36 - 12m + m^2$

   $$(6-m)^2 = 6^2 + 2 \cdot 6(-m) + m^2 = 36 - 12m + m^2$$

3. $(3b+2a)^2$

   answer: $9b^2 + 12ab + 4a^2$

   $$(3b+2a)^2 = (3b)^2 + 2(3b)(2a) + (2a)^2 = 9b^2 + 12ab + 4a^2$$

Special Product #3

$$(a+b)^3 = a^3 + 3a^2 b + 3ab^2 + b^3 \text{ and } (a-b)^3 = a^3 - 3a^2 b + 3ab^2 - b^3$$

There are methods for finding all powers of a binomial such as $(a+b)^4$, $(a+b)^7$, $(a-b)^{30}$, etc. You can use the Binomial Expansion and/or Pascal's Triangle, which are topics in an Algebra II course. Cubing a binomial is a common task, and the pattern in this special product helps. When the **sum** of two terms (a binomial) is cubed, there's a 1–3–3–1 pattern coupled with decreasing powers of the first term and increasing powers of the second term. In the answer, note whether instead of all positive signs, if there's a **difference** that's being cubed, then the only change in the basic pattern is that there are alternating signs.

Example Problems

These problems show the answers and solutions.

1. $(x+1)^3$

answer: $(x+1)^3 = x^3 + 3x^2 + 3x + 1$

Using the 1–3–3–1 pattern as the base, the powers of x—x^3, x^2, x^1, x^0—are placed with their decreasing powers, right after the numbers. Note that the x^0 isn't written that way, but as a 1, instead. Then the increasing powers of 1—$1^0, 1^1, 1^2, 1^3$—are placed with the numbers and x's. The terms are then simplified.

$$1 \cdot x^3 + 3 \cdot x^2 \cdot 1^1 + 3 \cdot x^1 \cdot 1^2 + 1 \cdot 1^3 = x^3 + 3x^2 + 3x + 1$$

2. $(y+2)^3$

answer: $y^3 + 6y^2 + 12y + 8$

This time the two terms have 0 exponents for emphasis—especially the number 2, which shows how the powers increase with each step. Now, simplify the expression.

$$1 \cdot y^3 \cdot 2^0 + 3 \cdot y^2 \cdot 2^1 + 3 \cdot y^1 \cdot 2^2 + 1 \cdot y^0 \cdot 2^3 = y^3 + 6y^2 + 12y + 8$$

As you can see, the 1–3–3–1 pattern has disappeared, but using it to build the product was simpler than multiplying this out the "long way" with distribution.

3. $(4-m)^3$

answer: $64 - 48m + 12m^2 - m^3$

Whenever the $-m$ is raised to an even power (0 power and second power), the term will be positive. Whenever the $-m$ is raised to an odd power (first power and third power), the term will be negative.

$$1 \cdot 4^3 \cdot (-m)^0 + 3 \cdot 4^2 \cdot (-m)^1 + 3 \cdot 4^1 \cdot (-m)^2 + 1 \cdot 4^0 \cdot (-m)^3 = 64 - 48m + 12m^2 - m^3$$

4. $(2a-c^2)^3$

answer: $8a^3 - 12a^2c^2 + 6ac^4 - c^6$

Here, this simplification is shown in two steps, first dealing with the powers of powers and then with the final products in each term.

$$1 \cdot 8a^3 \cdot 1 + 3 \cdot 4a^2 \cdot (-c^2) + 3 \cdot 2a^1 \cdot (c^4) + 1 \cdot 1 \cdot (-c^6) = 8a^3 - 12a^2c^2 + 6ac^4 - c^6$$

Special Product #4

$$(a+b)(a^2 - ab + b^2) = a^3 + b^3 \text{ and } (a-b)(a^2 + ab + b^2) = a^3 - b^3$$

These products are paired because of their results. These are very specific types of products. The terms in the binomial and trinomial have to be just so. You wouldn't necessarily expect to see such strange combinations of factors to be multiplied except that these wonderful results do occur. Also, these combinations of multipliers will show up again when factoring binomials.

The first multiplier is a binomial. The second multiplier is a trinomial with the squares of the two terms in the binomial in the first and third positions. The middle term of the trinomial multiplier is the opposite of the product of the two terms in the binomial. The result is always the sum or difference of two cubes, and the operation between the two terms in the answer is the same as the operation in the original binomial.

Example Problems

These problems show the answers and solutions.

1. $(x+2)(x^2-2x+4)$

   answer: x^3+8

   $$(x+2)(x^2-2x+4)=x^3+2^3=x^3+8$$

2. $(3+2z)(9-6z+4z^2)$

   answer: $27+8z^3$

   $$(3+2z)(9-6z+4z^2)=3^3+(2z)^3=27+8z^3$$

Work Problems

Use these problems to give yourself additional practice.

Find the product of each.

1. $(3a+4)(3a-4)$

2. $(2-3m)^2$

3. $(1+6az)^2$

4. $(a+4)^3$

5. $(2x-y)(4x^2+2xy+y^2)$

Worked Solutions

1. **$9a^2-16$**   These binomials represent the sum and difference of the same two numbers. Their product is the difference between their squares.

   $$(3a+4)(3a-4)=(3a)^2-4^2=9a^2-16$$

2. **$4-12m+9m^2$**   This is the square of the binomial. The result consists of the squares of the first and last terms, with twice the product of the two terms in between those squares.

   $$(2-3m)^2=2^2+2\cdot2\cdot(-3m)+(-3m)^2=4-12m+9m^2$$

3. $1 + 12az + 36a^2 z^2$ This is also the square of a binomial.

$$(1 + 6az)^2 = 1^2 + 2 \cdot 1 \cdot 6az + (6az)^2 = 1 + 12az + 36a^2 z^2$$

4. $a^3 + 12a^2 + 48a + 64$ When you cube a binomial, use the 1−3−3−1 pattern.

$$(a + 4)^3 = 1 \cdot a^3 \cdot 4^0 + 3 \cdot a^2 \cdot 4^1 + 3 \cdot a^1 \cdot 4^2 + 1 \cdot a^0 \cdot 4^3 = a^3 + 12a^2 + 48a + 64$$

5. $8x^3 - y^3$ This fits the pattern where the result is the difference between two cubes.

$$(2x - y)(4x^2 + 2xy + y^2) = (2x)^3 - (y)^3 = 8x^3 - y^3$$

Dividing Polynomials

The method used in the division of polynomials depends on the divisor—the term or terms that divide into the other expression. If the divisor has only one term, then you can divide by splitting up the dividend (the expression being divided into) into separate terms and making a fraction of each term in the dividend with the divisor in the denominator. If the divisor has more than one term, then you use long division. You can use a short cut with some special types of long division, but, in general, long division will always work.

Divisors with One Term

Where there's just one term in the divisor, split the problem up into fractions and simplify each term.

Example Problems

These problems show the answers and solutions.

1. Divide $(4x^3 - 3x^2 + 2x - 6) \div 2x$.

   answer: $2x^2 - \dfrac{3}{2}x + 1 - \dfrac{3}{x}$

   The divisor, $2x$, will be the denominator of each fraction.

   $$(4x^3 - 3x^2 + 2x - 6) \div 2x = \frac{4x^3}{2x} - \frac{3x^2}{2x} + \frac{2x}{2x} - \frac{6}{2x}$$
   $$= 2x^2 - \frac{3}{2}x + 1 - \frac{3}{x}$$

2. Divide $(15a^2 b^3 + 10ab^2 - 20a^3 b^4 + ab - 2) \div 10ab$.

   answer: $\dfrac{3}{2}ab^2 + b - 2a^2 b^3 + \dfrac{1}{10} - \dfrac{1}{5ab}$

   $$(15a^2 b^3 + 10ab^2 - 20a^3 b^4 + ab - 2) \div 10ab = \frac{15a^2 b^3}{10ab} + \frac{10ab^2}{10ab} - \frac{20a^3 b^4}{10ab} + \frac{ab}{10ab} - \frac{2}{10ab}$$
   $$= \frac{3}{2}ab^2 + b - 2a^2 b^3 + \frac{1}{10} - \frac{1}{5ab}$$

Work Problems

1. Divide $(16x^2 - 12x^3 + 20x^4) \div 4x$.

2. Divide $(15z^3 - 20z^2 - 35z - 15) \div 5z^2$.

3. Divide $(9abc + 18a^2 b + 27) \div 9ab$.

4. Divide $(6x^2 y - 8xy^2 + 12x^2 y^2) \div 4x^2 y^2$.

5. Divide $(100 - 25az + 10a^2) \div 10z$.

Worked Solutions

1. $4x - 3x^2 + 5x^3$

$$\frac{16x^2 - 12x^3 + 20x^4}{4x} = \frac{16x^2}{4x} - \frac{12x^3}{4x} + \frac{20x^4}{4x} = 4x - 3x^2 + 5x^3$$

2. $3z - 4 - \dfrac{7}{z} - \dfrac{3}{z^2}$

$$\frac{15z^3 - 20z^2 - 35z - 15}{5z^2} = \frac{15z^3}{5z^2} - \frac{20z^2}{5z^2} - \frac{35z}{5z^2} - \frac{15}{5z^2} = 3z - 4 - \frac{7}{z} - \frac{3}{z^2}$$

3. $c + 2a + \dfrac{3}{ab}$

$$\frac{9abc + 18a^2 b + 27}{9ab} = \frac{9abc}{9ab} + \frac{18a^2 b}{9ab} + \frac{27}{9ab} = c + 2a + \frac{3}{ab}$$

4. $\dfrac{3}{2y} - \dfrac{2}{x} + 3$

$$\frac{6x^2 y - 8xy^2 + 12x^2 y^2}{4x^2 y^2} = \frac{6x^2 y}{4x^2 y^2} - \frac{8xy^2}{4x^2 y^2} + \frac{12x^2 y^2}{4x^2 y^2} = \frac{3}{2y} - \frac{2}{x} + 3$$

5. $\dfrac{10}{z} - \dfrac{5a}{2} + \dfrac{a^2}{z}$

$$\frac{100 - 25az + 10a^2}{10z} = \frac{100}{10z} - \frac{25az}{10z} + \frac{10a^2}{10z} = \frac{10}{z} - \frac{5a}{2} + \frac{a^2}{z}$$

Long Division

When the divisor has two or more terms, you can do the operation with long division. This looks very much like the division done on whole numbers. It uses the same set up and operations.

Example Problems

These problems show the answers and solutions.

1. Divide $(5y^3 - 3y^2 + y - 1) \div (y + 2)$ using long division.

   answer: $5y^2 - 13y + 27 - \dfrac{55}{y + 2}$

Write the dividend in decreasing powers, leaving spaces for any missing terms (skipped powers).

$$y + 2 \overline{\smash{\big)}\, 5y^3 - 3y^2 + y - 1}$$

Focus on the first term in the divisor, the y. Then determine what must multiply that y so you get the first term in the dividend, the $5y^3$. Multiplying y by $5y^2$ will give you $5y^3$. Write the $5y^2$ above the $5y^3$.

$$\begin{array}{r} 5y^2 \\ y + 2 \overline{\smash{\big)}\, 5y^3 - 3y^2 + y - 1} \end{array}$$

Now multiply both terms in the divisor by $5y^2$ and put the results under the terms in the dividend that are alike.

$$\begin{array}{r} 5y^2 \\ y + 2 \overline{\smash{\big)}\, 5y^3 - 3y^2 + y - 1} \\ 5y^3 + 10y^2 \end{array}$$

Next, subtract. The easiest way is to change each term in the expression that you're subtracting to its opposite and add. As shown in Chapter 2, in the section "Subtraction of Signed Numbers," subtraction is changed to addition by changing the sign of the number being subtracted.

$$\begin{array}{r} 5y^2 \\ y + 2 \overline{\smash{\big)}\, 5y^3 - 3y^2 + y - 1} \\ \underline{-5y^3 - 10y^2 } \\ -13y^2 + y - 1 \end{array}$$

The remaining terms in the dividend are brought down and then the process is repeated. Determine what you have to multiply y by to get the new first term, $-13y^2$. Multiplying by $-13y$ will do it.

$$\begin{array}{r} 5y^2 - 13y \\ y + 2 \overline{\smash{\big)}\, 5y^3 - 3y^2 + y - 1} \\ \underline{-5y^3 - 10y^2 } \\ -13y^2 + y - 1 \\ -13y^2 - 26y \end{array}$$

Again, change the signs and add.

$$\begin{array}{r} 5y^2 - 13y \\ y + 2 \overline{\smash{\big)}\, 5y^3 - 3y^2 + y - 1} \\ \underline{-5y^3 - 10y^2 } \\ -13y^2 + y - 1 \\ \underline{-13y^2 + 26y } \\ 27y - 1 \end{array}$$

You get the last part of the answer by multiplying the y in the divisor by 27 to get the new first term, the $27y$.

$$\begin{array}{r} 5y^2 - 13y + 27 \\ y+2\overline{\smash{)}5y^3 - 3y^2 + y - 1} \\ \underline{-5y^3 - 10y^2} \\ -13y^2 + y - 1 \\ \underline{-13y^2 + 26y} \\ 27y - 1 \\ 27y + 54 \end{array}$$

Do the last subtraction. Whatever is left over is the remainder. The remainder is usually written as a fraction with the divisor in the denominator.

$$\begin{array}{r} 5y^2 - 13y + 27 \\ y+2\overline{\smash{)}5y^3 - 3y^2 + y - 1} \\ \underline{-5y^3 - 10y^2} \\ -13y^2 + y - 1 \\ \underline{-13y^2 + 26y} \\ 27y - 1 \\ \underline{-27y - 54} \\ -55 \end{array}$$

$$(5y^3 - 3y^2 + y - 1) \div (y + 2) = 5y^2 - 13y + 27 - \frac{55}{y+2}$$

2. Use long division to divide $(8m^4 - 3m^2 + m - 12) \div (2m - 1)$.

answer: $4m^3 + 2m^2 - \dfrac{1}{2}m + \dfrac{1}{4} - \dfrac{11\frac{3}{4}}{2m-1}$

In this problem, there are missing terms. The m^3 term is missing in the decreasing powers, so you need to insert a space or zero.

$$\begin{array}{r} 4m^3 + 2m^2 - \frac{1}{2}m + \frac{1}{4} \\ 2m-1\overline{\smash{)}8m^4 \qquad\quad - 3m^2 + m - 12} \\ \underline{8m^4 - 4m^3} \\ 4m^3 - 3m^2 + m - 12 \\ \underline{4m^3 - 2m^2} \\ -m^2 + m - 12 \\ \underline{-m^2 + \frac{1}{2}m} \\ \frac{1}{2}m - 12 \\ \underline{\frac{1}{2}m - \frac{1}{4}} \\ -11\frac{3}{4} \end{array}$$

$$(8m^4 - 3m^2 + m - 12) \div (2m - 1) = 4m^3 + 2m^2 - \frac{1}{2}m + \frac{1}{4} - \frac{11\frac{3}{4}}{2m-1}$$

Synthetic Division

Many division problems involving polynomials have special divisors of the form $x + a$ or $x - a$ where the coefficient of the variable is a 1. When this is the case, you can avoid long division and obtain the answer with a process called *synthetic division*. To do the problem $(3x^4 - 2x^3 + 91x + 11) \div (x + 3)$ using synthetic division, use the following steps:

1. Write just the coefficients of the terms in decreasing order, inserting 0s for any missing terms (powers).

$$3 \qquad -2 \qquad 0 \qquad 91 \qquad 11$$

2. Put the opposite of the constant in the divisor in front of the row of coefficients.

$$\underline{-3}\,|\,3 \qquad -2 \qquad 0 \qquad 91 \qquad 11$$

3. Draw a line about two spaces below the row of coefficients and drop the first coefficient to below the line.

$$
\begin{array}{r|rrrrr}
-3 & 3 & -2 & 0 & 91 & 11 \\
\hline
 & 3 & & & &
\end{array}
$$

4. Multiply the constant in front times the dropped number in the coefficient and place the result below the second coefficient in the row. Add the two numbers together and put the result below the line.

$$
\begin{array}{r|rrrrr}
-3 & 3 & -2 & 0 & 91 & 11 \\
 & & -9 & & & \\
\hline
 & 3 & -11 & & &
\end{array}
$$

5. Multiply that result by the constant in front and put the result below the third coefficient in the row. Add the two numbers and put the result below the line. Repeat this process until you run out of numbers in the row of coefficients.

$$
\begin{array}{r|rrrrr}
-3 & 3 & -2 & 0 & 91 & 11 \\
 & & -9 & 33 & -99 & 24 \\
\hline
 & 3 & -11 & 33 & -8 & 35
\end{array}
$$

6. Write the answer by using the new coefficients below the line and inserting powers of the variable, starting with a variable that has a power one less than the power of the dividend.

$$(3x^4 - 2x^3 + 91x + 11) \div (x + 3) = 3x^3 - 11x^2 + 33x - 8 + \frac{35}{x+3}$$

In general, synthetic division is much preferred over long division. It takes less time and less space, and you make fewer errors in signs and computations. It's always addition of the signed numbers, because the constant in the divisor is changed to its opposite right at the beginning, taking care of the subtraction process. This next example shows more on inserting the zeros for missed terms.

Example Problem

1. Divide $(4x^7 - 20x^4 - 3x^2 - 1) \div (x - 2)$.

   answer: $4x^6 + 8x^5 + 16x^4 + 12x^3 + 24x^2 + 45x + 90 + \dfrac{179}{x-2}$

   Set up the problem by writing just the coefficients of the terms in a row. Put in zeros for any terms that are missing in the list of decreasing powers of the variable.

   $$\begin{array}{r|rrrrrrrr} +2 & 4 & 0 & 0 & -20 & 0 & -3 & 0 & -1 \\ & & 8 & 16 & 32 & 24 & 48 & 90 & 180 \\ \hline & 4 & 8 & 16 & 12 & 24 & 45 & 90 & 179 \end{array}$$

   $(4x^7 - 2x^4 - 3x^2 - 1) \div (x - 2) = 4x^6 + 8x^5 + 16x^4 + 12x^3 + 24x^2 + 45x + 90 + \dfrac{179}{x-2}$

Work Problems

Use these problems to give yourself additional practice.

1. Use long division to divide $(8x^4 - 2x^2 + 18x + 6) \div (2x - 1)$.

2. Use long division to divide $(3x^3 + 22x^2 - 7x - 6) \div (3x - 2)$.

3. Use synthetic division to divide $(4x^4 - 2x^3 + 3x^2 - x - 8) \div (x + 2)$.

4. Use synthetic division to divide $(x^4 + 3x + 1) \div (x - 1)$.

5. Use synthetic division to divide $(x^9 + 27x^6 + x + 3) \div (x + 3)$.

Worked Solutions

1. $4x^3 + 2x^2 + 9 + \dfrac{15}{2x-1}$ Set up the long division problem. Focus on the first term of the divisor, the $2x$.

$$
\require{enclose}
\begin{array}{r}
4x^3 + 2x^2 + 9 + \frac{15}{2x-1} \\[2pt]
2x - 1 \enclose{longdiv}{8x^4 - 2x + 18x + 6} \\[2pt]
\underline{8x^4 - 4x^3} \\[2pt]
+4x^3 - 2x^2 + 18x + 6 \\[2pt]
\underline{4x^3 - 2x^2} \\[2pt]
18x + 6 \\[2pt]
\underline{18x - 9} \\[2pt]
15
\end{array}
$$

2. $x^2 + 8x + 3$

$$
\begin{array}{r}
x^2 + 8x + 3 \\
3x - 2 \overline{\smash{\big)}\, 3x^3 + 22x^2 - 7x - 6} \\
\underline{3x^3 - 2x^2} \\
24x^2 - 7x - 6 \\
\underline{24x^2 - 16x} \\
9x - 6 \\
\underline{9x - 6} \\
0
\end{array}
$$

In this case, there was no remainder. The binomial $3x - 2$ divided the expression evenly.

3. $4x^3 - 10x^2 + 23x - 47 + \dfrac{86}{x+2}$

$$
\begin{array}{r|rrrrr}
-2 & 4 & -2 & 3 & -1 & 8 \\
   &   & -8 & 20 & -46 & 94 \\
\hline
   & 4 & -10 & 23 & -47 & 86
\end{array}
$$

$$(4x^4 - 2x^3 + 3x^2 - x - 8) \div (x + 2) = 4x^3 - 10x^2 + 23x - 47 + \frac{86}{x+2}$$

4. $x^3 + x^2 + x + 4 + \dfrac{5}{x-1}$

$$
\begin{array}{r|rrrrr}
1 & 1 & 0 & 0 & 3 & 1 \\
  &   & 1 & 1 & 1 & 4 \\
\hline
  & 1 & 1 & 1 & 4 & 5
\end{array}
$$

$$(x^4 - 3x + 1) \div (x - 1) = x^3 + x^2 + x + 4 + \frac{5}{x-1}$$

5. $x^8 - 3x^7 + 9x^6 + 1$

$$
\begin{array}{r|rrrrrrrrr}
-3 & 1 & 0 & 0 & 27 & 0 & 0 & 0 & 0 & 1 & 3 \\
   &   & -3 & 9 & -27 & 0 & 0 & 0 & 0 & 0 & -3 \\
\hline
   & 1 & -3 & 9 & 0 & 0 & 0 & 0 & 0 & 1 & 0
\end{array}
$$

$$(x^9 + 27x^6 + x + 3) \div (x + 3) = x^8 - 3x^7 + 9x^6 + 1$$

Factoring

To factor an expression means to rewrite it as a big multiplication problem—all one term—instead of a string of several terms all added (or subtracted) together. For instance, the factored form for $x^4 + 3x^3 - 4x^2 - 12x$ is $x(x + 2)(x - 2)(x + 3)$. The first expression has 4 terms in it. The factored version has one term—four factors all multiplied together. The factored form has many uses in algebra. The factored form is needed in the numerator and denominator of a fraction, if it is to be reduced. The factored form is needed when solving most nonlinear equations. A factored form usually is more easily evaluated when substituting a number for the variable. For instance, in the two expressions mentioned earlier, let $x = 4$. When putting 4 in place of x in the first expression, you have to raise it to the fourth power, third power, and so on. You have to work with big numbers.

$$4^4 + 3(4)^3 - 4(4)^2 - 12(4) = 256 + 3(64) - 4(16) - 48 = 256 + 192 - 64 - 48 = 336.$$

I don't know about you, but I couldn't do that one in my head. Compare that to putting the 4 into the factored form: $x(x + 2)(x - 2)(x + 3) = 4(4 + 2)(4 - 2)(4 + 3) = 4(6)(2)(7) = 24 \cdot 14 = 336$. Okay, I couldn't do the last multiplication in my head either, but it was still easier!

Greatest Common Factor

Writing an expression in factored form seems almost like undoing distribution. The distributive property is covered in Chapter 1, if you need a review. Instead of multiplying every term by the same factor as done using the distributive property, factoring actually divides every term by the same factor and leaves the division result behind, inside the parentheses. The first step to factoring out the Greatest Common Factor (GCF) is to determine what the GCF is. In general, a common factor is a multiplier shared by every single term in an expression. The GCF is the largest possible factor—containing all of the smaller factors—that divides every single term in the expression evenly, leaving no remainders or fractions.

Example Problems

These problems show the answers and solutions.

1.   Factor out the GCF of $6x^2 y + 8x^3 y^2 + 4x^4$.

     answer: $2x^2(3y + 4xy^2 + 2x^2)$

     The GCF is $2x^2$ because each term is divisible evenly by 2 and x^2. There's no factor of y in the last term. The question now is $2x^2 (???) = 6x^2 y + 8x^3 y^2 + 4x^4$.

     Divide each term by $2x^2$ and put the result of each division in the parentheses.

     $$\frac{6x^2 y}{2x^2} = 3y, \frac{8x^3 y^2}{2x^2} = 4xy^2, \text{ and } \frac{4x^4}{2x^2} = 2x^2$$

     So, $6x^2 y + 8x^3 y^2 + 4x^4 = 2x^2(3y + 4xy^2 + 2x^2)$. This is completely factored. The terms in the parentheses have no common factor other than the number 1.

2.   Factor $-36y^5 + 24y^3 - 48y$.

     answer: $-12y(3y^4 - 2y^2 + 4)$

     The GCF is either $12y$ or $-12y$. Either divides evenly. It's often nicer to have a lead coefficient in the parentheses that's positive, so $-12y$ is a better choice. Divide each term by $-12y$.

     $$\frac{-36y^5}{-12y} = 3y^4, \frac{24y^3}{-12y} = -2y^2, \text{ and } \frac{-48y}{-12y} = 4$$

     So, $-36y^5 + 24y^3 - 48y = -12y(3y^4 - 2y^2 + 4)$

3. Factor $9x(y-3)^2 - 12x^2(y-3)$.

   answer: $3x(y-3)[3(y-3)-4x]$

   The two terms here have common factors of 3, x, and $(y-3)$, making the GCF equal to $3x(y-3)$.

$$\frac{9x(y-3)^2}{3x(y-3)} = 3(y-3) \text{ and } \frac{-12x^2(y-3)}{3x(y-3)} = -4x$$

   So, $9x(y-3)^2 - 12x^2(y-3) = 3x(y-3)[3(y-3)-4x]$. Because the terms in the brackets won't combine, even if you distribute the 3 over the terms in the parentheses, just leave it as is.

4. Factor $10a^{1/2}b^{-3/2} + 20a^{-1/2}b^{-1/2}$.

   answer: $10a^{-1/2}b^{-3/2}(a+2b)$

   The two terms have three factors in common. First, they are both divisible by 10. The other two factors are powers of a and b. To choose the powers correctly, determine which of the two powers of a is the smaller. The number $-\frac{1}{2}$ is smaller than $\frac{1}{2}$, so the common factor that can be taken out is $a^{-1/2}$. The two powers of b are $-\frac{3}{2}$ and $-\frac{1}{2}$. The smaller of those two is $-\frac{3}{2}$, so the common factor of the two preceding terms is $b^{-3/2}$. Putting them all together, the GCF of the two terms is $10a^{-1/2}b^{-3/2}$.

$$\frac{10a^{1/2}b^{-3/2}}{10a^{-1/2}b^{-3/2}} = a^{1/2-(-1/2)} = a^{1/2+1/2} = a^1 \text{ and } \frac{a^{-1/2}b^{-1/2}}{10a^{-1/2}b^{-3/2}} = 2b^{-1/2-(-3/2)} = 2b^{-1/2+3/2} = 2b^1$$

$$\text{So, } 10a^{1/2}b^{-3/2} + 20a^{-1/2}b^{-1/2} = 10a^{-1/2}b^{-3/2}(a+2b)$$

Work Problems

Use these problems to give yourself additional practice.

1. Factor $16y^4 - 12y^3 + 20y^2 + 8y$.

2. Factor $40a^2b^3 + 25a^3b^2 - 15ab^4$.

3. Factor $-3m^8 + 9m^6 - 6m^4 - 18m^2$.

4. Factor $-3(x+2)^4(y-1)^3 + 6(x+2)^3(y-1)^8$.

5. Factor $6p^{1/2}(t-1)^{-1} - 8p^{3/2}(t-1)^{-2}$.

Worked Solutions

1. $4y(4y^3 - 3y^2 + 5y + 2)$ The GCF is $4y$. Divide each term by $4y$:

$$\frac{16y^4}{4y} = 4y^3, \ \frac{-12y^3}{4y} = -3y^2, \ \frac{20y^2}{4y} = 5y, \text{ and } \frac{8y}{4y} = 2$$

So $16y^4 - 12y^3 + 20y^2 + 8y = 4y(4y^3 - 3y^2 + 5y + 2)$

2. $5ab^2(8ab + 5a^2 - 3b^2)$ The GCF is $5ab^2$. Divide each term by $5ab^2$:

$$\frac{40a^2 b^3}{5ab^2} = 8ab, \ \frac{25a^3 b^2}{5ab^2} = 5a^2 \text{ and } \frac{-15ab^4}{5ab^2} = -3b^2$$

So, $40a^2 b^3 + 25a^3 b^2 - 15ab^4 = 5ab^2(8ab + 5a^2 - 3b^2)$

3. $-3m^2(m^6 - 3m^4 + 2m^2 + 6)$ The GCF is $-3m^2$. Divide each term by $-3m^2$:

$$\frac{-3m^8}{-3m^2} = m^6, \ \frac{9m^6}{-3m^2} = -3m^4, \ \frac{-6m^4}{-3m^2} = 2m^2, \text{ and } \frac{-18m^2}{-3m^2} = 6$$

So, $-3m^8 + 9m^6 - 6m^4 - 18m^2 = -3m^2(m^6 - 3m^4 + 2m^2 + 6)$

Note that, if you chose the GCF to be $+3m^2$, then all of the terms in the parentheses would have the opposite signs of those in my answer.

4. $-3(x+2)^3(y-1)^3[(x+2) - 2(y-1)^5]$ The GCF is $-3(x+2)^3(y-1)^3$.

Divide each term by the GCF:

$$\frac{-3(x+2)^4(y-1)^3}{-3(x+2)^3(y-1)^3} = (x+2)^1 \text{ and } \frac{6(x+2)^3(y-1)^8}{-3(x+2)^3(y-1)^3} = -2(y-1)^5$$

So, $-3(x+2)^4(y-1)^3 + 6(x+2)^3(y-1)^8 = -3(x+2)^3(y-1)^3[(x+2) - 2(y-1)^5]$

5. $2p^{1/2}(t-1)^{-2}[3(t-1) - 4p]$ The GCF is $2p^{1/2}(t-1)^{-2}$. Divide each term by the GCF:

$$\frac{6p^{1/2}(t-1)^{-1}}{2p^{1/2}(t-1)^{-2}} = 3(t-1)^{-1-(-2)} \text{ and } \frac{-8p^{3/2}(t-1)^{-2}}{2p^{1/2}(t-1)^{-2}} = -4p^{3/2-1/2}$$

Simplify those results further: $3(t-1)^{-1-(-2)} = 3(t-1)^1$ and $-4p^{3/2-1/2} = -4p^1$.

So, $6p^{1/2}(t-1)^{-1} - 8p^{3/2}(t-1)^{-2} = 2p^{1/2}(t-1)^{-2}[3(t-1) - 4p]$.

Factoring Binomials

A binomial is an expression with two terms separated by either addition or subtraction. The goal is to make it all one term—with everything multiplied together. This is accomplished by factoring the two terms. You can use four basic methods to factor a binomial. If none of these methods works, the expression is considered to be **prime**—it cannot be factored.

The rules or patterns to use when doing the factoring are as follows:

Rule 1. Factoring out the Greatest Common Factor

$$ab + ac = a(b + c)$$

The common factor here is a; it's written as a multiplier of what's left after dividing it out of the terms.

Rule 2. Factoring using the pattern for the **difference of squares**

$$a^2 - b^2 = (a - b)(a + b)$$

Use one of the rules for special products of binomials (discussed earlier in the chapter); this is how the difference between two squares can be written as a product.

Rule 3. Factoring using the pattern for the **difference of cubes**

$$a^3 - b^3 = (a - b)(a^2 + ab + b^2)$$

This rule and the next use a similar pattern. There are two factors—a binomial and a trinomial. The binomial contains the two cube roots of the terms. The trinomial contains the squares of those roots in the first and third positions and then the opposite of the product of those two square roots as the middle term.

Rule 4. Factoring using the pattern for the **sum of cubes**

$$a^3 + b^3 = (a + b)(a^2 - ab + b^2)$$

Just like the preceding rule, two factors exist—a binomial and a trinomial. The binomial contains the two cube roots of the terms. The trinomial contains the squares of those roots in the first and third positions and then the opposite of the product of those two cube roots as the middle term.

The challenge will be in determining which factoring method to use. If you recognize that both terms are perfect squares and they're subtracted, then Rule 2 makes sense. If both terms are perfect cubes, then Rule 3 or 4 will work. If they have one or more factors in common, then use Rule 1. Sometimes, you get to use more than one rule to complete the job.

Example Problems

These problems show the answers and solutions.

1. Factor $4x^3 y^2 z + 6x^2 y^2 z^4$.

    answer: $2x^2 y^2 z(2x + 3z^3)$

    Rule 1 is used here. The two terms each have factors of 2, x^2, y^2, and z. That means that each of these factors, as well as the whole thing, $2x^2 y^2 z$, divides each of the terms evenly.

    $$\frac{4x^3 y^2 z}{2x^2 y^2 z} = 2x \text{ and } \frac{6x^2 y^2 z^4}{2x^2 y^2 z} = 3z^3$$

The common factor, $2x^2 y^2 z$, is written out in front of the parentheses containing the results of the divisions.

$$4x^3 y^2 z + 6x^2 y^2 z^4 = 2x^2 y^2 z(2x + 3z^3)$$

2. Factor $4x^2 - 9$.

   answer: $(2x - 3)(2x + 3)$

   Rule 2 is used because this expression is the difference between two perfect squares.

   $$4x^2 = (2x)^2 \text{ and } 9 = 3^2.$$
   $$\text{So } 4x^2 - 9 = (2x - 3)(2x + 3)$$

3. Factor $25a^2 y^2 - t^4$.

   answer: $(5ay - t^2)(5ay + t^2)$

   The term $25a^2 y^2$ is a perfect square; the rule involving exponents of products tells us that $(5ay)^2 = 5^2 a^2 y^2$ and $(t^2)^2 = t^4$. So $25a^2 y^2 - t^4 = (5ay - t^2)(5ay + t^2)$.

4. Factor $m^3 - 125$.

   answer: $(m - 5)(m^2 + 5m + 25)$

   Rule 3 is used here. The two terms are each perfect cubes. The rule for doing this factorization involves finding the two cube roots. If the two cubes are a^3 and b^3, then their roots are a and b, and those two roots are written in the first part of the factorization, $(a - b)$. Then a trinomial is built using the squares of those two roots and the **opposite of the product** of those same two roots. In this problem, m is the cube root of m^3, and -5 is the cube root of -125. The square of m is m^2, and the square of -5 is 25. The opposite of the product of those two cube roots is the opposite of $-5m$, which is $+5m$. Now, put everything into the pattern, $m^3 - 125 = (m - 5)(m^2 + 5m + 25)$.

5. Factor $y^6 + 8z^3$.

   answer: $(y^2 + 2z)(y^4 - 2y^2 z + 4z^2)$

   The two terms are each perfect cubes and, because they're added, Rule 4 is used. $y^6 = (y^2)^3$, so y^2 is the cube root of y^6, and $8z^3 = (2z)^3$, so $2z$ is the cube root of $8z^3$. Using the same pattern as in the previous example, the two cube roots are written in the binomial, and their squares are the first and last terms in the trinomial. The opposite of the product of these two cube roots is $-2y^2 z$, so it's written as the middle term in the trinomial. $y^6 + 8z^3 = (y^2 + 2z)(y^4 - 2y^2 z + 4z^2)$

6. Factor $ab^3 x^2 - 9ab^3$.

   answer: $ab^3 (x - 3)(x + 3)$

   Even though squares and cubes are in both terms, this expression doesn't fit any of the patterns for factoring differences of cubes or squares. A common factor exists in the two terms, though. It's ab^3. Factoring that out, $ab^3 x^2 - 9ab^3 = ab^3 (x^2 - 9)$. Now, something nice has happened. The expression in the parentheses can be factored using the difference of squares. So, factor again, $ab^3 (x^2 - 9) = ab^3 (x - 3)(x + 3)$.

Work Problems

Use these problems to give yourself additional practice.

Factor each expression.

1. $x^3 + 1$

2. $16x^2 - 81$

3. $x^2 y^2 - 4x^3 y^3$

4. $27 - 8a^6$

5. $25a^3 n^2 + 25n^2 z^3$

Worked Solutions

1. **$(x + 1)(x^2 - x + 1)$**   This is the sum of two cubes. Use the pattern in which the binomial contains the cube roots of the two terms and the trinomial contains the squares of the roots and the opposite of the product of the roots.

$$x^3 + 1 = (x + 1)(x^2 - x + 1)$$

2. **$(4x - 9)(4x + 9)$**   This is the difference of two squares. Use the rule in which the factored form has two binomials that are the sum and difference of the roots of those squares.

$$16x^2 - 81 = (4x - 9)(4x + 9)$$

3. **$x^2 y^2 (1 - 4xy)$**   Even though this problem contains squares and cubes, this doesn't fit any of the last three rules. The Greatest Common Factor is $x^2 y^2$, so factor it out, $x^2 y^2 - 4x^3 y^3 = x^2 y^2 (1 - 4xy)$.

4. **$(3 - 2a^2)(9 + 6a^2 + 4a^4)$**   This is the difference between two cubes. Use the following pattern:

$$27 - 8a^6 = (3 - 2a^2)(9 + 6a^2 + 4a^4)$$

5. **$25n^2 (a + z)(a^2 - az + z^2)$**   First, factoring out the common factor of $25n^2$, you get $25a^3 n^2 + 25n^2 z^3 = 25n^2 (a^3 + z^3)$.

The binomial in the parentheses is the sum of two perfect cubes, so it can be factored.

$$25n^2 (a^3 + z^3) = 25n^2 (a + z)(a^2 - az + z^2)$$

Factoring Trinomials

Trinomials can be factored in two ways—either by finding a Greatest Common Factor or by un-FOILing (or both). Factoring out the GCF is discussed in the previous section. The other option, un-FOILing, involves determining what two binomials were multiplied together to get the trinomial. Sometimes, both methods will need to be used. There can be a common factor **and** there can be two binomials whose product is the trinomial. In those cases, it's best to take out the common factor first, to make the numbers smaller.

If you need to review multiplying binomials together using the FOIL method, refer to the section earlier in this chapter. Just as a quick review/reference, consider multiplying the two binomials $(a + b)(c + d)$. When using FOIL, you add the products of the First terms, Outer terms, Inner terms, and Last terms. So, $(a + b)(c + d) = a \cdot c + a \cdot d + b \cdot c + b \cdot d$.

Consider the general trinomial $ax^2 \pm bx \pm c$. The $\pm$ means that there can be either an add or a subtract in that position. Using this trinomial, the procedure for factoring the trinomial by un-FOILing is:

> First, find two terms whose product is the first term in the trinomial, ax^2. These two terms will be in the First positions in the two binomials of the factorization.
>
> Second, find two terms whose product is the last term in the trinomial, c. These two terms will be in the Last position in the binomials.
>
> If c is **positive**, arrange the terms determined previously so that the Outer and Inner products will have a **sum** equal to bx, the middle term in the trinomial.
>
> If c is **negative**, arrange the terms determined previously so that the Outer and Inner products will have a **difference** of bx, the middle term. The sign of b is what is considered when you assign $+$ and $-$ signs to make the product correct.
>
> If a is negative, the easiest thing to do is to first factor out -1 from each term and work with the trinomial in the parenthesis.

There may be several choices for the First position terms and Last position terms. You may have to try more than one arrangement.

Example Problems

These problems show the answers and solutions.

1. Factor $x^2 - 2x - 8$.

   answer: $(x + 2)(x - 4)$

   The two terms whose product is x^2 are x and x. You have two choices for the product of 8: $1 \cdot 8$ and $2 \cdot 4$. The possible arrangements for these terms are $(x\ 1)(x\ 8)$ or $(x\ 2)(x\ 4)$. Because 8 is negative, the Outer and Inner products need to have a difference of 2, the coefficient of the middle term. That makes the choice of factors $(x\ 2)(x\ 4)$. There will have to be a $+$ in one of the binomials and a $-$ in the other, so that the Last product will be negative. The arrangement that gives the correct product with the correct sign on the middle term is $(x + 2)(x - 4)$.

2. Factor $2x^2 + 5x - 12$.

   answer: $(2x - 3)(x + 4)$

   The two terms whose product is $2x^2$ are $2x$ and x. You have three different choices for the product of 12: $1 \cdot 12$, $2 \cdot 6$, and $3 \cdot 4$. The possible arrangements for these terms are

   $(2x\ 1)(x\ 12)$ or $(2x\ 12)(x\ 1)$ or $(2x\ 2)(x\ 6)$ or $(2x\ 6)(x\ 2)$ or $(2x\ 3)(x\ 4)$ or $(2x\ 4)(x\ 4)$

Because the 12 is negative, the Outer and Inner products need to have a difference of 5, the coefficient of the middle term. The choice that accomplishes this is $(2x\ 3)(x\ 4)$. The Outer product is $8x$, and the Inner product is $3x$. Their difference is $5x$. To make the signs of the product come out correctly, put in the + and − as follows: $(2x - 3)(x + 4)$.

3. Factor $9y^2 - 24y + 16$.

    answer: $(3y - 4)(3y - 4)$

    You have two choices for the terms needed to get $9y^2$. They are $9y \cdot y$ and $3y \cdot 3y$.

    You have three choices for the 16: $1 \cdot 16$, $2 \cdot 8$, and $4 \cdot 4$. The possible arrangements are

$(9y\ 1)(y\ 16)$ or $(9y\ 16)(y\ 1)$ or $(9y\ 2)(y\ 8)$ or $(9y\ 8)(y\ 2)$ or $(9y\ 4)(y\ 4)$ or $(3y\ 1)(3y\ 16)$ or
$(3y\ 2)(3y\ 8)$ or $(3y\ 4)(3y\ 4)$

One thing that can help is if you notice that the first and last terms in the trinomial are perfect squares. Earlier in this chapter, under "Special Products," you learned the pattern for when a binomial is squared. When this happens, the first and last terms are perfect squares, and the middle term is twice the product of those original factors. This appears to be the situation with this problem. The choice of $(3y\ 4)(3y\ 4)$ works. And, because the last term is positive and the middle term is negative, both of the operations in the binomials are negative. The factorization is $(3y - 4)(3y - 4)$.

4. Factor $15y^2 - 26y + 8$.

    answer: $(5y - 2)(3y - 4)$

    This time you need to consider several choices and combinations of factors.

    $15y^2$ can be written as $15y \cdot y$ or $5y \cdot 3y$. 8 can be written as $8 \cdot 1$ or $4 \cdot 2$. Instead of writing down all of the choices and arrangements, look at the trinomial and determine whether it's a sum or difference of products that has to be chosen. Because the 8 is positive, the Outer and Inner products will need to be either both positive or both negative, so it'll be a sum of products. Which combination of factors will give products that have a sum of 26?

    Just considering the numbers, it's the $5 \cdot 3$ and $4 \cdot 2$ combinations that will work. Multiplying the 5 and 4 gives you 20, and multiplying the 3 and 2 gives you 6. Now, the numbers have to be arranged in the parentheses so that the correct products are obtained.

    $(5y\ 2)(3y\ 4)$ is what works. Both of the operations will be minus, so the solution is $(5y - 2)(3y - 4)$.

    Sometimes, it takes a while to find the correct combination. When you have many choices to work with and lots of arrangements to consider, it's a good idea to go through them systematically, trying all the arrangements with one pair of factors and then moving on to the next until you find the one you need. This way, you won't miss any and won't repeat yourself.

5. Factor $16x^2 + 108x + 72$.

    answer: $4(4x + 3)(x + 6)$

Before factoring this trinomial using unFOIL, divide out the common factor. Each term is divisible by 4. So, $16x^2 + 108x + 72 = 4(4x^2 + 27x + 18)$. That leaves a trinomial to be factored whose First term, $4x^2$, can be $4x \cdot x$ or $2x \cdot 2x$; Last term, 18, can be $18 \cdot 1$, $9 \cdot 2$, or $6 \cdot 3$. Consider what the choices would be if you didn't factor the 4 out first. You'd have $16x^2$ can be $16x \cdot x$, $8x \cdot 2x$, or $4x \cdot 4x$. Seventy-two can be $72 \cdot 1$, $36 \cdot 2$, $24 \cdot 3$, $18 \cdot 4$, $12 \cdot 6$, or $9 \cdot 8$. The last term is positive, so the Outer and Inner products need to have a sum of 27. The combination of factors that will work is $(4x \quad 3)(x \quad 6)$. Putting addition signs for the operations and putting the 4 that was factored out in the factorization, results in $16x^2 + 108x + 72 = 4(4x + 3)(x + 6)$.

Work Problems

Use these problems to give yourself additional practice.

1. Factor $y^2 + 3y - 10$.

2. Factor $2a^2 + 14a + 24$.

3. Factor $12x^2 - 25x + 12$.

4. Factor $12n^2 - 18n - 120$.

5. Factor $4x^2 + 20x + 25$.

Worked Solutions

1. **$(y - 2)(y + 5)$**   The First term, y^2, can be written as $y \cdot y$, only. The Last term, 10, can be written as $10 \cdot 1$ or $5 \cdot 2$. Because the 10 is negative, you want the **difference** of the Outer and Inner products to be $3y$.

    $$y^2 + 3y - 10 = (y - 2)(y + 5)$$

2. **$2(a + 3)(a + 4)$**   The first thing to do is factor a 2 out of each term.

    $$2a^2 + 14a + 24 = 2(a^2 + 7a + 12)$$

    Now, in the new trinomial, the First term, a^2, can be written as $a \cdot a$, only. The Last term, 12, can be written as $12 \cdot 1$, $6 \cdot 2$, or $4 \cdot 3$. Because the last term is positive, you want the **sum** of the Outer and Inner products to be $7a$. Because the middle term is also positive, both of the operation signs will be $+$. $2a^2 + 14a + 24 = 2(a^2 + 7a + 12) = 2(a + 3)(a + 4)$

3. **$(3x - 4)(4x - 3)$**   The First term, $12x^2$, can be written as $12x \cdot x$, $6x \cdot 2x$, or $4x \cdot 3x$. The Last term, another 12, can be written as $12 \cdot 1$, $6 \cdot 2$, or $4 \cdot 3$. Because the last term is positive, you want the **sum** of the Outer and Inner products to be $17x$. Because the middle term is negative, both of the operation signs will be negative. $12x^2 - 25x + 12 = (3x - 4)(4x - 3)$

4. **$6(2n + 5)(n - 4)$**   Here's another trinomial that can be factored first, by dividing by 6.

    $$12n^2 - 18n - 120 = 6(2n^2 - 3n - 20)$$

The First term in the new trinomial can be written as $2n \cdot n$.

The Last term in that trinomial can be written as $20 \cdot 1$, $10 \cdot 2$, or $5 \cdot 4$.

Because the last term is negative, you want the *difference* of the Outer and Inner products to be $3n$. The operations in the binomials will be one of each, $+$ and $-$.

$$12n^2 - 18n - 120 = 6(2n^2 - 3n - 20) = 6(2n + 5)(n - 4)$$

5. **$(2x + 5)(2x + 5)$**   The First and Last terms are both perfect squares. Check first to see whether this is a perfect square trinomial—if the middle term is twice the product of the roots.

   The binomials will both be $2x + 5$, the two roots of the first and last terms.

   $$(2x + 5)(2x + 5) = 4x^2 + 10x + 10x + 25 = 4x^2 + 20x + 25$$

   It worked! Be careful, though. Not all trinomials with perfect squares in the first and last terms are perfect square trinomials. Be sure to check carefully.

Factoring Other Polynomials

Two processes considered in this section are

- ❑ Factoring quadratic-like expressions
- ❑ Factoring by grouping

Quadratic-Like Expressions

A quadratic-like expression is one of the form $ax^{2n} + bx^n + c$ in which there's a trinomial with a term raised to an even power, a term raised to half that even power, and a constant. Some examples are as follows: $y^4 + 6y^2 + 5$, $6a^6 - a^3 - 2$, or $9z^{-2} - 30z^{-1} + 25$. These can be factored using the same unFOIL process that is used to factor quadratics such as $y^2 - 9y - 10$. The same First, Outer, Inner, and Last pattern exists. Go back to products of binomials and/or factoring trinomials for more information on this.

Example Problems

These problems show the answers and solutions.

1. Factor $y^4 + 6y^2 + 5$.

   answer: $(y^2 + 5)(y^2 + 1)$

   The First term can be written as $y^2 \cdot y^2$. The Last term can be written as $5 \cdot 1$. Because the last term is positive, you're looking for the **sum** of the Outer and Inner products to be equal to the middle term, the $6y^2$. The middle term is positive, so both of the operations in the binomials will be positive.

   $$y^4 + 6y^2 + 5 = (y^2 + 5)(y^2 + 1)$$

2. Factor $6a^6 - a^3 - 2$.

   answer: $(3a^3 - 2)(2a^3 + 1)$

   The First term can be written as $6a^3 \cdot a^3$ or $3a^3 \cdot 2a^3$. The Last term can be written as $2 \cdot 1$. The last term is negative, so you're looking for the **difference** of the Outer and Inner products to be equal to the middle term, a^3. The middle term is negative, so there will be one negative operation and one positive operation.

   $$6a^6 - a^3 - 2 = (3a^3 - 2)(2a^3 + 1)$$

3. Factor $9z^{-2} - 30z^{-1} + 25$.

   answer: $(3z^{-1} - 5)(3z^{-1} - 5)$

   The First term can be written as $9z^{-1} \cdot z^{-1}$ or $3z^{-1} \cdot 3z^{-1}$. The Last term can be written as $25 \cdot 1$ or $5 \cdot 5$. Notice that these first and last terms are both perfect squares. This could be a perfect square trinomial. Try this out: $(3z^{-1}\ 5)(3z^{-1}\ 5)$. The Outer and Inner products are each $15z^{-1}$, and their sum is equal to the middle term.

   $$\text{So, } 9z^{-2} - 30z^{-1} + 25 = (3z^{-1} - 5)(3z^{-1} - 5)$$

4. Factor $n^4 - 8n^2 - 9$.

   answer: $(n - 3)(n + 3)(n^2 + 1)$

   The First term can be written as $n^2 \cdot n^2$. The Last term can be written as $9 \cdot 1$ or $3 \cdot 3$.

   The last term is negative, so you're looking for a **difference** of the Outer and Inner products of $8n^2$. The operations will be one of each. $n^4 - 8n^2 - 9 = (n^2 - 9)(n^2 + 1)$. Note, now, that the first factor is the difference of two perfect squares. In the previous section, the factorization of this type of binomial is discussed. Factoring the first factor into the sum and difference of the same two roots, the factorization now becomes:

   $$(n^2 - 9)(n^2 + 1) = (n - 3)(n + 3)(n^2 + 1)$$

Factoring by Grouping

Factoring by grouping is a method that is usually done on expressions with four, six, eight, or higher numbers of terms—usually an even number. Typically, no common factor exists for all of the terms in the expression, but common factors do exist in pairs of the terms. The common factor is identified for each pair or grouping, the factoring is done, and then the new expression, with new terms, is checked to see whether there is a common factor throughout. If the factoring of separate groups doesn't result in a common factor in the new terms, try rearranging the terms into different groups. It could be that the expression just can't be factored, but different arrangements should be tried first.

Example Problems

These problems show the answers and solutions.

1. Factor $x^2 y - 2x^2 + 3y - 6$ by grouping.

    answer: $(y - 2)(x^3 + 3)$

    As you can see, no factor is common to all four terms at once. The first two terms have a common factor of x^2, and the last two terms have a common factor of 3.

    $$x^2 y - 2x^2 = x^2(y - 2)$$
    $$3y - 6 = 3(y - 2)$$

    Put these together, and $x^2 y - 2x^2 + 3y - 6 = x^2(y - 2) + 3(y - 2)$. Now you have two terms instead of four. The two terms each have a common factor of $(y - 2)$. Factor that out of the two terms, $x^2(y - 2) + 3(y - 2) = (y - 2)(x^3 + 3)$.

2. Factor $a^5 - 8a^3 - 6a^2 + 48$ by grouping.

    answer: $(a^2 - 8)(a^3 - 6)$

    Again, you have no factor common to all four terms at once. The first two terms have a common factor of a^3, and the last two terms have a common factor of 6 or −6.

    $$a^5 - 8a^3 = a^3(a^2 - 8)$$
    $$-6a^2 + 48 = 6(-a^2 + 8)$$

    Put these together, $a^5 - 8a^3 - 6a^2 + 48 = a^3(a^2 - 8) + 6(-a^2 + 8)$. This time, the two new terms don't have a common factor. The two factors in the parentheses aren't exactly alike. They are opposite in sign, though, so the factorization of those last two terms should be revisited. Factor out a −6 instead of 6.

    $$-6a^2 + 48 = -6(a^2 - 8)$$
    $$\text{So, } a^5 - 8a^3 - 6a^2 + 48 = a^3(a^2 - 8) - 6(a^2 - 8).$$

    Now the two terms have the common factor of $(a^2 - 8)$.

    $$a^5 - 8a^3 - 6a^2 + 48 = a^3(a^2 - 8) - 6(a^2 - 8) = (a^2 - 8)(a^3 - 6)$$

3. Factor $n^5 - 4n^3 + n^2 - 4$ using grouping.

    answer: $(n + 1)(n^2 - n + 1)(n - 2)(n + 2)$

    The first two terms have a common factor of n^3, and the second two terms only have a common factor of 1. Actually, it would work to break the terms up this way but rearrange them to create common factors in both groups. Rewrite the expression as $n^5 + n^2 - 4n^3 - 4$. Now factor n^2 out of the first two terms and −4 out of the second two terms. $n^5 + n^2 - 4n^3 - 4 = n^2(n^3 + 1) - 4(n^3 + 1)$. The two new terms each have the factor $(n^3 + 1)$. Factor that out, $n^2(n^3 + 1) - 4(n^3 + 1) = (n^3 + 1)(n^2 - 4)$. This is now factored by grouping, but it's not completely factored. Each of the binomials can be factored further. The first binomial is the sum of two perfect cubes. The second binomial is the difference of two squares. In the section "Factoring Binomials," each of these factorizations is discussed. To complete the factorization,

$$(n^3 + 1)(n^2 - 4) = (n + 1)(n^2 - n + 1)(n - 2)(n + 2)$$

4. Factor $2x^2 a + 6x^2 - ax - 3x - 3a - 9$ using grouping.

answer: $(a + 3)(2x - 3)(x + 1)$

No factor is common to all six terms. The first two terms have a common factor of $2x^2$; the second two terms have a common factor of $-x$; and the last two terms have a common factor of -3.

$$2x^2 a + 6x^2 - ax - 3x - 3a - 9 = 2x^2(a + 3) - x(a + 3) - 3(a + 3)$$

Now there are three terms instead of six, and each has the common factor of $(a + 3)$.

Factor that out, $2x^2(a + 3) - x(a + 3) - 3(a + 3) = (a + 3)(2x^2 - x - 3)$.

And, just when you think you're finished, you notice that the trinomial in this result can be factored. Using unFOIL, the final factorization is $(a + 3)(2x^2 - x - 3) = (a + 3)(2x - 3)(x + 1)$.

The preceding example could have been factored, by grouping, using groups of three instead of two. With some rearranging and getting the three terms with a in them first, the last three terms then have a common factor of 3. It was just as easy to do it in pairs of factors. There's usually more than one way to group and factor these problems, so you can't go wrong, as long as you follow the rules.

Work Problems

Use these problems to give yourself additional practice.

1. Factor $3x^8 - 2x^4 - 16$.

2. Factor $y^6 - 13y^3 + 40$.

3. Factor $x^4 + x^3 + 3x + 3$.

4. Factor $6m^5 + 15m^3 - 12m^2 - 30$.

5. Factor $2ace + 2ade + 2bce + 2bde - acf - adf - bcf - bdf$.

Worked Solutions

1. $(3x^4 - 8)(x^4 + 2)$ This is quadratic-like, so you can look for two binomials whose product is this expression.

    The First term can be written $3x^4 \cdot x^4$.

    The Last term can be written $16 \cdot 1$, $8 \cdot 2$, or $4 \cdot 4$.

The last term is negative, so you'll be looking for the **difference** between the Outer and Inner products to be $-2x^4$. The operations in the binomials will be one of each.

$$3x^8 - 2x^4 - 16 = (3x^4 - 8)(x^4 + 2)$$

2. **$(y-2)(y^4+2y+4)(y^3-5)$**   This is quadratic-like.

The First term can be written $y^3 \cdot y^3$.

The Last term can be written $40 \cdot 1$, $20 \cdot 2$, $10 \cdot 4$, or $8 \cdot 5$.

The last term is positive, so you'll be looking for the **sum** of the Outer and Inner products to be $13y^3$. The middle term is negative, so both operations in the binomials will be negative.

$$y^6 - 13y^3 + 40 = (y^3 - 8)(y^3 - 5).$$

The first binomial is the difference between two cubes, so it can be factored using the rules in the section "Factoring Binomials."

$$(y^3 - 8)(y^3 - 5) = (y - 2)(y^4 + 2y + 4)(y^3 - 5)$$

3. **$(x+1)(x^3+3)$**   This will be factored by grouping. The first two terms have a common factor of x^3, and the last two terms have a common factor of 3.

$$x^4 + x^3 + 3x + 3 = x^3(x+1) + 3(x+1)$$

Now the two new terms each have a common factor of $(x+1)$.

$$x^3(x+1) + 3(x+1) = (x+1)(x^3+3)$$

4. **$(2m^2+5)(3m^3-6)$**   This will be factored by grouping. The first two terms have a common factor of $3m^3$, and the last two terms have a common factor of -6.

$$6m^5 + 15m^3 - 12m^2 - 30 = 3m^3(2m^2 + 5) - 6(2m^2 + 5)$$

Now the two new terms have a common factor of $(2m^2 + 5)$.

$$3m^3(2m^2 + 5) - 6(2m^2 + 5) = (2m^2 + 5)(3m^3 - 6)$$

5. **$(c+d)(a+b)(2e-f)$**   This will be factored using grouping. The first two terms have a common factor of $2ae$; the second two terms have a common factor of $2be$; the third two terms have a common factor of $-af$; and the last two terms have a common factor of $-bf$. Factoring the pairs of terms,

$$2ace + 2ade + 2bce + 2bde - acf - adf - bcf - bdf = 2ae(c+d) + 2be(c+d) - af(c+d) - bf(c+d)$$

The four terms in the new expression each have a common factor of $(c+d)$. Factoring that out,

$$2ae(c+d) + 2be(c+d) - af(c+d) - bf(c+d) = (c+d)(2ae + 2be - af - bf)$$

The factored form can, again, be factored. The expression with four terms, in the parenthesis on the right, can be factored by grouping. The first two terms have a common factor of 2e, and the last two terms have a common factor of −f.

$$(c + d)(2ae + 2be - af - bf) = (c + d)[2e(a + b) - f(a + b)]$$

The two terms in the brackets have a common factor of $(a + b)$.

$$(c + d)[2e(a + b) - f(a + b)] = (c + d)[(a + b)(2e - f)]$$

Chapter Problems and Solutions

Problems

1. Distribute $4y(6y^2 + 2y - x + 1)$.

2. Multiply $(2x + 3)(3x - 1)$.

3. Multiply $(z - 7)(z^2 + 2z - 1)$.

4. Multiply $(5a - 2)(5a + 2)$.

5. Multiply $(6 + c)^2$.

6. Multiply $(2x - 3)^3$.

7. Multiply $4(y - 3)^2$.

8. Divide $(27x^3 - 8) \div (3x - 2)$.

9. Divide $(2x^4 - 3x^3 + 5x^2 - 4x + 1) \div (x - 3)$.

10. Factor out the Greatest Common Factor in $8a^2 b^3 + 12a^3 b^2 - 32ab^4$.

11. Factor out the Greatest Common Factor in $y^4(x^2 + 1)^3 + y^5(x^2 + 1)^2$.

Problems 12 through 25, factor completely.

12. $x^2 - 16$

13. $81 - 36y^2$

14. $a^3 + 27$

15. $8 - 125b^3$

16. $x^4 - 81$

17. $x^2 + 5x - 24$

18. $y^2 - 12y + 35$

19. $6z^2 + z - 1$

20. $8w^2 + 6w - 9$

21. $15t^2 + 57t - 12$

22. $a^6 + 7a^3 - 44$

23. $4y^{-6} - 13y^{-3} + 3$

24. $a^2x - 2a^2y + bx - 2by$

25. $4x^5 - 9x^3 + 4x^2 - 9$

Answers and Solutions

1.　**Answer: $24y^3 + 8y^2 - 4xy + 4y$**　Multiply each term in the parenthesis by $4y$.

2.　**Answer: $6x^2 + 7x - 3$**　Use FOIL to find the products of the First terms, Outer terms, Inner terms, and Last terms. Simplify the answer.

3.　**Answer: $z^3 - 5z^2 - 15z + 7$**　Distribute the binomial over the trinomial.
$(z - 7) \cdot z^2 + (z - 7) \cdot 2z + (z - 7)(- 1)$. Multiply to get $z^3 - 7z^2 + 2z^2 - 14z - z + 7$. Then simplify to get the answer.

4.　**Answer: $25a^2 - 4$**　Use the special product for the sum and difference of the same two numbers.

5.　**Answer: $36 + 12c + c^2$**　Use the special product squaring a binomial and getting a perfect square trinomial.

6.　**Answer: $8x^3 - 36x^2 + 54x - 27$**　Use the 1–3–3–1 pattern and decreasing and increasing powers to find the cube of the binomial.
$1 (2x)^3 (- 3)^0 + 3 (2x)^2 (- 3)^1 + 3 (2x)^1 (- 3)^2 + 1 (2x)^0 (- 3)^3$ is then simplified.

7.　**Answer: $4y^2 - 24y + 36$**　First square the binomial using the special product pattern. Then multiply each term by 4.

8.　**Answer: $9x^2 + 6x + 4$**　Use long division. There's no remainder.

$$
\begin{array}{r}
9x^2 + 6x + 4 \\
3x - 2 \overline{\smash{\big)}\, 27x^3 \qquad\qquad -8} \\
\underline{-27x^3 + 18x^2} \\
18x^2 \qquad -8 \\
\underline{-18x^2 + 12x} \\
12x - 8 \\
\underline{12x - 8}
\end{array}
$$

9. **Answer:** $2x^3 + 3x^2 + 14x + 38 + \dfrac{115}{x-3}$ Use synthetic division.

$$
\begin{array}{r|rrrrr}
3 & 2 & -3 & 5 & -4 & 1 \\
  &   & 6 & 9 & 42 & 114 \\
\hline
  & 2 & 3 & 14 & 38 & 115
\end{array}
$$

10. **Answer:** $4ab^2(2ab + 3a^2 - 8b^2)$ The GCF is $4ab^2$. Divide each term by that and write the results in the parentheses.

11. **Answer:** $y^4(x^2+1)^2\left[(x^2+1)+y\right]$ The binomial is part of the GCF.

12. **Answer:** $(x-4)(x+4)$ The difference of two squares factors into the difference and sum of the roots.

13. **Answer:** $9(3-2y)(3+2y)$ First factor out the GCF, 9. Then use the pattern for factoring the difference of two squares.

14. **Answer:** $(a+3)(a^2-3a+9)$ The sum of two cubes factors into a binomial times a trinomial.

15. **Answer:** $(2-5b)(4+10b+25b^2)$ The difference between two cubes factors into a binomial times a trinomial.

16. **Answer:** $(x-3)(x+3)(x^2+9)$ First factor the binomial into the difference and sum of the roots, $(x^2-9)(x^2+9)$. Then factor the first binomial into the difference and sum of the roots.

17. **Answer:** $(x+8)(x-3)$ Use unFOIL. Because the last term is negative, the Outer and Inner products need to have a difference of $5x$.

18. **Answer:** $(y-5)(y-7)$ Use unFOIL. Because the last term is positive, the Outer and Inner products need to have a sum of $-12y$.

19. **Answer:** $(2z+1)(3z-1)$ Use unFOIL.

20. **Answer:** $(4w-3)(2w+3)$ Use unFOIL.

21. **Answer:** $3(5t-1)(t+4)$ First factor out the GCF, 3. Then use unFOIL on the trinomial $5t^2 + 19t - 4$.

22. **Answer:** $(a^3-4)(a^3+11)$ Quadratic-like trinomial factor using unFOIL.

23. **Answer:** $(y^{-3}-3)(4y^{-3}-1)$ Quadratic-like trinomial factor using unFOIL.

24. **Answer:** $(x-2y)(a^2+b)$ Use grouping. Factor a^2 out of the first two terms and b out of the second two terms to get $a^2(x-2y) + b(x-2y)$. Now, the binomial has a GCF of $x-2y$ that can be factored out.

25. **Answer:** $(x+1)(x^2-x+1)(2x-3)(2x+3)$ First, rearrange the terms to read $4x^5 + 4x^2 - 9x^3 - 9$. Then use grouping to factor $4x^2$ out of the first two terms and -9 out of the second two terms to get $4x^2(x^3+1) - 9(x^3+1)$. The common factor is x^3+1. Factoring this out, you get $(x^3+1)(4x^2-9)$. Each of these factors also factors. The first is the sum of two cubes, and the second is the difference of two squares.

Supplemental Chapter Problems

Problems

1.　Distribute $-x^3(4x^2 - 3x - 2)$.

2.　Multiply $(4y - 5)(2y + 7)$.

3.　Multiply $(2w + 1)(3w^2 - w + 2)$.

4.　Multiply $(4a - b)^2$.

5.　Multiply $(16 - m)(16 + m)$.

6.　Multiply $(z + 5)^3$.

7.　Multiply $2(r - 2t)^2$.

8.　Divide $(3x^4 - 11x^2 - 4) \div (x + 2)$.

9.　Divide $(5x^5 - 3x^3 + x - 8) \div (x + 2)$.

10.　Factor out the GCF of $15m^2n^3 - 21m^3n^4 + 33m^2n^5$.

11.　Factor out the GCF of $ab(c - 1) + ad(c - 1)^2 + a(c - 1)^3$.

Problems 12 through 25, factor completely.

12.　$4z^2 - 100$

13.　$49 - w^2$

14.　$d^3 - 1$

15.　$6e^3 + 48$

16.　$y^8 - 25$

17.　$t^2 + 9t + 14$

18.　$z^2 + 2z - 80$

19.　$20p^2 + 3p - 2$

20.　$12m^2 - 8m - 15$

21.　$10x^2 - 15x - 45$

22.　$y^8 + 2y - 3$

23.　$3x^{-4} + 5x^{-2} - 12$

24.　$3x^2 + mx^2 - 15 - 5m$

25.　$200 - 8y^2 - 25y^3 + y^5$

Answers

1. $-4x^5 + 3x^4 + 2x^3$ (Multiplying Monomials, p. 145)

2. $8y^2 + 18y - 35$ (Multiplying Polynomials, p. 148)

3. $6w^3 + w^2 + 3w + 2$ (Multiplying Polynomials, p. 148)

4. $16a^2 - 8ab + b^2$ (Special Products, p. 151)

5. $256 - m^2$ (Special Products, p. 151)

6. $z^3 + 15z^2 + 75z + 125$ (Special Products, p. 151)

7. $2r^2 - 8rt + 8t^2$ (Special Products, p. 151)

8. $3x^3 - 6x^2 + x - 2$ (Dividing Polynomials, p. 156)

9. $5x^4 - 10x^3 + 17x^2 - 34x + 69 - \dfrac{146}{x+2}$ (Synthetic Division, p. 160)

10. $3m^2 n^3 (5 - 7mn + 11n^2)$ (Greatest Common Factor, p. 163)

11. $a(c-1)\left[b + d(c-1) + (c-1)^2\right]$ (Greatest Common Factor, p. 163)

12. $4(z-5)(z+5)$ (Factoring Binomials, p. 165)

13. $(7-w)(7+w)$ (Factoring Binomials, p. 165)

14. $(d-1)(d^2 + d + 1)$ (Factoring Binomials, p. 165)

15. $6(e+2)(e^2 - 2e + 4)$ (Factoring Binomials, p. 165)

16. $(y^4 - 5)(y^4 + 5)$ (Factoring Binomials, p. 165)

17. $(t+2)(t+7)$ (Factoring Trinomials, p. 168)

18. $(z-8)(z+10)$ (Factoring Trinomials, p. 168)

19. $(4p-1)(5p+2)$ (Factoring Trinomials, p. 168)

20. $(6m+5)(2m-3)$ (Factoring Trinomials, p. 168)

21. $5(x-3)(2x+3)$ (Factoring Trinomials, p. 168)

22. $(y^4 + 3)(y-1)(y+1)(y^2 + 1)$ (Quadratic-Like Expressions, p. 172)

23. $(x^{-2} + 3)(3x^{-2} - 4)$ (Quadratic-Like Expressions, p. 172)

24. $(3+m)(x^2 - 5)$ (Factoring by Grouping, p. 173)

25. $(2-y)(4 + 2y + y^2)(5-y)(5+y)$ (Factoring by Grouping, p. 173)

Chapter 5

Inequalities, Absolute Value Inequalities, and Radicals

This chapter presents more information about the relationship of **inequality** and two operations, **absolute value** and **square root.** These topics are covered in Chapters 1 and 2, but now they're used in statements that can be right or wrong. A statement that is right or true has the correct value or values for the variable. That's what solving these statements is all about. Inequalities have a couple of peculiarities about them. Operations on certain numbers cause a reversal of their sense. Solving absolute value and radical equations is made simpler by converting them into nicer forms, but you must observe some cautions.

Inequalities

An inequality is a statement saying that one thing is greater than, >; greater than or equal to, ≥; less than, <; or less than or equal to, ≤, something else.

The symbols ≥ or ≤ mean that one number is greater (or less) than another **or** the number is equal to that number. It really can't be both at the same time.

To ride on the roller coaster, your height has to be greater than or equal to 4 feet, $h \geq 4$. To avoid paying income tax, your income has to be less than $2,000, $i < 2000$. Statements can be written about inequalities, and the inequalities can be solved much like linear equations. See Chapter 3 for information on solving linear equations. The procedures that are used to solve linear equations, such as adding the same thing to each side, and so on, can be used to solve inequalities, too. But caution! Two notable exceptions to the rules involve both multiplication and division. In the following rules for operating on inequalities, the letters a, b, and c represent any real number. The rules applying to the inequality > also apply to ≥.

Rules for Operating on Inequalities

❑ If $a > b$, then $a + c > b + c$; if $a < b$, then $a + c < b + c$.

> You can add the same number to each side of an inequality and preserve the truth of the statement.

❑ If $a > b$, then $a - c > b - c$; if $a < b$, then $a - c < b - c$.

> You can subtract the same number from each side of an inequality and preserve the truth of the statement.

❑ If $a > b$, and if c is **positive**, then $a \cdot c > b \cdot c$; if $a < b$, and if c is **positive**, then $a \cdot c < b \cdot c$.

You can multiply the same positive number to each side of an inequality and preserve the truth of the statement.

❑ If $a > b$, and if c is **negative**, then $a \cdot c < b \cdot c$; if $a < b$, and if c is **negative**, then $a \cdot c > b \cdot c$.

If you multiply each side of an inequality by a negative number, the sense or inequality is reversed to make it true.

❑ If $a > b$, and if c is **positive**, then $\frac{a}{c} > \frac{b}{c}$; if $a < b$, and if c is **positive**, then $\frac{a}{c} < \frac{b}{c}$.

You can divide each side of an inequality by a positive number and preserve the truth of the statement.

❑ If $a > b$, and if c is **negative**, then $\frac{a}{c} < \frac{b}{c}$; if $a < b$, and if c is **negative**, then $\frac{a}{c} > \frac{b}{c}$.

If you divide each side of an inequality by a negative number, the sense is reversed to make it true.

❑ If $a > b$ and $b > c$, then $a > c$; if $a < b$ and $b < c$, then $a < c$.

This is the Transitive Law.

❑ If $a > b > c$, then $c < b < a$.

If you reverse the order of the sections in an inequality, you must reverse all of the senses.

❑ If $\frac{a}{b} > \frac{c}{d}$, then $\frac{b}{a} < \frac{d}{c}$ when a, b, c, and d are all positive.

When two fractions are written in an inequality, flipping the fractions reverses the sense.

Here are some examples using inequality signs. To solve an inequality, the variable should be alone on one side, and a number or some value should be on the other side.

Example Problems

These problems show the answers and solutions.

1. Solve for x in $3x + 2 > 5$.

   answer: $x > 1$

   Add -2 to each side.

$$3x + 2 > 5$$
$$\underline{-2 \quad -2}$$
$$3x \qquad > 3$$

   Now divide each side by 3.

$$\frac{3x}{3} > \frac{3}{3}$$
$$x > 1$$

   The sense (inequality symbol) didn't reverse in direction, because the 3 that divided each side was positive.

2. Solve for y. $9 - 2y \le y + 6$

answer: $y \ge 1$

Add -9 to each side. Then subtract y from each side.

$$
\begin{array}{r}
9 - 2y \le y + 6 \\
\underline{-9 \qquad\qquad -9} \\
-2y \le y - 3 \\
\underline{-y \qquad -y} \\
-3y \le -3
\end{array}
$$

Divide by -3. Because this is dividing by a negative number, the inequality symbol gets reversed.

$$\frac{-3y}{-3} \ge \frac{-3}{-3}$$
$$y \ge 1$$

3. Solve for z. $1 - \dfrac{2z}{5} < z - 6$

answer: $z > 5$

Subtracting 1 from each side and then subtracting z from each side doesn't affect the sense. Subtracting $-z$ from $-\dfrac{2z}{5}$ involves fractions and changes the first term to one with a denominator of 5. $-z - \dfrac{2z}{5} = -\dfrac{5z}{5} - \dfrac{2z}{5} = -\dfrac{7z}{5}$.

$$
\begin{array}{r}
1 - \dfrac{2z}{5} < z - 6 \\
\underline{-1 \qquad\qquad -1} \\
-\dfrac{2z}{5} < z - 7 \\
\underline{-z \quad -z} \\
-\dfrac{7z}{5} < -7
\end{array}
$$

To solve for z, you can do it in two steps or one. The two steps would be to multiply each side by 5 and divide each side by -7. One step that does both is to multiply each side by the reciprocal of $-\dfrac{7}{5}$ which is $-\dfrac{5}{7}$. Because that's a negative value, the sense (inequality sign) gets reversed.

$$-\frac{5}{7}\left(-\frac{7z}{5}\right) > -7\left(-\frac{5}{7}\right)$$
$$z > 5$$

4. Solve for t. $\frac{3}{t} > \frac{15}{2}$

    answer: $t < \frac{2}{5}$ where t is positive.

    One of the rules for inequalities says that you can flip both of the fractions if you reverse the inequality sign. That will be convenient, because it'll put the variable, t, in the numerator, which makes the statement easier to solve.

    Flip the inequality: $\frac{t}{3} < \frac{2}{15}$.

    Now, multiply each side by 3 to finish the job.

    $$\frac{\cancel{3}}{1} \cdot \frac{t}{\cancel{3}} < \frac{2}{\cancel{15}_5} \cdot \frac{\cancel{3}}{1}$$

    $$t < \frac{2}{5}$$

5. Solve for z. $-3 < 2z + 1 \le 5$

    answer: $-2 < z \le 2$

    To solve an inequality problem with multiple inequality signs, whatever is done to one section must be done to all of the sections. To solve this, first add -1 to each section. The goal is to get the z alone between those two inequality symbols.

    $$
    \begin{array}{rcrcr}
    -3 & < & 2z+1 & \le & 5 \\
    -1 & & -1 & & -1 \\
    \hline
    -4 & < & 2z & \le & 4
    \end{array}
    $$

    Now divide each section by 2. Because 2 is positive, the inequality symbols stay in that direction.

    $$\frac{-4}{2} < \frac{\cancel{2}z}{\cancel{2}} \le \frac{4}{2}$$

    $$-2 < z \le 2$$

    This solution says that z is some number that's strictly bigger than -2 **and** smaller than or equal to 2 at the same time. It's between the two numbers.

Work Problems

Use these problems to give yourself additional practice.

1. Solve for x. $2x - 1 \ge 3x + 5$

2. Solve for y. $\frac{3y + 21}{-4} < y$

3. Solve for z. $4 \le 9 - 5z < 19$

4. Solve for w. $\frac{5w}{2} + \frac{1}{3} > w - \frac{7}{6}$

5. Solve for u. $\frac{4}{3u} \le \frac{6}{u + 14}$

Worked Solutions

1. **$x \leq -6$**   To get x alone on the left, first add 1 to each side. Then subtract $3x$ from each side.

$$
\begin{array}{rl}
2x - 1 & \geq 3x + 5 \\
\underline{ +1 \quad\;\; +1} & \\
2x & \geq 3x + 6 \\
\underline{-3x \qquad -3x} & \\
-x & \geq 6
\end{array}
$$

To solve for x, multiply each side by -1, which means that the sense (inequality symbol) gets reversed.

$$
\begin{array}{c}
-x \geq 6 \\
-1(-x) \leq 6(-1) \\
x \leq -6
\end{array}
$$

2. **$y < -3$**   Multiplying each side by -4 will reverse the sense. Then, rather than subtracting 21 from each side and then subtracting y from each side, do just one operation—subtract $3y$ from each side.

$$
\begin{array}{c}
-\dfrac{\cancel{4}}{1} \cdot \dfrac{3y + 21}{-\cancel{4}} > y(-4) \\[2mm]
3y + 21 > -4y \\
\underline{-3y \qquad\quad -3y} \\
21 > -7y
\end{array}
$$

Next, divide each side by -7, which reverses the sense.

$$
\begin{array}{c}
\dfrac{21}{-7} < \dfrac{-7y}{-7} \\[2mm]
-3 < y
\end{array}
$$

To write this with the y on the left, it can be switched if the sense is also switched/reversed.

$$
y > -3
$$

3. **$-2 < z \leq 1$**   First add -9 to each section. Then, when dividing each section by -5, the inequality symbols will both get reversed.

$$
\begin{array}{c}
4 \leq 9 - 5z < 19 \\
\underline{-9 \;\; -9 \qquad\;\; -9} \\
-5 \leq \;\; -5z < 10 \\
\dfrac{-5}{-5} \geq \;\; \dfrac{-5z}{-5} > \dfrac{10}{-5} \\[2mm]
1 \geq \quad z \;\; > -2
\end{array}
$$

This solution has the numbers reversed from their order on the number line, where the smaller numbers are to the left of the larger numbers—the negatives are to the left of the positives. To reverse the numbers, everything gets flipped—the numbers and the inequality symbols.

$$-2 < z \leq 1$$

4. **$w > -1$** The first thing to do here is get rid of the fractions. This can be done by multiplying every term by 6, which is the common denominator of all of the fractions. The number 6 is positive, so the inequality symbol stays the way it is.

$$\frac{\overset{3}{\cancel{6}}}{1} \cdot \frac{5w}{\cancel{2}} + \frac{\overset{2}{\cancel{6}}}{1} \cdot \frac{1}{\cancel{3}} > 6 \cdot w - \frac{\cancel{6}}{1} \cdot \frac{7}{\cancel{6}}$$

$$15w + 2 > 6w - 7$$

Now, to solve for w, subtract 2 from each side and subtract $6w$ from each side.

$$
\begin{array}{rcl}
15w + 2 & > & 6w - 7 \\
-2 & & -2 \\
\hline
15w & > & 6w - 9 \\
-6w & & -6w \\
\hline
9w & > & -9
\end{array}
$$

Dividing each side by 9 solves for w.

$$\frac{\cancel{9}w}{\cancel{9}} > \frac{-9}{9}$$

$$w > -1$$

5. **$u \geq 4$** You can rewrite this by flipping the fractions and reversing the inequality symbol.

$$\frac{3u}{4} \geq \frac{u + 14}{6}$$

Now, multiply each side by 12 to get rid of the fractions.

$$\overset{3}{\cancel{12}} \cdot \frac{3u}{\cancel{4}} \geq \frac{u + 14}{\cancel{6}} \cdot \cancel{12}^{2}$$

$$9u \geq (u + 14) \cdot 2$$

$$9u \geq 2u + 28$$

Subtract $2u$ from each side.

$$
\begin{array}{rcl}
9u & \geq & 2u + 28 \\
-2u & & -2u \\
\hline
7u & \geq & 28
\end{array}
$$

And now, divide each side by 7.

$$\frac{7u}{7} \ge \frac{28}{7}$$

$$u \ge 4$$

Solving Inequalities by Graphing on a Number Line

The solutions to inequalities can be expressed in several ways. You can use inequality notation, interval notation, or a shaded number line. Interval notation is discussed in Chapter 2 in the section, "Number Lines." Each of these ways of describing a solution has its place in different applications. None is better than another. They're just different and used for different purposes.

The number line is very helpful when solving multiple inequalities. It's an essential part of solving absolute value inequalities, which come later in this chapter. Two key words in the solutions of these problems are **and** and **or**. In mathematics, these two words have very specific meanings. The word **and** means that **both** requirements must be met. The word **or** means that one requirement **or** the other **or** both must be met.

Example Problems
These problems show the answers and solutions.

1. Solve for the values of x that satisfy the statement that $x > 3$ and $x \le 8$.

    answer: $3 < x \le 8$

    Graphing the values of x that satisfy both of these statements:

    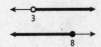

    You can put these together on the same number line by looking at where their graphs share solutions. The two number lines that follow are both shaded between the 3 and the 8.

    The combined graph would look like ⟵○———●⟶.
                                       3   8

    From this, the common solutions are seen easily. Any number between 3 and 8, including the 8, are solutions of the graph. The same is true of the original inequality statements: $x > 3$ and $x \le 8$.

    The inequality statements can be combined to reflect what the combined graph looks like.

    $x > 3$ and $x \le 8$ are combined to read $3 < x \le 8$.

This may seem rather trivial, but awareness of how the inequalities work and combine is very important for writing inequality statements that are correct and make sense. For instance, what would the common solution of the two inequalities, $x < -3$ and $x > -1$ look like? You might be tempted to write $-1 < x < -3$, because it puts the x between the two values. But look at how the statements graph on the number line.

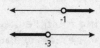

You can put these together on the same number line by looking at where their graphs share solutions.

These solutions go in completely different directions. They don't have anything in common. So the solution should read $x < -3$ or $x > -1$. There aren't any numbers that can be smaller than -3 and bigger than -1 at the same time.

2. Solve the inequalities and find their common solutions—what numbers work for both at the same time. $4x - 1 > x + 8$ and $1 < x + 2 \leq 7$.

answer: $3 < x \leq 5$

First, the inequality $4x - 1 > x + 8$

Add 1 to both sides and subtract x from both sides.

$$
\begin{array}{r}
4x - 1 > x + 8 \\
\underline{+1 \quad\quad +1} \\
4x \quad > x + 9 \\
\underline{-x \quad\quad -x} \\
3x \quad > \quad 9
\end{array}
$$

Now, divide each side by 3 and graph the result.

$$\frac{\cancel{3}x}{\cancel{3}} > \frac{9}{3}$$

$$x > 3$$

To solve the other inequality, $1 < x + 2 \leq 7$, subtract 2 from each section.

$$
\begin{array}{r}
1 < x + 2 \leq 7 \\
\underline{-2 \quad -2 -2} \\
-1 < x \quad \leq 5
\end{array}
$$

The graph of this solution is 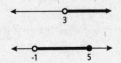 .

The original statement says **and**, so the solution will include all values that satisfy both of the inequalities at the same time. Looking at their graphs together will help determine this solution.

The places where both graphs are shaded at the same time seem to be between 3 and 5, including the 5 but not the 3. The solution to the entire statement $4x - 1 > x + 8$ **and** $1 < x + 2 \leq 7$ would be written $3 < x \leq 5$.

3.   Illustrate on the number line the solution to $-2 < x \leq 4$ or $x > 6$.

   answer:

   Graph the two inequalities separately:

$$-2 < x \leq 4$$

$$x > 6$$

   As you can see, there's no overlap. The statements have to be kept separate. The combined graph is

Work Problems

Use these problems to give yourself additional practice.

1.   If possible, combine the inequalities into one. $x > 3$, $x < -2$

2.   If possible, combine the inequalities into one. $x > -2$, $x < 3$

3.   Find the common solution for $0 < x + 5 \leq 9$, $2x - 1 < 5$.

4.   Write the inequality that matches the graph.

5.   Write the inequality that matches the graph.

Worked Solutions

1. $x > 3$ or $x < -2$ First, look at their graphs.

$$\longleftarrow\!\!\!\underset{3}{\circ}\!\!\!-\!\!\!\bullet \qquad x > 3$$

$$\bullet\!\!\!-\!\!\!\underset{-2}{\circ}\!\!\!\longrightarrow \qquad x < -2$$

There's no overlap, so the inequalities have to be written separately, with an **or** between them.

2. **−2 < x < 3**   Their graphs are

$$\longleftarrow\underset{-2}{\circ}\!\!\!-\!\!\!-\!\!\!-\!\!\!\longrightarrow \qquad x > -2$$

$$\longleftarrow\!\!\!-\!\!\!-\!\!\!\underset{3}{\circ}\!\!\!\longrightarrow \qquad x < 3$$

These graphs overlap for all the numbers between −2 and 3.

3. **−5 < x < 3**   Solving the first inequality, subtract 5 from each section.

$$0 < x + 5 \le 9$$
$$\underline{-5 \qquad -5 \ -5}$$
$$-5 < x \le 4$$

Solving the second inequality, add 1 to each side and then divide each side by 2.

$$2x - 1 < 5$$
$$\underline{\qquad +1 +1}$$
$$2x \quad < 6$$
$$\frac{2x}{2} < \frac{6}{2}$$
$$x < 3$$

Now look at the graphs of these two solutions.

$$\longleftarrow\underset{-5}{\circ}\!\!\!-\!\!\!\underset{4}{\bullet} \qquad -5 < x \le 4$$

$$\longleftarrow\!\!\!-\!\!\!\underset{3}{\circ}\!\!\!\longrightarrow \qquad x < 3$$

As you can see, there's an overlap between −5 and 3.

4. **−2 ≤ x < 3**   The −2 is included, but the 3 is not.

5. **−1 < x < 3 or 4 ≤ x < 8**   The only endpoint included is the 4. These two parts of the number line are completely disconnected, so an **or** is used between the inequalities.

Absolute Value Equations

The absolute value equations covered in this section are the linear equations. Absolute value equations with quadratics and fractions are discussed in the next chapter, after the section on quadratic equations.

The absolute value operation takes a number written between the two absolute value bars and makes it a positive value. The absolute value of a number is the distance of that number from 0 on the number line. The numbers 4 and −4 are the same distance from 0 on the number line. They're just in opposite directions. The absolute value pays no attention to what direction the number goes. For more on absolute values, see Chapter 1, "The Basics."

Because a variable can be either positive or negative, and you can't tell which it is from looking at the variable, both situations have to be considered when solving absolute value equations. The rules for solving absolute value equations are

❑ If $|x| = k$, then either $x = k$ or $x = -k$.

   You can't tell, by looking at the x, whether it represents a positive number or a negative number, so you have to account for both possibilities.

❑ If $|ax + b| = k$, then either $ax + b = k$ or $ax + b = -k$.

Example Problems

These problems show the answers and solutions.

1. Solve $|3x - 1| = 5$.

    answer: $x = 2$, $x = -\dfrac{4}{3}$

    Rewrite this as two different equations without the absolute value operation. Then solve both equations. Using the rule for absolute value, this inequality is equivalent to $3x - 1 = 5$ or $3x - 1 = -5$.

    Solve the first equation, $3x - 1 = 5$.

    Add 1 to each side and divide each side by 3.

    $$3x - 1 = 5$$
    $$\underline{+1 \quad +1}$$
    $$\frac{3x}{3} = \frac{6}{3}$$
    $$x = 2$$

    Solve the second equation, $3x - 1 = -5$.

    Add 1 to each side and divide each side by 3.

    $$3x - 1 = -5$$
    $$\underline{+1 \quad +1}$$
    $$\frac{3x}{3} = \frac{-4}{3}$$
    $$x = \frac{-4}{3}$$

    Do both of these answers really work?

    Place the answer $x = 2$ into the original equation, $|3x - 1| = 5$.

$$|3(2) - 1| = 5$$
$$|6 - 1| = 5$$
$$|5| = 5$$

Now, put the answer $x = -\dfrac{4}{3}$ into the original equation, $|3x - 1| = 5$.

$$\left|3\left(-\frac{4}{3}\right) - 1\right| = 5$$
$$|-4 - 1| = 5$$
$$|-5| = 5$$

This method produced both the positive and negative value whose absolute values are 5. So, both answers work.

2.   Solve $|5 - 2y| + 1 = 3$.

   answer: $y = \dfrac{3}{2}$, $y = \dfrac{7}{2}$

Before applying the rule for rewriting this problem without the absolute value signs, the equation has to be put in the exact form used by the rule. The 1 can't be on the same side of the equation as the absolute value term. Subtract 1 from each side.

$$|5 - 2y| + 1 = 3$$
$$\underline{ -1 \; -1}$$
$$|5 - 2y| = 2$$

Now you can apply the rule and determine the two linear equations. The two equations to solve are $5 - 2y = 2$ and $5 - 2y = -2$.

Solving the equation $5 - 2y = 2$, subtract 5 from each side and divide by -2.

$$5 - 2y = 2$$
$$\underline{-5 \qquad -5}$$
$$\frac{-2}{-2}y = \frac{-3}{-2}$$
$$y = \frac{3}{2}$$

Solving the equation $5 - 2y = -2$, subtract 5 from each side and divide by -2.

$$5 - 2y = -2$$
$$\underline{-5 \qquad -5}$$
$$\frac{-2}{-2}y = \frac{-7}{-2}$$
$$y = \frac{7}{2}$$

3.   Solve $3|5w - 1| - 4 = 8$.

   answer: $w = 1$, $w = -\dfrac{3}{5}$

Before applying the rule for rewriting this without the absolute value signs, you have to put the equation in the exact form used by the rule. The 3 shouldn't be multiplied by the absolute value term, and the 4 can't be subtracted. Add 4 to each side and divide each side by 3.

$$3|5w - 1| - 4 = 8$$
$$\underline{+4 \quad +4}$$
$$\frac{3|5w - 1|}{3} = \frac{12}{3}$$
$$|5w - 1| = 4$$

Now the absolute value equation can be written as two linear equations, $5w - 1 = 4$ and $5w - 1 = -4$.

Solving the first equation, $5w - 1 = 4$, add 1 to each side and divide by 5.

$$5w - 1 = 4$$
$$\underline{+1 \quad +1}$$
$$\frac{5w}{5} = \frac{5}{5}$$
$$w = 1$$

Solving the second equation, $5w - 1 = -4$, add 1 to each side and divide by 5.

$$5w - 1 = -4$$
$$\underline{+1 \quad +1}$$
$$\frac{5w}{5} = \frac{-3}{5}$$
$$w = -\frac{3}{5}$$

4. Solve $|9 - 2t| = -5$.

answer: No answer is possible.

This problem can catch you unaware, if you don't notice that it can't be solved. Look carefully. The equation says that the absolute value of some number is equal to a negative number and that contradicts the definition of absolute value. The answer to this problem is that there is no answer. But, what happens if you don't notice the negative and try to solve it anyway?

If you write it as two linear equations, changing the −5 to 5 in the second one, you get $9 - 2t = -5$ or $9 - 2t = 5$.

Solving the first equation, subtract 9 from each side and divide by −2.

$$9 - 2t = -5$$
$$\underline{-9 -9}$$
$$\frac{-2t}{-2} = \frac{-14}{-2}$$
$$t = 7$$

Solving the second equation, subtract 9 from each side and divide by −2.

$$9 - 2t = 5$$
$$\underline{-9 \qquad -9}$$
$$\frac{-2t}{-2} = \frac{-4}{-2}$$
$$t = 2$$

These are two very nice answers, but do they work? Try putting them back into the original equation to check.

$$|9 - 2t| = -5. \text{ Letting } t = 7,$$
$$|9 - 2(7)| = -5$$
$$|9 - 14| = -5$$
$$|-5| = -5$$

The last equation is wrong! Not so!

$$\text{Letting } t = 2,$$
$$|9 - 2(2)| = -5$$
$$|9 - 4| = -5$$
$$|5| = -5$$

This also is wrong!

If you're not paying attention, you can get a nonsense answer. Just be careful when using rules that change the original form of an equation to something more convenient. This sort of thing happens in several instances in algebra. The methods are the nicest to use, but you have to be wary of false solutions.

Work Problems

Use these problems to give yourself additional practice.

1. Solve $|4x - 5| = 3$.

2. Solve $3|3y + 8| = 6$.

3. Solve $4 - |5 + 2y| = -11$.

4. Solve $6 + |y + 1| = 2$.

5. Solve $5|3y + 1| = 5$.

Worked Solutions

1. $x = 2, x = \frac{1}{2}$ Change the absolute value equation into the two corresponding linear equations.

    $|4x - 5| = 3$ is equivalent to $4x - 5 = 3$ or $4x - 5 = -3$.

    Solving the first equation, add 5 to each side and divide by 4.

$$4x - 5 = 3$$
$$\underline{+5 \quad +5}$$
$$\frac{4x}{4} = \frac{8}{4}$$
$$x = 2$$

Solving the second equation, add 5 to each side and divide by 4.

$$4x - 5 = -3$$
$$\underline{+5 \quad +5}$$
$$\frac{4x}{4} = \frac{2}{4}$$
$$x = \frac{1}{2}$$

2. **$y = -2$, $y = -\dfrac{10}{3}$** Before changing this to the two linear equations, first divide each side by 3.

$$\frac{3\left|3y + 8\right|}{3} = \frac{6}{3}$$
$$\left|3y + 8\right| = 2$$

Now the absolute value equation can be written as the linear equations $3y + 8 = 2$ and $3y + 8 = -2$.

Solving the first equation, subtract 8 from each side and divide by 3.

$$3y + 8 = 2$$
$$\underline{-8 \quad -8}$$
$$\frac{3y}{3} = \frac{-6}{3}$$
$$y = -2$$

Solving the second equation, subtract 8 from each side and divide by 3.

$$3y + 8 = -2$$
$$\underline{-8 \quad -8}$$
$$\frac{3y}{3} = \frac{-10}{3}$$
$$y = -\frac{10}{3}$$

3. **$y = 5$, $y = -10$** Before changing this into the two linear equations, first subtract 4 from each side and then multiply each side by -1.

$$4 - \left|5 + 2y\right| = -11$$
$$\underline{-4 \qquad\qquad -4}$$
$$-\left|5 + 2y\right| = -15$$
$$-1\left(-\left|5 + 2y\right|\right) = -1\left(-15\right)$$
$$\left|5 + 2y\right| = 15$$

It looked, at first, as if there would be no solution. The original absolute value equation had a negative sign in front of the absolute value, and it was equal to a negative value. It looks okay, now, but still check your answers at the end.

The two linear equations to solve are $5 + 2y = 15$ and $5 + 2y = -15$.

Solving $5 + 2y = 15$, first subtract 5 from each side and then divide each side by 2.

$$\begin{array}{r} 5 + 2y = 15 \\ \underline{-5 \qquad -5} \\ \dfrac{2y}{2} = \dfrac{10}{2} \\ y = 5 \end{array}$$

Solving $5 + 2y = -15$, first subtract 5 from each side and then divide each side by 2.

$$\begin{array}{r} 5 + 2y = -15 \\ \underline{-5 \qquad -5} \\ \dfrac{2y}{2} = \dfrac{-10}{2} \\ y = -5 \end{array}$$

4. **No answer**   First, subtract 6 from each side.

$$\begin{array}{r} 6 + |y + 1| = 2 \\ \underline{-6 \qquad\qquad -6} \\ |y + 1| = -4 \end{array}$$

You can stop right there. The absolute value cannot equal a negative number. This equation has no solution.

5. $y = 0$, $y = -\dfrac{2}{3}$ First, divide each side by 5.

$$\dfrac{5|3y + 1|}{5} = \dfrac{5}{5}$$
$$|3y + 1| = 1$$

The two equations to solve are $3y + 1 = 1$ and $3y + 1 = -1$.

Solving $3y + 1 = 1$, subtract 1 from each side and then divide by 3.

$$\begin{array}{r} 3y + 1 = 1 \\ \underline{-1 \quad -1} \\ \dfrac{3y}{3} = \dfrac{0}{3} \\ y = 0 \end{array}$$

Solving $3y + 1 = -1$, subtract 1 from each side and then divide by 3.

$$3y + 1 = -1$$
$$\underline{\quad -1 \quad -1 \quad}$$
$$\frac{3y}{3} = \frac{-2}{3}$$
$$y = -\frac{2}{3}$$

Absolute Value Inequalities

When absolute values and inequalities get mixed together, the goal is to rewrite the statement in a form that's recognizable and solvable **without** the absolute value symbol. The new, rewritten statement has to have the same solution as the original absolute value statement; the new one will just be in a more convenient form. The new form then can be solved with the inequality rules, which are found at the beginning of this chapter. The main challenge here is to correctly rewrite the statement. There's a separate rule for each type of inequality, determined by the direction that the inequality symbol is facing. The rules are

❑ If $|ax + b| > k$, then either $ax + b > k$ or $ax + b < -k$.

   Notice that you have to switch the sense around and change the constant to a negative in the second equation. This is very much like the absolute value equations in the previous section. Also, this same rule applies if it's $\geq$ rather than $>$.

❑ If $|ax + b| < k$, then $-k < ax + b < k$.

   This time, the expression inside the absolute value is sandwiched between the number and its opposite.

Example Problems

These problems show the answers and solutions.

1. Solve $|3x - 1| > 4$.

    answer: $x > \frac{5}{3}$ or $x < -1$

    Rewrite this problem as $3x - 1 > 4$ or $3x - 1 < -4$. Solving the first inequality, add 1 to each side and then divide by 3.

$$3x - 1 > 4$$
$$\underline{\quad +1 \quad +1 \quad}$$
$$\frac{3x}{3} > \frac{5}{3}$$
$$x > \frac{5}{3}$$

    Solving the second inequality, add 1 to each side and then divide by 3.

$$3x - 1 < -4$$
$$\underline{\quad +1 \quad +1 \quad}$$
$$\frac{3x}{3} < \frac{-3}{3}$$
$$x < -1$$

These two solutions go in completely different directions. There's no overlap. You write the solution, therefore, as $x > \frac{5}{3}$ or $x < -1$. To write this solution using interval notation, you'd write $(-\infty, -1) \cup \left(\frac{5}{3}, \infty\right)$. The $\cup$ is the symbol for "union," which means to include everything from both parts. For more on interval notation, see "Number Lines," in Chapter 2. Graphing this on a number line, you would get .

2. Solve $|9 - 2x| \le 11$.

   answer: $-1 \le x \le 10$

   Rewrite this as $-11 \le 9 - 2x \le 11$.

   To solve this problem, subtract 9 from each section and then divide each section by -2. Don't forget that, when dividing an inequality by a negative number, you need to reverse the sense (inequality symbol).

$$-11 \le 9 - 2x \le 11$$
$$\underline{-9 \quad -9 \qquad -9}$$
$$-20 \le \quad -2x \le 2$$
$$\frac{-20}{-2} \ge \frac{-2x}{-2} \ge \frac{2}{-2}$$
$$10 \ge x \ge -1$$

   This makes more sense if it's written with the smaller number to the left and the larger to the right. That means switching the sense again, if everything is getting flipped. $-1 \le x \le 10$ written in interval notation is $[-1, 10]$, and graphed on a number line looks like this: .

Work Problems

Use these problems to give yourself additional practice.

1. Solve $|x + 2| \ge 5$. Write the answer in inequality notation, interval notation, and graph the solution on a number line.

2. Solve $|x - 9| < 15$. Write the answer in inequality notation, interval notation, and graph the solution on a number line.

3. Solve $|3x + 1| - 4 > 3$.

4. Solve $2|5x - 3| \le 26$.

5. Solve $5|3 - 4x| + 7 > 3$.

Worked Solutions

1. **$x \ge 3$ or $x \le -7$**   Rewrite the absolute inequality as $x + 2 \ge 5$ or $x + 2 \le -5$.

   Solving $x + 2 \ge 5$, subtract 2 from each side.

$$x + 2 \geq 5$$
$$\underline{-2 \quad -2}$$
$$x \qquad \geq 3$$

Solving $x + 2 \leq -5$, subtract 2 from each side.

$$x + 2 \leq -5$$
$$\underline{-2 \quad -2}$$
$$x \qquad \leq -7$$

The solution is that $x \geq 3$ or $x \leq -7$. In interval notation, that's $(-\infty, -7] \cup [3, \infty)$. The graph looks like ◄━━━●━━━●━━━►.

-7   3

2. **$-6 < x < 24$** Rewrite the absolute inequality as $-15 < x - 9 < 15$. Add 9 to each section of the inequality.

$$-15 < x - 9 < 15$$
$$\underline{+9 \quad +9 \quad +9}$$
$$-6 < x \qquad < 24$$

In interval notation, that's $(-6, 24)$, which looks like the coordinates of a point in a graph. The only way to tell that it's an interval on a number line is the context. The graph looks like ◄━━○━━━○━━►.

-6   24

3. **$x > 2$ or $x < -\dfrac{8}{3}$** Before rewriting this without the absolute value symbols, add 4 to each side.

$$|3x + 1| - 4 > 3$$
$$\underline{+4 \quad +4}$$
$$|3x + 1| > 7$$

Now, rewrite without the absolute value signs, $3x + 1 > 7$ or $3x + 1 < -7$.

Solving $3x + 1 > 7$, subtract 1 from each side and then divide by 3.

$$3x + 1 > 7$$
$$\underline{-1 \quad -1}$$
$$\frac{3x}{3} \quad > \frac{6}{3}$$
$$x > 2$$

Solving $3x + 1 < -7$, subtract 1 from each side and then divide by 3.

$$3x + 1 < -7$$
$$\underline{-1 \quad -1}$$
$$\frac{3x}{3} \quad < \frac{-8}{3}$$
$$x < -\frac{8}{3}$$

The solution reads that $x > 2$ or $x < -\dfrac{8}{3}$.

4. $-2 \le x \le \dfrac{16}{5}$ First, divide each side by 2.

$$\frac{2|5x-3|}{2} \le \frac{26}{2}$$
$$|5x-3| \le 13$$

Rewrite the absolute value inequality without the absolute value.

$$-13 \le 5x - 3 \le 13$$

Add 3 to each section and then divide each section by 5.

$$-13 \le 5x - 3 \le 13$$
$$\underline{ +3 +3 +3}$$
$$\frac{-10}{5} \le \frac{5x}{5} \le \frac{16}{5}$$
$$-2 \le x \le \frac{16}{5}$$

5. $(-\infty, \infty)$ Before rewriting, first subtract 7 from each side and then divide by 5.

$$5|3-4x| + 7 > 3$$
$$\underline{ -7 -7}$$
$$\frac{5|3-4x|}{5} > \frac{-4}{5}$$
$$|3-4x| > \frac{-4}{5}$$

This now states that the absolute value of something is bigger than a negative number. By the definition of absolute value, you always get a positive number, so it seems that you can't go wrong. Rewrite this using the rule: $3 - 4x > \dfrac{-4}{5}$ or $3 - 4x < \dfrac{4}{5}$.

Solving $3 - 4x > \dfrac{-4}{5}$, subtract 3 from each side and then divide by -4. Remember to reverse the inequality sign, which is what you do when multiplying or dividing an inequality by a negative number.

$$3 - 4x > \frac{-4}{5}$$
$$\underline{-3 -3}$$
$$-4x > -\frac{19}{5}$$
$$\frac{-4x}{-4} < -\frac{19}{5} \div -4$$
$$x < -\frac{19}{5}\left(-\frac{1}{4}\right)$$
$$x < \frac{19}{20}$$

Now, solving the other inequality, $3 - 4x < \frac{4}{5}$, subtract 3 from each side and then divide by −4. Remember to reverse the inequality sign, which is what you do when multiplying or dividing an inequality by a negative number.

$$3 - 4x < \frac{4}{5}$$
$$\underline{-3 \qquad\quad -3}$$
$$-4x < -\frac{11}{5}$$
$$\frac{-4x}{-4} > -\frac{11}{5} \div -4$$
$$x > -\frac{11}{5} \cdot -\frac{1}{4}$$
$$x > \frac{11}{20}$$

The solutions, put together, say that $x < \frac{19}{12}$ or $x > \frac{11}{12}$. What does this look like on a number line? The statement is that one solution **or** the other will work.

You can use all of the solutions in either statement, so the combination of the two would be **everything** on the number line, from negative infinity to positive infinity, $(-\infty, \infty)$.

Simplifying Square Roots

When you're working with the square roots of numbers, you can either find an exact value, if it's a perfect square, or sometimes you can simplify the root into a multiple of a number times a square root. One other option is to estimate what the value of the square root is, to a specified number of decimal places. You can get more detail on this in the section "Square Roots and Cube Roots," in Chapter 1. When you are working with variables and algebraic expressions with square roots in them, you use all the same rules that apply to square roots of numbers. Here are some guidelines for working with the square roots of variables.

$\sqrt{x}$ means to find the square root of some number, represented by x. The value of x has to be nonnegative, because no real number times itself will give you a negative result.

$\sqrt{x^2}$ is equal to $|x|$. This happens because the x is squared under the radical, which allows you to have x represent a negative number. But, the answer has to be positive; the square root of the positive square is a positive number. So, by putting the absolute value symbol around the variable, you're assured of always having a positive result.

$\sqrt{4x}$ is equal to $2\sqrt{x}$. The expression is $\sqrt{4x} = \sqrt{4}\sqrt{x} = 2\sqrt{x}$.

Simplifying square roots refers to a very specific process. Simplifying here means you need to factor what's under the radical into a perfect square factor that can be separated and computed and leave the rest of the factors under the radical. This allows you to combine the expression with others and do other algebraic manipulations. For more information, refer to the section, "Simplifying Square Roots" in Chapter 1.

In the following examples, to simplify matters, just assume that the variables are positive.

Example Problems

These problems show the answers and solutions.

1. Simplify $\sqrt{16x^4 y}$.

   answer: $4x^2 \sqrt{y}$

   Write this as a product of two radicals—one with perfect squares under it and the other with what's left of the expression that wasn't a perfect square.

   $$\sqrt{16x^4 y} = \sqrt{16x^4} \sqrt{y} = 4x^2 \sqrt{y}$$

2. Simplify $\sqrt{90a^3 b^2}$.

   answer: $3ab \sqrt{10a}$

   There are three perfect square factors, 9, a^2, and b^2.

   $$\sqrt{90a^3 b^2} = \sqrt{9a^2 b^2} \sqrt{10a} = 3ab \sqrt{10a}$$

3. Simplify $\sqrt{28m^5 (n-1)^7}$.

   answer: $2m^2 (n-1)^3 \sqrt{7m(n-1)}$

   The binomial can be treated like any other variable factor.

   $$a\sqrt{28m^5 (n-1)^7} = \sqrt{4m^4 (n-1)^6} \sqrt{7m(n-1)} = 2m^2 (n-1)^3 \sqrt{7m(n-1)}$$

4. Multiply and simplify $\sqrt{5xy^3} \sqrt{10x^3 y^4}$.

   answer: $5x^2 y^3 \sqrt{2y}$

   There are some perfect square factors under these individual radicals, but it's more efficient to multiply first and then simplify.

   $$\sqrt{5xy^3} \sqrt{10x^3 y^4} = \sqrt{50x^4 y^7} = \sqrt{25x^4 y^6} \sqrt{2y} = 5x^2 y^3 \sqrt{2y}$$

5. Simplify $\sqrt{12a^3 b^9} - \sqrt{27a^3 b^9} + ab\sqrt{75ab^7}$.

   answer: $4ab^4 \sqrt{3ab}$

   Subtraction and addition are indicated, but these terms can't be combined unless exactly the same term is under the radical and exactly the same variables are multiplying the radical. Simplify each term, first, to see whether any subtraction or addition can be performed.

   $$\sqrt{12a^3 b^9} - \sqrt{27a^3 b^9} + ab\sqrt{75ab^7} = \sqrt{4a^2 b^8} \sqrt{3ab} - \sqrt{9a^2 b^8} \sqrt{3ab} + ab\sqrt{25b^6} \sqrt{3ab}$$

   $$= 2ab^4 \sqrt{3ab} - 3ab^4 \sqrt{3ab} + ab(5b^3)\sqrt{3ab}$$

   $$= 2ab^4 \sqrt{3ab} - 3ab^4 \sqrt{3ab} + 5ab^4 \sqrt{3ab}$$

The resulting terms are exactly the same, except for the coefficient—the numerical part. Combining them, you get $2ab^4\sqrt{3ab} - 3ab^4\sqrt{3ab} + 5ab^4\sqrt{3ab} = 4ab^4\sqrt{3ab}$.

Work Problems

Use these problems to give yourself additional practice.

1. Simplify $\sqrt{20a^3bc^5}$.

2. Simplify $\sqrt{200x^4(x^2+3)^5}$.

3. Multiply and simplify $\sqrt{21y^3z^2}\sqrt{7y^{10}z^5}$.

4. Multiply and simplify $3\sqrt{2a^3}\sqrt{6ab}$.

5. Simplify $6\sqrt{8b^2z^3} - 4z\sqrt{18b^2z} - b\sqrt{144z^3}$.

Worked Solutions

1. $2ac^2\sqrt{5abc}$ Determine which factors of the expression under the radical are perfect squares.

$$\sqrt{20a^3bc^5} = \sqrt{4a^2c^4}\sqrt{5abc} = 2ac^2\sqrt{5abc}$$

2. $10x^2(x^2+3)^2\sqrt{2(x+3)}$

$$\sqrt{200x^4(x^2+3)^5} = \sqrt{100x^4(x^2+3)^4}\sqrt{2(x+3)} = 10x^2(x^2+3)^2\sqrt{2(x+3)}$$

3. $7y^6z^3\sqrt{3yz}$ First multiply the factors together under the radical. Then find any perfect square factors.

$$\sqrt{21y^3z^2}\sqrt{7y^{10}z^5} = \sqrt{147y^{13}z^7} = \sqrt{49y^{12}z^6}\sqrt{3yz} = 7y^6z^3\sqrt{3yz}$$

4. $6a^2\sqrt{3b}$ First multiply the values under the radical. Then look for perfect square factors.

$$3\sqrt{2a^3}\sqrt{6ab} = 3\sqrt{12a^4b} = 3\sqrt{4a^4}\sqrt{3b} = 3(2a^2)\sqrt{3b} = 6a^2\sqrt{3b}$$

5. $-12bz\sqrt{z}$ Simplify each term first to see whether any can be combined.

$$6\sqrt{8b^2z^3} - 4z\sqrt{18b^2z} - b\sqrt{144z^3} = 6\sqrt{4b^2z^2}\sqrt{2z} - 4z\sqrt{9b^2}\sqrt{2z} - b\sqrt{144z^2}\sqrt{z}$$

$$= 6(2bz)\sqrt{2z} - 4z(3b)\sqrt{2z} - b(12z)\sqrt{z}$$

$$= 12bz\sqrt{2z} - 12bz\sqrt{2z} - 12bz\sqrt{z}$$

The first two terms are opposites of one another, so they'll give a result of 0. The last term isn't the same as the first two; the value under the radical is different. So, it doesn't combine with the others.

$$12bz\sqrt{2z} - 12bz\sqrt{2z} - 12bz\sqrt{z} = 0 - 12bz\sqrt{z}$$
$$= -12bz\sqrt{z}$$

Simplifying Other Roots

Square roots are probably more popular algebraic manipulations than other roots, but cube roots, fourth roots, and so on play an important part in equations and solutions of equations and need to be discussed. The process for simplifying other roots is much like that of square roots, but attention has to be paid to the powers of the variables and numbers and which root is being applied.

First, a quick listing of some powers is shown in the following table. It helps when you recognize that a number is a power of another. The higher powers of the larger numbers aren't given, because they're used infrequently.

Powers										
1	2	3	4	5	6	7	8	9	10	11
2	4	9	16	25	36	49	64	81	100	121
3	8	27	64	125	216	343	512	729	1000	1331
4	16	81	256	625	1296					
5	32	243	1024	3125						
6	64	729								
7	128									

When you take the root of a variable with a power, the power of the variable has to divide evenly by the root.

These equations show you how to compute roots using the radical symbols or the equivalent fractional exponents and their rules:

$$\sqrt[3]{x^6} = x^{6 \div 3} = x^2 \text{ or } \sqrt[3]{x^6} = \left(x^6\right)^{1/3} = x^{6 \cdot \frac{1}{3}} = x^2$$

$$\sqrt[4]{x^{20}} = x^5 \qquad \sqrt[7]{y^{14}} = y^2 \qquad \sqrt[3]{27z^6} = \sqrt[3]{3^3 z^6} = 3z^2$$

$$\sqrt[3]{x^{30}} = x^{10} \qquad \sqrt[5]{y^{15}} = y^3 \qquad \sqrt[5]{32m^{10}z^{25}} = \sqrt[5]{2^5 m^{10} z^{25}} = 2m^2 z^5$$

If the variable has a power that isn't evenly divisible by the root, then the expression can still be simplified. Using a technique similar to simplifying square roots, the expression is broken into two factors—one that is a perfect root and the other with what's left over.

Example Problems

These problems show the answers and solutions.

1. Simplify $\sqrt[3]{x^7}$.

   answer: $x^2 \sqrt[3]{x}$

   Break the expression into two factors—one with the prefect cube and the other with the rest.

   $$\sqrt[3]{x^7} = \sqrt[3]{x^6}\sqrt[3]{x} = x^2\sqrt[3]{x}$$

2. Simplify $\sqrt[4]{32a^4 b^6 c^{13}}$.

   answer: $2abc^3 \sqrt[4]{2b^2 c}$

   This time, look for factors that are something raised to the fourth power.

   $$\sqrt[4]{32a^4 b^6 c^{13}} = \sqrt[4]{16a^4 b^4 c^{12}}\sqrt[4]{2b^2 c} = 2abc^3\sqrt[4]{2b^2 c}$$

Work Problems

Use these problems to give yourself additional practice.

1. Simplify $\sqrt[3]{8a^3 b^6 c^{31} d}$.

2. Simplify $\sqrt[6]{64x^6 y^{66}}$.

3. Simplify $\sqrt[4]{2m^5 n^6 p^8}$.

4. Simplify $\sqrt[3]{125x^6 y^{19}}$.

5. Simplify $\sqrt[5]{320r^4 s^{10} t^{23}}$.

Worked Solutions

1. $2ab^2 c^{10}\sqrt[3]{cd}$ Write two radical factors—one with perfect cube factors and the other with what's left.

   $$\sqrt[3]{8a^3 b^6 c^{31} d} = \sqrt[3]{2^3 a^3 b^6 c^{30}}\sqrt[3]{cd} = 2ab^2 c^{10}\sqrt[3]{cd}$$

2. $2xy^{11}$ All of the factors are the sixth power of some value.

   $$\sqrt[6]{64x^6 y^{66}} = \sqrt[6]{2^6 x^6 y^{66}} = 2xy^{11}$$

3. $mnp^2 \sqrt[4]{2mn^2}$ Separate the factors that are fourth powers from the others.

$$\sqrt[4]{2m^5 n^6 p^8} = \sqrt[4]{m^4 n^4 p^8} \sqrt[4]{2mn^2} = mnp^2 \sqrt[4]{2mn^2}$$

4. $5x^2 y^6 \sqrt[3]{y}$ Separate the factors that are perfect cubes from the others.

$$\sqrt[3]{125x^6 y^{19}} = \sqrt[3]{5^3 x^6 y^{18}} \sqrt[3]{y} = 5x^2 y^6 \sqrt[3]{y}$$

5. $2s^2 t^4 \sqrt[5]{10r^4 t^3}$ Only the variable r doesn't have a fifth power factor.

$$\sqrt[5]{320r^4 s^{10} t^{23}} = \sqrt[5]{32s^{10} t^{20}} \sqrt[5]{10r^4 t^3} = 2s^2 t^4 \sqrt[5]{10r^4 t^3}$$

Radical Equations

A radical equation is one with a radical term on one or both sides of the equal sign. The most efficient way of solving radical equations is to get rid of the radical. It's not as hard as it sounds. If each side of the equation is raised to a power that gets rid of the radical, then the new equation can be solved. The only hitch to this is that the new equation, although solvable, may have extraneous solutions. Extraneous in this case means that, even though it's a solution of the new equation, it doesn't work in the original. This type of situation is also encountered when changing absolute value equations and inequalities into solvable linear equations. It comes up later, in other algebra situations. Whenever the original equation is changed, there's a possibility for solutions that are not really solutions. Just be aware and on the lookout. Even though this is a potential problem with the method, it's still worth using because it's the easiest. Just check your work.

Example Problems

These problems show the answers and solutions.

1.   Solve for m in $\sqrt{2m - 3} = 7$.

answer: $m = 26$

Square both sides of the equation.

$$\left(\sqrt{2m - 3}\right)^2 = 7^2$$
$$2m - 3 = 49$$

Now add 3 to each side and divide each side by 2.

$$2m - 3 = 49$$
$$\underline{ + 3 \quad +3}$$
$$\frac{2m}{2} = \frac{52}{2}$$
$$m = 26$$

Because the format of the equation was changed by squaring both sides, check to be sure that this solution works in the original equation.

$$\sqrt{2(26)-3}=7$$
$$\sqrt{52-3}=7$$
$$\sqrt{49}=7$$

It works!

2. Solve for x in $\sqrt{5-4x}-2=9$.

answer: $x=-29$

First add 2 to each side.

$$\sqrt{5-4x}-2=9$$
$$\underline{\phantom{\sqrt{5-4x}}+2+2}$$
$$\sqrt{5-4x}=11$$

Now square both sides.

$$\left(\sqrt{5-4x}\right)^2=11^2$$
$$5-4x=121$$

Subtract 5 from each side and then divide each side by −4.

$$5-4x=121$$
$$\underline{-5-5}$$
$$\frac{-4x}{-4}=\frac{116}{-4}$$
$$x=-29$$

This needs to be checked in the original equation to see whether it is really the solution.

$$\sqrt{5-4(-29)}-2=9$$
$$\sqrt{5+116}-2=9$$
$$\sqrt{121}-2=9$$
$$11-2=9$$

Yes, it works!

3. Solve for x in $\sqrt{6x+1}+3=1$.

answer: No answer is possible.

First subtract 3 from each side. Then square each side.

$$\sqrt{6x+1}+3=1$$
$$\underline{\phantom{\sqrt{6x+1}}-3-3}$$
$$\sqrt{6x+1}=-2$$
$$\left(\sqrt{6x+1}\right)^2=(-2)^2$$
$$6x+1=4$$

Now subtract 1 from each side and divide by 6.

Check this to see whether it's a solution of the original equation.

$$\sqrt{6\left(\frac{1}{2}\right)+1}+3=1$$
$$\sqrt{3+1}+3=1$$
$$\sqrt{4}+3=1$$
$$2+3=1$$

No, this one didn't work. There isn't a solution, because the only one that was found using this method didn't satisfy the equation.

4. Solve for z in $\sqrt{5z-1}=\sqrt{25-8z}$.

    answer: $z=2$

    Square both sides of the equation.

    $$\left(\sqrt{5z-1}\right)^2=\left(\sqrt{25-8z}\right)^2$$
    $$5z-1=25-8z$$

    Now add $8z$ to each side and then add 1 to each side.

    $$5z-1=25-8z$$
    $$\underline{8z \qquad\qquad +8z}$$
    $$13z-1=25$$
    $$\underline{\qquad +1 \ +1}$$
    $$13z \quad =26$$

    Now divide by 2 and check the answer.

    $$\frac{13z}{13}=\frac{26}{13}$$
    $$z=2$$

    Checking:

    $$\sqrt{5(2)-1}=\sqrt{25-8(2)}$$
    $$\sqrt{10-1}=\sqrt{25-16}$$
    $$\sqrt{9}=\sqrt{9}$$

    It works.

5. Solve for x in $\sqrt{4x+3}-\sqrt{6x-3}=0$.

    answer: $x=3$

First, move the second radical over to the right by adding it to each side. Then square both sides.

$$\sqrt{4x+3} - \sqrt{6x-3} = 0$$
$$\underline{+\sqrt{6x-3} \quad +\sqrt{6x-3}}$$
$$\sqrt{4x+3} \qquad = \sqrt{6x-3}$$
$$\left(\sqrt{4x+3}\right)^2 = \left(\sqrt{6x-3}\right)^2$$
$$4x+3 = 6x-3$$

Now subtract 6x from each side and subtract 3 from each side.

$$4x+3 = 6x-3$$
$$\underline{-6x \qquad -6x}$$
$$-2x+3 = \qquad -3$$
$$\underline{\qquad -3 \qquad -3}$$
$$-2x \qquad = -6$$

Now divide each side by −2 and check the answer in the original equation.

$$\frac{-2x}{-2} = \frac{-6}{-2}$$
$$x = 3$$

Checking:

$$\sqrt{4(3)+3} - \sqrt{6(3)-3} = 0$$
$$\sqrt{12+3} - \sqrt{18-3} = 0$$
$$\sqrt{15} - \sqrt{15} = 0$$

It works.

Work Problems

Use these problems to give yourself additional practice.

1. Solve for n in $\sqrt{5n+9} = 2$.

2. Solve for p in $10 - \sqrt{9-5p} = 3$.

3. Solve for t in $\sqrt{2t-3} + 4 = 1$.

4. Solve for x in $\sqrt{5x-2} = \sqrt{x+2}$.

5. Solve for w in $\sqrt{6w+7} - \sqrt{4+3w} = 0$.

Worked Solutions

1. **n = −1**   Square both sides of the equation.

$$\left(\sqrt{5n+9}\right)^2 = 2^2$$
$$5n+9 = 4$$

Now subtract 9 from each side and then divide each side by 5.

$$5n+9 = 4$$
$$\underline{\quad -9 \quad -9 \quad}$$
$$\frac{5n}{5} = \frac{-5}{5}$$
$$n = -1$$

The answer needs to be checked in the original equation.

$$\sqrt{5(-1)+9} = 2$$
$$\sqrt{-5+9} = 2$$
$$\sqrt{4} = 2$$

The answer works.

2. **p = −8**   First subtract 10 from each side. Then square both sides.

$$10-\sqrt{9-5p} = 3$$
$$\underline{-10 \qquad\qquad\quad -10}$$
$$\left(-\sqrt{9-5p}\right)^2 = (-7)^2$$
$$9-5p = 49$$

Now subtract 9 from each side. Then divide each side by −5.

$$9-5p = 49$$
$$\underline{-9 \qquad\quad -9}$$
$$\frac{-5p}{-5} = \frac{40}{-5}$$
$$p = -8$$

Check the answer in the original equation:

$$10-\sqrt{9-5(-8)} = 3$$
$$10-\sqrt{9+40} = 3$$
$$10-\sqrt{49} = 3$$
$$10-7 = 3$$

It worked.

3. **No answer** First subtract 4 from each side. Then square both sides.

$$\sqrt{2t-3}+4=1$$
$$\underline{\qquad -4 \quad -4\qquad}$$
$$\left(\sqrt{2t-3}\right)^2 = (-3)^2$$
$$2t-3 \;=9$$

Now add 3 to each side and then divide each side by 2.

$$2t-3=9$$
$$\underline{\qquad +3+3\qquad}$$
$$\frac{2t}{2}=\frac{12}{2}$$
$$t=6$$

Check this in the original equation:

$$\sqrt{2(6)-3}+4=1$$
$$\sqrt{12-3}+4=1$$
$$\sqrt{9}+4=1$$
$$3+4=1$$

This is false. This isn't the solution; this equation has no solution.

4. $x = 1$　Square both sides of the equation.

$$\left(\sqrt{5x-2}\right)^2=\left(\sqrt{x+2}\right)^2$$
$$5x-2=x+2$$

Subtract x from each side and then add 2 to each side.

$$5x-2=x+2$$
$$\underline{-x \qquad\quad -x\qquad}$$
$$4x-2=2$$
$$\underline{\qquad +2 \quad +2\qquad}$$
$$4x \;=4$$

Divide each side by 4 to get $x = 1$. Now check that answer:

$$\sqrt{5(1)-2}=\sqrt{(1)+2}$$
$$\sqrt{5-2}=\sqrt{1+2}$$
$$\sqrt{3}=\sqrt{3}$$

It worked.

5. $w = -1$　First move the second radical to the other side by adding that radical to each side. Then square both sides of the equation.

$$\sqrt{6w+7} - \sqrt{4+3w} = 0$$

$$\frac{+\sqrt{4+3w} \quad +\sqrt{4+3w}}{\left(\sqrt{6w+7}\right)^2 \quad = \left(\sqrt{4+3w}\right)^2}$$

$$6w+7 \quad = \quad 4+3w$$

Subtract 3w from each side and subtract 7 from each side.

$$6w+7 = 4+3w$$

$$\frac{-3w \qquad -3w}{3w+7 = 4}$$

$$\frac{-7 \; -7}{3w \quad = -3}$$

Now divide each side by 3 and check the answer.

$$\frac{3w}{3} = \frac{-3}{3}$$

$$w = -1$$

$$\sqrt{6(-1)+7} - \sqrt{4+3(-1)} = 0$$

$$\sqrt{-6+7} - \sqrt{4-3} = 0$$

$$\sqrt{1} - \sqrt{1} = 0$$

It works.

Chapter Problems and Solutions

Problems

For problems 1 through 4, solve for the value(s) of the variable.

1. $4x - 3 < 8x + 7$

2. $5(3 - y) \geq 1 - 3y$

3. $-3 < 3z + 4 \leq 5$

4. $\dfrac{5}{w+1} > \dfrac{3}{2w-1}$

For problems 5 through 7, find the common solution for the two inequalities.

5. $x \geq 3$ and $x < 7$

6. $y < -2$ and $y \geq 2$

7. $0 < x + 1 \leq 5$ and $x - 1 > 1$

For problems 8 through 13, solve for the value(s) of the variable.

8. $|3z - 1| = 7$

9. $4|x - 3| - 2 = 18$

10. $|4 - 3y| + 5 = 0$

11. $|x - 1| \geq 3$

12. $|4 - y| < 6$

13. $3|3x + 4| > 15$

For problems 14 through 20, simplify the expression.

14. $\sqrt{300x^4 y^5 z^6}$

15. $\sqrt{40a^3 b^7 c^9}$

16. $4\sqrt{6xz}\sqrt{12x^2 yz}$

17. $8\sqrt{2a^3 b^2} - ab\sqrt{18a}$

18. $\sqrt[3]{27m^3 n^4 z^6}$

19. $\sqrt[4]{32a^8 b^{10}}$

20. $\sqrt[5]{4x^2 y^3}\sqrt[5]{16x^3 y^3}$

For problems 21 through 25, solve for the value(s) of the variable.

21. $\sqrt{2z - 1} = 4$

22. $\sqrt{3x + 2} - 1 = 3$

23. $\sqrt{2m - 1} = \sqrt{5 - m}$

24. $3\sqrt{1 - 4x} + 4 = 1$

25. $\sqrt[3]{5x + 1} = 6$

Answers and Solutions

1. **Answer:** $x > -\dfrac{5}{2}$ Subtract $8x$ from each side and add 3 to each side.

$$4x - 3 < 8x + 7$$
$$\underline{-8x + 3 \quad -8x + 3}$$
$$-4x \quad < \quad 10$$

Divide each side by −4, which means you turn the sense around.

$\frac{-4x}{-4} > \frac{10}{-4}$, which means that $x > -\frac{5}{2}$.

2. **Answer: $y \le 7$**   Distribute the 5 on the left. Then add 5y to each side and subtract 1 from each side.

$5(3 - y) \ge 1 - 3y$

$15 - 5y \ge 1 - 3y$

$\underline{-1 + 5y \;\; -1 + 5y}$ Divide each side by 2 . For the more conventional way of writing
$14 \qquad \ge 2y$ the answer, reverse the values and the inequality.

3. **Answer: $-\frac{7}{3} < z \le \frac{1}{3}$**   Subtract 4 from each section; then divide by 3.

$$-3 < 3z + 4 \le 5$$
$$\underline{-4 \qquad -4 \;\; -4}$$
$$\frac{-7}{3} < \frac{3z}{3} \quad \le \frac{1}{3}$$

4. **Answer: $w > \frac{8}{7}$**   Flip the fractions by reversing the inequality sense. Then multiply both sides by 15, the common denominator.

$$\frac{5}{w + 1} > \frac{3}{2w - 1}$$
$$\frac{w + 1}{5} < \frac{2w - 1}{3}$$
$$^3\cancel{15} \cdot \frac{w + 1}{\cancel{5}} < \frac{2w - 1}{\cancel{3}} \cdot \cancel{15}^5$$
$$3(w + 1) < (2w - 1)5$$
$$3w + 3 < 10w - 5$$

Subtract 3w from each side and add 5 to each side.

$3w + 3 < 10w - 5$

$\underline{-3w + 5 \;\; -3w + 5}$ Now divide each side by 7. Reversing the values and the sense puts
$8 < 7w$ the answer in a more conventional form, with the variable first.

5. **Answer: $3 \le x < 7$**   Look at the graphs of the inequalities and see where they overlap.

$x \ge 3$ and $x < 7$

6. **Answer: no answer**   The graphs of these inequalities don't have any common values or overlap.

$y < -2$ and $y \ge 2$

7. **Answer: $2 < x \le 4$**   Look at the graphs and see where they overlap.

$$0 < x + 1 \le 5 \text{ and } x - 1 > 1$$

$0 < x + 1 \le 5$ is equivalent to $-1 < x \le 4$.

$x - 1 > 1$ is equivalent to $x > 2$.

8. **Answer $\frac{8}{3}, -2$**   Solve the two equations, $3z - 1 = 7$ and $3z - 1 = -7$.

9. **Answer: 8,–2**   First, add 2 to each side and then divide each side by 4.

$$4|x - 3| - 2 = 18$$
$$ \underline{+2 \quad +2}$$
$$\frac{4|x - 3|}{4} = \frac{20}{4}$$
$$|x - 3| = 5$$

Now solve the two equations $x - 3 = 5$ and $x - 3 = -5$.

10. **Answer: No answer possible**   When subtracting 5 from each side, the equation has the absolute value equal to a negative number, which cannot be.

11. **Answer: $x \le -2$ or $x \ge 4$**   Solve the two inequalities, $x - 1 \ge 3$ and $x - 1 \le -3$.

12. **Answer: $-2 < y < 10$**   Solve the inequality, $-6 < 4 - y < 6$. The answer first comes out to be $-10 < -y < 2$. Multiplying through by -1 results in $10 > y > -2$. Switch the positions of the values and the senses to get the final answer.

13. **Answer: $x < -3$ or $x > \frac{1}{3}$**   First divide both sides by 3. Then solve the two inequalities $3x + 4 > 5$ and $3x + 4 < -5$.

14. **Answer: $10x^2 y^2 z^3 \sqrt{3y}$**   The perfect square factors are $100, x^4, y^4, z^6$.

15. **Answer: $2ab^3 c^4 \sqrt{10abc}$**   The perfect square factors are $4, a^2, b^6, c^8$.

16. **Answer: $24xz \sqrt{2xy}$**   Multiply the two radicals together to get $4\sqrt{72x^3 yz^2}$. Then break up the radical into two radicals—one with all the perfect square factors. $4\sqrt{36x^2 z^2}\sqrt{2xy} = 4 \cdot 6xz \sqrt{2xy}$. Multiply the 4 and 6.

17. **Answer: $5ab \sqrt{2a}$**   First, simplify the radicals.

$$8\sqrt{2a^3 b^2} - ab\sqrt{18a} = 8\sqrt{a^2 b^2}\sqrt{2a} - ab\sqrt{9}\sqrt{2a}$$
$$= 8ab\sqrt{2a} - 3ab\sqrt{2a}$$

Now the two terms can be subtracted.

18. **Answer: $3mnz^2 \sqrt[3]{n}$**   The perfect cube factors, which can be written in their own radical, are $27, m^3, n^3, z^6$.

19. **Answer: $2a^2 b^2 \sqrt[4]{2b^2}$**   The fourth power factors that go in their own radical are $16, a^8$, and b^8.

20. **Answer: $2xy\sqrt[5]{2y}$**   First, multiply the two radicals together. $\sqrt[5]{64x^5y^6}$

    Then separate the fifth degree factors from the other. $\sqrt[5]{64x^5y^6} = \sqrt[5]{32x^5y^5}\sqrt[5]{2y}$. Take the roots of the first radical.

21. **Answer: $z = \dfrac{17}{2}$**   Square both sides; then solve the equation $2z - 1 = 16$.

22. **Answer: $x = \dfrac{14}{3}$**   First, add 1 to each side. Then square both sides, and you'll have the linear equation, $3x + 2 = 16$, to solve.

23. **Answer: $m = 2$**   Square both sides of the equation. Then solve the equation $2m - 1 = 5 - m$.

24. **Answer: No answer possible**   When 4 is subtracted from each side (and both sides are divided by 3), the radical then is set equal to a negative number. The radical must be positive, though.

25. **Answer: $x = 43$**   Raise each side to the third power. Then solve the linear equation, $5x + 1 = 216$.

Supplemental Chapter Problems

Problems

For problems 1 through 4, solve for the value(s) of the variable.

1. $5x + 1 \geq 2x - 11$

2. $\dfrac{y+3}{5} < \dfrac{y-2}{7}$

3. $3 < 5 - 2z < 9$

4. $\dfrac{1}{3t-8} > \dfrac{3}{4t+1}$

In problems 5 through 7, find the common solution for the two inequalities.

5. $y < -3$ and $y \geq -5$

6. $3x + 1 < 8$ and $1 - 2x \geq 5$

7. $-1 \leq x < 4$ and $3x + 1 > 4$

For problems 8 through 14, solve for the value(s) that make the statement true.

8. $|5 - 2x| = 3$

9. $|4 + 3x| = -1$

10. $2|x - 7| - 3 = 11$

11. $|3x + 2| \le 5$

12. $|4x - 1| > 3$

13. $3|2x - 1| < 9$

14. $|5x + 5| + 5 > 10$

In problems 15 through 20, simplify.

15. $\sqrt{48a^2 (b+c)^3}$

16. $\sqrt{50t^4 z^{10}}$

17. $9\sqrt{5a^2 b}\,\sqrt{10a^2 b}$

18. $3\sqrt{6x^2 y} - x\sqrt{24y}$

19. $\sqrt[4]{243t^4 z^4}$

20. $\sqrt[5]{64m^4 n^3}$

For problems 21 through 25, solve for the value(s) of the variable.

21. $\sqrt{9 - 4z} = 7$

22. $\sqrt{8 + x} = \sqrt{3x + 6}$

23. $4 + \sqrt{y - 2} = 1$

24. $2\sqrt{3x + 11} = 4$

25. $\sqrt[4]{x + 6} = 1$

Answers

1. $x \ge -4$ (Inequalities, p. 183)

2. $y < -\frac{31}{2}$ (Inequalities, p. 183)

3. $-2 < z < 1$ (Inequalities, p. 183)

4. $t < 5$ and $t > \frac{8}{3}$ to keep it positive.   (Inequalities, p. 183)

5. $-5 \le y < -3$ (Solving Inequalities by Graphing on a Number Line, p. 189)

6. $x \le -2$ (Solving Inequalities by Graphing on a Number Line, p. 189)

7. $1 < x < 4$　(Solving Inequalities by Graphing on a Number Line, p. 189)

8. $x = 1, x = 4$　(Absolute Value Equations, p. 192)

9. No answer possible　(Absolute Value Equations, p. 192)

10. $x = 14, x = 0$　(Absolute Value Equations, p. 192)

11. $-\frac{7}{3} \leq x \leq 1$　(Absolute Value Inequalities, p. 192)

12. $x < -\frac{1}{2}$ or $x > 1$　(Absolute Value Inequalities, p. 192)

13. $-1 < x < 2$　(Absolute Value Inequalities, p. 192)

14. $x < -2$ or $x > 0$　(Absolute Value Inequalities, p. 192)

15. $4a(b+c)\sqrt{3(b+c)}$　(Simplifying Square Roots, p. 203)

16. $5t^2 z^5 \sqrt{2}$　(Simplifying Square Roots, p. 203)

17. $45a^2 b \sqrt{2}$　(Simplifying Square Roots, p. 203)

18. $x\sqrt{6y}$　(Simplifying Square Roots, p. 203)

19. $3tz\sqrt[4]{3}$　(Simplifying Other Roots, p. 206)

20. $2\sqrt[5]{2m^4 n^3}$　(Simplifying Other Roots, p. 206)

21. $z = -10$　(Radical Equations, p. 208)

22. $x = 1$　(Radical Equations, p. 208)

23. No answer possible　(Radical Equations, p. 208)

24. $x = -\frac{7}{3}$　(Radical Equations, p. 208)

25. $x = -5$　(Radical Equations, p. 208)

Chapter 6
Introducing Quadratic Equations—Testing Solutions

A quadratic equation is a statement with a variable in it that's raised to the second power, or squared, and there aren't any variables in the statement that have a higher power than that. Having the variable squared opens up all sorts of possibilities for the solutions to the equation. If the equation is quadratic, it may have two solutions, one solution, or no solution. The "two solutions" are two distinct (different) numbers. One solution means that an answer has been repeated—the same thing appears twice. No solution means that there's no real number that makes the equation true. You can solve a quadratic equation with four possible methods:

❏ BGBG: By Guess or By Golly

This isn't very efficient, but sometimes solutions seem obvious, and there's no harm in recognizing or **knowing** the solution!

❏ Factoring

This is probably the most efficient way of solving a quadratic equation—as long as the solutions are rational numbers, which means it's factorable. The factored form and the Multiplication Property of Zero (see the "Properties of Algebraic Expressions," section in Chapter 1 for more on that) join to make this a preferred method.

❏ Quadratic Formula

The quadratic formula **always** works. Whether an equation factors or not, you can always use the quadratic formula to get the answers. Why isn't it **always** used then? The quadratic formula can be rather messy and cumbersome, and you can make errors more easily using this method than by factoring. This formula is your second resort.

❏ Completing the Square

This method should **never** be used to solve a quadratic equation. However, students are asked to do just that. Completing the square is an important technique used to rewrite equations of circles, parabolas, hyperbolas, and ellipses in their standard form—a more useful form. By learning completing the square as a method of solving quadratics, you get practice for later. And, this method does give the answers to the problems.

The last three techniques listed here will each be dealt with in their own sections—the next three sections in this chapter. The focus here will be on checking or testing solutions to be sure that they satisfy the quadratic equation. This way, if you've made an error in your algebra or if the solutions you've found are extraneous, you'll be able to catch the discrepancies.

Example Problems

These problems show the answers and solutions.

1. Check to see whether $x = 2$ and $x = -3$ make the equation $x^2 + x - 6 = 0$ true.

   answer: Yes, they both work.

   First, let $x = 2$; then, put it into the equation, $2^2 + 2 - 6 = 4 + 2 - 6 = 6 - 6 = 0$, so 2 is a solution.

   Next, let $x = -3$; then, put it into the equation, $(-3)^2 + (-3) - 6 = 9 - 3 - 6 = 6 - 6 = 0$, so -3 is a solution.

2. Check to see whether $x = -2$ makes the equation $x^2 + 4 = 0$ true.

   answer: No, it doesn't work.

   Substitute -2 for x in the left-hand side of the equation, $(-2)^2 + 4 = 0$, $4 + 4 = 0$.

   This is not true. Actually, no real solution exists for this quadratic equation.

3. Check to see whether $y = \sqrt{3}$ and $y = -\sqrt{3}$ work in $y^2 - 3 = 0$.

   answer: Yes, they both work.

   First, let $y = \sqrt{3}$; then, put it into the equation, $\left(\sqrt{3}\right)^2 - 3 = 0$, $3 - 3 = 0$.

   Next, let $y = -\sqrt{3}$; then, put it into the equation, $\left(-\sqrt{3}\right)^2 - 3 = 0$, $3 - 3 = 0$.

4. Check to see whether $z = 2 + \sqrt{10}$ and $2 - \sqrt{10}$ work in the quadratic equation $z^2 - 4z - 6 = 0$.

   answer: Yes, both work.

   First let $z = 2 + \sqrt{10}$. Put it into the equation.

   $$\left(2 + \sqrt{10}\right)^2 - 4\left(2 + \sqrt{10}\right) - 6 = \left(2 + \sqrt{10}\right)\left(2 + \sqrt{10}\right) - 8 - 4\sqrt{10} - 6$$
   $$= 4 + 2\sqrt{10} + 2\sqrt{10} + 10 - 14 - 4\sqrt{10}$$
   $$= 14 - 14 + 4\sqrt{10} - 4\sqrt{10}$$
   $$= 0$$

   Then, let $z = 2 - \sqrt{10}$ and put it into the equation.

$$\left(2-\sqrt{10}\right)^2 - 4\left(2-\sqrt{10}\right) - 6 = \left(2-\sqrt{10}\right)\left(2-\sqrt{10}\right) - 8 + 4\sqrt{10} - 6$$

$$= 4 - 2\sqrt{10} - 2\sqrt{10} + 10 - 14 + 4\sqrt{10}$$

$$= 14 - 14 - 4\sqrt{10} + 4\sqrt{10}$$

$$= 0$$

The numbers that have been substituted into the quadratic equations are the types of results that you can expect to find by factoring and using the quadratic formula to get the solutions. In some instances, you have to check the solutions carefully. These mainly arise in problems in which the original equation has absolute value or a radical or fractions, and they've been changed into a more convenient form to solve. Extraneous roots can occur. Even though a particular solution may work for the quadratic equation that resulted in the change, the solution may not work for the original equation. Checking is essential.

Work Problems

Use these problems to give yourself additional practice.

In each case, determine whether or not the values are solutions of the quadratic equation.

1. $x = \dfrac{3}{2}$ and $x = 1$ in $2x^2 - 5x + 3 = 0$

2. $y = 4$ and $y = 9$ in $y^2 - 36 = 0$

3. $z = 5$ in $z^2 - 10x + 25 = 0$

4. $w = \sqrt{7}$ and $w = -\sqrt{7}$ in $w^2 - 7 = 0$

5. $t = -3 + 3\sqrt{2}$ and $t = -3 - 3\sqrt{2}$ in $t^2 + 6t - 9 = 0$

Worked Solutions

1. **Both work.**   First, let $x = \dfrac{3}{2}$. Put it into the equation,

$$2\left(\frac{3}{2}\right)^2 - 5\left(\frac{3}{2}\right) + 3 = 2\left(\frac{9}{4}\right) - \frac{15}{2} + 3$$

$$= \frac{18}{4} - \frac{15}{2} + 3 = \frac{9}{2} - \frac{15}{2} + \frac{6}{2} = \frac{15}{2} - \frac{15}{2} = 0$$

Next, let $x = 1$.

$$2(1)^2 - 5(1) + 3 = 2(1) - 5 + 3$$

$$= 2 - 5 + 3 = 5 - 5 = 0$$

2. **Neither works.** First, let $y = 4$, $\begin{matrix} 4^2 = 36 \\ 16 \neq 36 \end{matrix}$. The 4 doesn't work.

   Next, let $y = 9$, $\begin{matrix} 9^2 = 36 \\ 81 \neq 36 \end{matrix}$. The 9 doesn't work.

3. **It works.** Let $z = 5$, $(5)^2 - 10(5) + 25 = 25 - 50 + 25 = 50 - 50 = 0$.

   This value works. This is a case where there's a double root. The number 5 is the only solution. This situation occurs when the quadratic factors into the square of a single binomial.

4. **Both work.** First, let $w = \sqrt{7}$, $\left(\sqrt{7}\right)^2 - 7 = 7 - 7 = 0$.

   Next, let $w = -\sqrt{7}$, $\left(-\sqrt{7}\right)^2 - 7 = 7 - 7 = 0$.

5. **Both work.** First, let $t = -3 + 3\sqrt{2}$.

$$\left(-3 + 3\sqrt{2}\right)^2 + 6\left(-3 + 3\sqrt{2}\right) - 9 = \left(-3 + 3\sqrt{2}\right)\left(-3 + 3\sqrt{2}\right) - 18 + 18\sqrt{2} - 9$$
$$= 9 - 9\sqrt{2} - 9\sqrt{2} + 9 \cdot 2 - 18 + 18\sqrt{2} - 9$$
$$= 9 + 18 - 18 - 9 - 18\sqrt{2} + 18\sqrt{2} = 0$$

Next, let $t = -3 - 3\sqrt{2}$.

$$\left(-3 - 3\sqrt{2}\right)^2 + 6\left(-3 - 3\sqrt{2}\right) - 9 = \left(-3 - 3\sqrt{2}\right)\left(-3 - 3\sqrt{2}\right) - 18 - 18\sqrt{2} - 9$$
$$= 9 + 9\sqrt{2} + 9\sqrt{2} + 9 \cdot 2 - 18 - 18\sqrt{2} - 9$$
$$= 9 + 18 - 18 - 9 + 18\sqrt{2} - 18\sqrt{2} = 0$$

Solving Quadratic Equations by Factoring

When solving a quadratic equation by factoring, you're taking advantage of the Multiplication Property of Zero. See "Properties of Algebraic Expressions," in Chapter 1 for more on that property. The Multiplication Property of Zero says that if the product of two values is equal to zero, then one or the other of the two values must be zero. This works with factoring the quadratic, because when the factored expression is set equal to zero, each of the factors can be set equal to zero, independently, to see what value of the variable would make it so. For a review on factoring quadratic expressions, see "Factoring Trinomials" in Chapter 4.

Example Problems

These problems show the answers and solutions.

1. Solve for x by factoring $x^2 + 6x - 7 = 0$.

   answer: $x = -7$ and $x = 1$

   The trinomial factors, giving $(x + 7)(x - 1) = 0$.

The Multiplication Property of Zero says that one or the other of the factors must be equal to zero, so

$$x + 7 = 0 \text{ or } x - 1 = 0$$
$$\text{If } x + 7 = 0, \text{ then } x = -7.$$
$$\text{If } x - 1 = 0, \text{ then } x = 1.$$

The two solutions to this quadratic equation are $x = -7$ and $x = 1$. If you put them into the original equation, both will work.

2. Solve for y by factoring $3y^2 - 11y = 4$.

   answer: $y = -\frac{1}{3}$ and $y = 4$

   This equation must first be set equal to 0. Subtract 4 from each side.

$$\begin{array}{r} 3y^2 - 11y = 4 \\ \underline{-4 \qquad\quad -4} \\ 3y^2 - 11y - 4 = 0 \end{array}$$

   Now the trinomial can be factored.

$$3y^2 - 11y - 4 = (3y + 1)(y - 4) = 0$$

   One or the other of the factors must equal 0.

   If $3y + 1 = 0$, then $3y = -1$, and $y = -\frac{1}{3}$.

   If $y - 4 = 0$, then $y = 4$.

3. Solve for z by factoring $11z = z^2$.

   answer: $z = 0$ and $z = 11$

   A common error in solving this equation is to divide each side by z. As a rule, don't divide by the variable—you'll lose a solution. It's okay to divide each side by a constant number but not a variable. Set the equation equal to zero by subtracting $11z$ from each side.

$$11z = z^2$$
$$0 = z^2 - 11z$$

   Now factor out the z.

$$0 = z(z - 11)$$

   One of the factors must be equal to 0.

$$z = 0 \text{ or } z - 11 = 0$$

   As you see, one of the solutions just pops up. If $z - 11 = 0$, then $z = 11$, so the two solutions are $z = 0$ and $z = 11$.

4. Solve for w in $w^2 = 25$.

   answer: $w = 5$ and $w = -5$

   This can actually be done in one of two ways, by factoring or by taking the square root of each side.

   Using factoring, first set the equation equal to 0, and then factor and solve for the solutions.

   $$w^2 - 25 = 0$$
   $$(w - 5)(w + 5) = 0$$

   If $w - 5 = 0$, then $w = 5$. If $w + 5 = 0$, then $w = -5$. The two solutions are $w = 5$ and $w = -5$.

   Take the square root of each side.

   $$w^2 = 25$$
   $$\sqrt{w^2} = \sqrt{25}$$

   In Chapter 5, it's explained that $\sqrt{w^2} = |w|$, because w could be positive or negative. Because w can be positive or negative, both of these have to be considered when taking the square root of both sides. When doing this, write

   $$\sqrt{w^2} = \sqrt{25}$$
   $$w = \pm 5$$

   That takes care of the absolute value and accounts for both solutions. Just as with the factoring, the solutions are $w = 5$ and $w = -5$.

5. Solve for t by factoring $4t^2 + 12t + 9 = 0$.

   answer: $t = -\dfrac{3}{2}$

   $$4t^2 + 12t + 9 = (2t + 3)(2t + 3) = 0$$

   Set the first factor equal to 0.

   $$2t + 3 = 0$$
   $$2t = -3$$
   $$t = -\dfrac{3}{2}$$

   The same thing happens with the second factor. This is a case of a quadratic equation having a double root—two roots that are exactly the same. This happens when the quadratic is a perfect square trinomial.

Work Problems

Use these problems to give yourself additional practice.

Solve each quadratic equation by factoring.

1. $x^2 - 13x + 40 = 0$

2. $6y^2 = 7y + 3$

3. $m^2 - 14m + 49 = 0$

4. $14p^2 = 56$

5. $9 - 27z - 36z^2 = 0$

Worked Solutions

1. **$x = 5$ and $x = 8$** $x^2 - 13x + 40 = (x - 5)(x - 8) = 0$

    If $x - 5 = 0$, then $x = 5$. If $x - 8 = 0$, then $x = 8$.

2. **$y = -\frac{1}{3}$ and $y = \frac{3}{2}$**   First, set the equation equal to 0 by subtracting the two terms on the right from each side.

    $$6y^2 = 7y + 3$$
    $$6y^2 - 7y - 3 = 0$$

    Now factor: $6y^2 - 7y - 3 = (3y + 1)(2y - 3) = 0$.

    Setting each factor equal to 0, first, if $3y + 1 = 0$:

    $$3y = -1$$
    $$y = -\frac{1}{3}$$

    Next, if $2y - 3 = 0$:

    $$2y = 3$$
    $$y = \frac{3}{2}$$

3. **$m = 7$** $m^2 - 14m + 49 = (m - 7)(m - 7) = 0$

    This is a double root. Both solutions are the same.

    If $m - 7 = 0$, then $m = 7$.

4. **$p = 2$ and $p = -2$**   First subtract 56 from each side. Then take out the common factor of 14.

$$14p^2 = 56$$
$$14p^2 - 56 = 0$$
$$14(p^2 - 4) = 0$$

Now the quadratic can be factored.

$$14(p - 2)(p + 2) = 0$$

Setting the factors equal to 0, the first factor does not qualify. The factor 14 is never equal to 0. Only the factors with variables in them need be set equal to 0.

If $p - 2 = 0$, then $p = 2$. If $p + 2 = 0$, then $p = -2$.

5. **$z = \frac{1}{4}$ and $z = -1$**   This will be easier to factor if it's written with the squared term first, and if the squared term is positive. Add the opposite of each term to both sides to get $0 = 36z^2 + 27z - 9$.

Now factor.

$$36z^2 + 27z - 9 = 9(4z^2 + 3z - 1) = 9(4z - 1)(z + 1) = 0$$

Set the two factors with variables equal to zero.

$$\text{If } 4z - 1 = 0, \text{ then } 4z = 1, z = \frac{1}{4}.$$
$$\text{If } z + 1 = 0, \text{ then } z = -1.$$

Solving Quadratic Equations with the Quadratic Formula

The quadratic formula is just that. It's a formula used to solve for the solutions of a quadratic equation. Many quadratic equations can't be factored, because not all solutions are nice integers or fractions. When the solutions are irrational numbers, with radicals, or if the factoring isn't easy to do, then the quadratic formula is the way to go. In the next section, on completing the square, the quadratic formula is developed—you'll see where it came from. For now, the formula is given, and its use is demonstrated.

Given the standard form of a quadratic equation, $ax^2 + bx + c = 0$, the solutions of the equation are

$$x = \frac{-b \pm \sqrt{b^2 - 4ac}}{2a}$$

Example Problems

These problems show the answers and solutions.

1. Solve $x^2 - 8x + 7 = 0$ using the quadratic formula.

   answer: $x = 7$ and $x = 1$

The trinomial actually can be factored easily, but the quadratic formula works for factorable equations, also.

From the standard form of the quadratic equation, in this equation $a = 1$, $b = -8$, and $c = 7$.

Put those values into the quadratic formula:

$$x = \frac{8 \pm \sqrt{(-8)^2 - 4(1)(7)}}{2(1)}$$

Notice that, since $b = -8$, the first term in the fraction is the opposite of -8, or $+8$.

$$x = \frac{8 \pm \sqrt{64 - 28}}{2} = \frac{8 \pm \sqrt{36}}{2} = \frac{8 \pm 6}{2}$$

Break this up into two fractions, $x = \frac{8+6}{2}$ or $x = \frac{8-6}{2}$, so $x = \frac{14}{2} = 7$ or $x = \frac{2}{2} = 1$.

2. Use the quadratic formula to solve $2y^2 + 4y - 9 = 0$.

    answer: $x = \frac{-2 + \sqrt{22}}{2}$ and $x = \frac{-2 - \sqrt{22}}{2}$

    The values of the coefficients are $a = 2$, $b = 4$, and $c = -9$.

    Put these into the formula.

    $$x = \frac{-4 \pm \sqrt{4^2 - 4(2)(-9)}}{2(2)} = \frac{-4 \pm \sqrt{16 - (-72)}}{4} = \frac{-4 \pm \sqrt{88}}{4}$$

    The radical can be simplified and the fraction reduced.

    $$\frac{-4 \pm \sqrt{88}}{4} = \frac{-4 \pm \sqrt{4}\sqrt{22}}{4} = \frac{-4 \pm 2\sqrt{22}}{4} = \frac{\cancel{2}\left(-2 \pm \sqrt{22}\right)}{\cancel{4}_2} = \frac{-2 \pm \sqrt{22}}{2}$$

    The fraction then is broken up into the two answers, $x = \frac{-2 + \sqrt{22}}{2}$ and $x = \frac{-2 - \sqrt{22}}{2}$.

3. Use the quadratic formula to solve $3z^2 - 2z + 5 = 0$.

    answer: There's no real answer.

    The coefficients are $a = 3$, $b = -2$, and $c = 5$.

    Put them into the formula.

    $$x = \frac{2 \pm \sqrt{(-2)^2 - 4(3)(5)}}{2(3)} = \frac{2 \pm \sqrt{4 - 60}}{6} = \frac{2 \pm \sqrt{-56}}{6}$$

    You can stop right there. You can't take the square root of a negative number and get a real answer, so this equation has no real solution.

4. Use the quadratic formula to solve $21x^2 - 110x + 125 = 0$.

    answer: $x = \frac{25}{7}$ or $x = \frac{5}{3}$

Even though this could be factored, the numbers are a bit unwieldy. The numbers in the quadratic formula will be large, too, but it's still probably the easier way to go.

The coefficients are $a = 21$, $b = -110$, and $c = 125$.

Put them into the formula.

$$x = \frac{110 \pm \sqrt{(-110)^2 - 4(21)(125)}}{2(21)} = \frac{110 \pm \sqrt{12100 - 10500}}{42} = \frac{110 \pm \sqrt{1600}}{42} = \frac{110 \pm 40}{42}$$

Break it up into two fractions.

$$x = \frac{110 + 40}{42} = \frac{150}{42} = \frac{25}{7} \text{ or } x = \frac{110 - 40}{42} = \frac{70}{42} = \frac{5}{3}$$

Work Problems

Use these problems to give yourself additional practice.

Use the quadratic formula to solve each of the following.

1. $x^2 - 3x - 18 = 0$

2. $10y^2 - 13y = 3$

3. $3w^2 - 50w + 75 = 0$

4. $4x^2 - 9 = 0$

5. $z^2 + 8z - 10 = 0$

Worked Solutions

1. **$x = 6$, $x = -3$** The coefficients are $a = 1$, $b = -3$, and $c = -18$.

   Put them into the formula.

   $$x = \frac{3 \pm \sqrt{(-3)^2 - 4(1)(-18)}}{2(1)} = \frac{3 \pm \sqrt{9 - (-72)}}{2} = \frac{3 \pm \sqrt{81}}{2} = \frac{3 \pm 9}{2}$$

   Break up the fraction, $x = \frac{3+9}{2} = \frac{12}{2} = 6$ or $x = \frac{3-9}{2} = \frac{-6}{2} = -3$.

2. **$y = \frac{3}{2}$, $y = -\frac{1}{5}$** First, write the equation in standard form by subtracting 3 from each side.

   $$10y^2 - 13y - 3 = 0$$

   The coefficients are $a = 10$, $b = -13$, and $c = -3$.

   Put them into the formula.

$$y = \frac{13 \pm \sqrt{(-13)^2 - 4(10)(-3)}}{2(10)} = \frac{13 \pm \sqrt{169 - (-120)}}{20} = \frac{13 \pm \sqrt{289}}{20} = \frac{13 \pm 17}{20}$$

Break up the fraction, $y = \frac{13+17}{20} = \frac{30}{20} = \frac{3}{2}$ or $y = \frac{13-17}{20} = \frac{-4}{20} = -\frac{1}{5}$.

3. **$w = 15$, $w = \frac{5}{3}$**   The coefficients are $a = 3$, $b = -50$, and $c = 75$.

    Put them into the formula.

    $$w = \frac{50 \pm \sqrt{(-50)^2 - 4(3)(75)}}{2(3)} = \frac{50 \pm \sqrt{2500 - 900}}{6} = \frac{50 \pm \sqrt{1600}}{6} = \frac{50 \pm 40}{6}.$$

    Break up the fraction, $w = \frac{50+40}{6} = \frac{90}{6} = 15$ or $w = \frac{50-40}{6} = \frac{10}{6} = \frac{5}{3}$.

4. **$x = \frac{3}{2}$, $x = -\frac{3}{2}$**   Only two terms are in this quadratic. The middle term is missing, so the coefficient is equal to 0.

    The coefficients are $a = 4$, $b = 0$, and $c = -9$.

    Put them into the formula.

    $$x = \frac{0 \pm \sqrt{0^2 - 4(4)(-9)}}{2(4)} = \frac{\pm\sqrt{144}}{8} = \frac{\pm 12}{8}$$

    Break up the fraction, $x = \frac{+12}{8} = \frac{3}{2}$ or $x = \frac{-12}{8} = -\frac{3}{2}$.

    As you can see, this would have been much easier to do by factoring.

5. **$z = -4 + \sqrt{26}$ and $z = -4 - \sqrt{26}$**   The coefficients are $a = 1$, $b = 8$, and $c = -10$.

    Put them into the formula.

$$z = \frac{-8 \pm \sqrt{8^2 - 4(1)(-10)}}{2(1)} = \frac{-8 \pm \sqrt{64 - (-40)}}{2} = \frac{-8 \pm \sqrt{104}}{2} = \frac{-8 \pm \sqrt{4}\sqrt{26}}{2} = \frac{-8 \pm 2\sqrt{26}}{2} = -4 \pm \sqrt{26}$$

Solving Quadratic Equations by Completing the Square

As mentioned in the introductory section of this chapter, you'll never choose to use completing the square to solve a quadratic equation if you can factor it or use the quadratic formula. But completing the square is used when rewriting equations of conic sections (circles, ellipses, hyperbolas, and parabolas) in a form that can be easily analyzed and graphed. By using completing the square to solve a quadratic equation, you'll get the same solutions you'd get using other methods, so at least you'll have an answer for all your efforts.

The procedure used for completing the square on the quadratic equation $ax^2 + bx + c = 0$ is as follows:

A. If a isn't 1, then divide every term in the equation by a.

B. Subtract c (or add the opposite of c) from each side of the equation. (Subtract $\frac{c}{a}$ if you divided through by a.)

C. Determine what new constant to add to each side of the equation so that the left side will become a perfect square trinomial. To find that constant, divide the coefficient of the x term by 2 and square that result. The coefficient of the x term is either b or $\frac{b}{a}$. Add the square to each side.

D. Factor the left side (it's now a perfect square trinomial).

E. Take the square root of each side.

F. Solve for the variable.

Example Problems

These problems show the answers and solutions.

1. Solve $x^2 + 6x - 7 = 0$ using completing the square.

    answer: $x = 1$, $x = -7$

    A. Since a is 1, you don't need to divide by anything.

    B. Add 7 to each side.

    $$x^2 + 6x - 7 = 0$$
    $$\underline{ +7 +7}$$
    $$x^2 + 6x = 7$$

    C. Complete the square by dividing 6 by 2, squaring that, and adding it to each side.

    $$\left(\frac{6}{2}\right)^2 = 3^2 = 9$$
    $$x^2 + 6x + 9 = 7 + 9$$
    $$x^2 + 6x + 9 = 16$$

    D. Factor the left side.

    $$x^2 + 6x + 9 = (x + 3)(x + 3) = (x + 3)^2$$

    So, $(x + 3)^2 = 16$.

    E. Take the square root of each side.

    Don't forget the $\pm$ signs. See "Solving Quadratic Equations by Factoring," earlier in this chapter for more on the $\pm$ sign.

    $$x + 3 = \pm 4$$

    F. Solve for x by subtracting 3 from each side.

    $$x = -3 \pm 4$$
    $$x = -3 + 4 = 1 \text{ or } x = -3 - 4 = -7$$

Yes, this would have been much easier and quicker using factoring, but it's a good example to start with when explaining completing the square.

2. Solve $3y^2 - 8y - 1 = 0$ using completing the square.

answer: $y = \dfrac{4 + \sqrt{19}}{3}$ or $y = \dfrac{4 - \sqrt{19}}{3}$

A. Since a is 3, divide each term by 3.

$$\frac{3y^2}{3} - \frac{8y}{3} - \frac{1}{3} = \frac{0}{3}$$

$$y^2 - \frac{8}{3}y - \frac{1}{3} = 0$$

B. Add $\frac{1}{3}$ to each side.

$$y^2 - \frac{8}{3}y - \frac{1}{3} = 0$$

$$\underline{\phantom{y^2 - \frac{8}{3}y} + \frac{1}{3} + \frac{1}{3}}$$

$$y^2 - \frac{8}{3}y = \frac{1}{3}$$

C. Dividing $-\frac{8}{3}$ by 2 is the same as multiplying by $\frac{1}{2}$.

$$-\frac{8}{3} \cdot \frac{1}{2} = -\frac{8}{6} = -\frac{4}{3}$$

Now, $-\frac{4}{3}$ is squared and added to each side of the equation.

$$\left(-\frac{4}{3}\right)^2 = \frac{16}{9}$$

$$y^2 - \frac{8}{3}y + \frac{16}{9} = \frac{1}{3} + \frac{16}{9}$$

$$y^2 - \frac{8}{3}y + \frac{16}{9} = \frac{3}{9} + \frac{16}{9}$$

$$y^2 - \frac{8}{3}y + \frac{16}{9} = \frac{19}{9}$$

D. Factor the left side of the equation.

$$\left(y - \frac{4}{3}\right)^2 = \frac{19}{9}$$

E. Take the square root of each side, $\sqrt{\left(y - \frac{4}{3}\right)^2} = \sqrt{\frac{19}{9}}$.

$$y - \frac{4}{3} = \pm\sqrt{\frac{19}{9}} = \pm\frac{\sqrt{19}}{3}$$

F. To solve for y, add $\frac{4}{3}$ to each side.

$$y = \frac{4}{3} \pm \frac{\sqrt{19}}{3}$$

So $y = \frac{4 + \sqrt{19}}{3}$ or $y = \frac{4 - \sqrt{19}}{3}$.

3.   Prove the quadratic formula by completing the square on the standard form of the quadratic equation, $ax^2 + bx + c = 0$.

answer: $x = \dfrac{-b \pm \sqrt{b^2 - 4ac}}{2a}$

A. Divide every term by a.

$$\frac{\cancel{a}x^2}{\cancel{a}} + \frac{bx}{a} + \frac{c}{a} = x^2 + \frac{b}{a}x + \frac{c}{a} = 0$$

B. Subtract $\frac{c}{a}$ from each side.

$$x^2 + \frac{b}{a}x = -\frac{c}{a}$$

C. Add $\dfrac{b^2}{4a^2}$ to each side. This comes from taking half of $\frac{b}{a}$ and squaring it.

$$\frac{1}{2} \cdot \frac{b}{a} = \frac{b}{2a}, \left(\frac{b}{2a}\right)^2 = \frac{b^2}{4a^2}$$

$$x^2 + \frac{b}{a}x + \frac{b^2}{4a^2} = -\frac{c}{a} + \frac{b^2}{4a^2} = -\frac{c}{a} \cdot \frac{4a}{4a} + \frac{b^2}{4a^2} = \frac{-4ac}{4a^2} + \frac{b^2}{4a^2} = \frac{b^2 - 4ac}{4a^2}$$

$$\text{or } x^2 + \frac{b}{a}x + \frac{b^2}{4a^2} = \frac{b^2 - 4ac}{4a^2}$$

D. Factor the left side.

$$\left(x + \frac{b}{2a}\right)^2 = \frac{b^2 - 4ac}{4a^2}$$

E. Take the square root of each side.

$$\sqrt{\left(x + \frac{b}{2a}\right)^2} = \sqrt{\frac{b^2 - 4ac}{4a^2}}$$

$$x + \frac{b}{2a} = \pm\sqrt{\frac{b^2 - 4ac}{4a^2}} = \pm\frac{\sqrt{b^2 - 4ac}}{2a}$$

F. Solve for x by subtracting $\frac{b}{2a}$ from each side.

$$x = -\frac{b}{2a} \pm \frac{\sqrt{b^2 - 4ac}}{2a} = \frac{-b \pm \sqrt{b^2 - 4ac}}{2a}$$

Work Problems

Use these problems to give yourself additional practice.

Solve each of the quadratic equations using completing the square.

1. $y^2 + 3y - 4 = 0$

2. $2z^2 - 11z + 12 = 0$

3. $m^2 + 5m - 2 = 0$

4. $3x^2 - 7 = 0$

5. $y^2 - 6y = 0$

Worked Solutions

1. **$y = 1, y = -4$**   Add 4 to each side. $y^2 + 3y = 4$

   Add the square of half the coefficient of y to each side. $y^2 + 3y + \left(\frac{3}{2}\right)^2 = 4 + \left(\frac{3}{2}\right)^2$

   $$y^2 + 3y + \frac{9}{4} = 4 + \frac{9}{4} = \frac{16}{4} + \frac{9}{4} = \frac{25}{4}$$

   Factor the left side $\left(y + \frac{3}{2}\right)^2 = \frac{25}{4}$.

   Take the square root of each side. $\sqrt{\left(y + \frac{3}{2}\right)^2} = \sqrt{\frac{25}{4}}$

   $$y + \frac{3}{2} = \pm\sqrt{\frac{25}{4}} = \pm\frac{5}{2}$$

   Subtract $\frac{3}{2}$ from each side. $y = -\frac{3}{2} \pm \frac{5}{2} = \frac{-3 \pm 5}{2}$

   So, $y = \frac{-3 + 5}{2} = \frac{2}{2} = 1$ or $y = \frac{-3 - 5}{2} = \frac{-8}{2} = -4$.

2. **$z = 4, z = \frac{3}{2}$** $\frac{2z^2}{2} - \frac{11z}{2} + \frac{12}{2} = z^2 - \frac{11}{2}z + 6 = 0$

   $$z^2 - \frac{11}{2}z = -6$$

   $$z^2 - \frac{11}{2}z + \left(\frac{11}{4}\right)^2 = -6 + \left(\frac{11}{4}\right)^2$$

   $$z^2 - \frac{11}{2}z + \frac{121}{16} = \frac{-96}{16} + \frac{121}{16} = \frac{25}{16}$$

   $$\left(z - \frac{11}{4}\right)^2 = \frac{25}{16}$$

$$\sqrt{\left(z - \frac{11}{4}\right)^2} = \sqrt{\frac{25}{16}}$$

$$z - \frac{11}{4} = \pm\sqrt{\frac{25}{16}} = \pm\frac{5}{4}$$

$$z = \frac{11}{4} \pm \frac{5}{4} = \frac{11 \pm 5}{4}$$

$$\text{So } z = \frac{11 + 5}{4} = \frac{16}{4} = 4 \text{ or } z = \frac{11 - 5}{4} = \frac{6}{4} = \frac{3}{2}.$$

3. $m = \dfrac{-5 + \sqrt{33}}{2}, \ m = \dfrac{-5 - \sqrt{33}}{2}$

$$m^2 + 5m + \left(\frac{5}{2}\right)^2 = 2 + \left(\frac{5}{2}\right)^2$$

$$m^2 + 5m + \frac{25}{4} = \frac{8}{4} + \frac{25}{4} = \frac{33}{4}$$

$$\left(m + \frac{5}{2}\right)^2 = \frac{33}{4}$$

$$\sqrt{\left(m + \frac{5}{2}\right)^2} = \sqrt{\frac{33}{4}}$$

$$m + \frac{5}{2} = \pm\sqrt{\frac{33}{4}} = \pm\frac{\sqrt{33}}{2}$$

$$m = -\frac{5}{2} \pm \frac{\sqrt{33}}{2} = \frac{-5 \pm \sqrt{33}}{2}$$

4. $x = \sqrt{\dfrac{7}{3}}, \ x = -\sqrt{\dfrac{7}{3}}$

$$\frac{3x^2}{3} - \frac{7}{3} = x^2 - \frac{7}{3} = 0$$

$$x^2 = \frac{7}{3}$$

In this case, you don't need to add anything to each side to "complete the square;" the left side is already a square, so that part is done. Now, just take the square root of each side.

$$\sqrt{x^2} = \sqrt{\frac{7}{3}}$$

$$x = \pm\sqrt{\frac{7}{3}}$$

5. **$y = 6, y = 0$**   There is no constant to move over to the right, so completing the square can proceed from here.

$$y^2 - 6y + (3)^2 = y^2 - 6y + 9 = 9$$

Factor the left side.

$$(y-3)^2 = 9$$
$$\sqrt{(y-3)^2} = \sqrt{9}$$
$$y - 3 = \pm 3$$
$$y = 3 \pm 3$$

So, $y = 3 + 3 = 6$ or $y = 3 - 3 = 0$.

Solving Quadratic-Like Equations

In Chapter 4 in the section "Factoring Other Polynomials," there's a section on factoring quadratic-like expressions. The description and explanation of this type of algebraic arrangement is discussed there. In this section, solving equations that contain quadratic-like expressions is covered. Solving some higher-order equations (those with exponents greater than 2) and some equations with negative or fractional exponents can be accomplished fairly easily if the terms in the equation are quadratic-like. The same techniques that are applied when solving quadratic equations can be used to help solve quadratic-like equations. There's just an extra step or two at the end, after applying the factoring, to finish the particular type of equation.

Some quadratic-like equations are

$$x^4 - 4x^2 - 21 = 0$$
$$8y^6 - 9y^3 + 1 = 0$$
$$z^{1/2} - 5z^{1/4} + 4 = 0$$

Notice that they're each a trinomial, that the exponent/power on the first term is twice that of the middle term, and that the last term is a constant. They're all of the form $ax^{2n} + bx^n + c = 0$. To solve them, either factor using the same patterns as in factoring quadratic equations from Chapter 5 or use the quadratic formula (from a section earlier in this chapter). Then, to finish it off, solve for the specific variable.

Example Problems

These problems show the answers and solutions.

1. Solve for x in $x^4 - 4x^2 - 21 = 0$.

   answer: $x = \sqrt{7}$ and $x = -\sqrt{7}$

   Factor this into the product of two binomials using unFOIL.

   $$(x^2 - 7)(x^2 + 3) = 0$$

   Apply the Multiplication Property of Zero (see Chapter 1).

   $$x^2 - 7 = 0 \text{ or } x^2 + 3 = 0$$

   Solve $x^2 - 7 = 0$, which gives $x^2 = 7$ and $x = \pm\sqrt{7}$.

$x^2 + 3 = 0$ has no real solutions, because $\sqrt{x^2} = \sqrt{-3}$ asks for the square root of a negative number.

The only two solutions of this equation are $x = \sqrt{7}$ and $x = -\sqrt{7}$.

2. Solve for y in $8y^6 - 9y^3 + 1 = 0$.

    answer: $y = \dfrac{1}{2}$ and $y = 1$

    Factor the trinomial into the product of two binomials.

    $$(8y^3 - 1)(y^3 - 1) = 0$$

    Set the two factors equal to 0. First, $8y^3 - 1 = 0$ means $y^3 = \dfrac{1}{8}$. Take the cube root of each side, $\sqrt[3]{y^3} = \sqrt[3]{\dfrac{1}{8}}$, $y = \dfrac{1}{2}$. Then, $y^3 - 1 = 0$ means that $y^3 = 1$. Take the cube root of each side, $\sqrt[3]{y^3} = \sqrt[3]{1}$, $y = 1$.

    So the two solutions are $y = \dfrac{1}{2}$ and $y = 1$.

3. Solve for z in $z^{1/2} - 5z^{1/4} + 4 = 0$.

    answer: $z = 256$ and $z = 1$

    Factor, $(z^{1/4} - 4)(z^{1/4} - 1) = 0$.

    Now, set the two binomials equal to 0, if $z^{1/4} - 4 = 0$ and then $z^{1/4} = 4$. This time, to solve for z, raise each side to the fourth power.

    $$\left(z^{1/4} \right)^4 = (4)^4, z = 256$$

    If $z^{1/4} - 1 = 0$, then $z^{1/4} = 1$. Raise each side to the fourth power. $\left(z^{1/4} \right)^4 = (1)^4$, $z = 1$

    The two solutions are $z = 256$ and $z = 1$.

4. Solve for w in $w^{-2} + 2w^{-1} - 15 = 0$.

    answer: $w = -\dfrac{1}{5}$ or $w = \dfrac{1}{3}$

    You can factor this into the product of two binomials. Sometimes, the exponents get a bit distracting. You can see that they fit the pattern, but it's hard to tell how to factor them. If that happens, just rewrite the equation as a standard quadratic equation. For instance, rewrite this problem as $x^2 + 2x - 15 = 0$ and let x^1 replace w^{-1}. This new equation factors into $(x + 5)(x - 3) = 0$. Now apply this same pattern on the equation with the negative exponents.

    $$w^{-2} + 2w^{-1} - 15 = (w^{-1} + 5)(w^{-1} - 3) = 0$$

    If the first factor equals 0, $w^{-1} + 5 = 0$ means $w^{-1} = -5$. Write both sides as fractions, $\dfrac{1}{w} = \dfrac{-5}{1}$

Now, solve for w, $\dfrac{w}{1} = \dfrac{1}{-5}$ or $w = -\dfrac{1}{5}$.

If the second factor equals 0, $w^{-1} - 3 = 0$ means $w^{-1} = 3$. Write both sides as fractions, $\dfrac{1}{w} = \dfrac{3}{1}$. Now, solve for w, $\dfrac{w}{1} = \dfrac{1}{3}$ or $w = \dfrac{1}{3}$.

5. Solve for t in $t^4 + 3t^2 - 1 = 0$.

   answer: $t = +\sqrt{\dfrac{-3 + \sqrt{13}}{2}}$, $t = -\sqrt{\dfrac{-3 + \sqrt{13}}{2}}$

   This equation doesn't factor. The quadratic formula can be applied if a substitution is done, somewhat like in the preceding example.

   Let y replace the t^2. Then, if $y = t^2$, $y^2 = t^4$. So the t's can be replaced.

   The new equation reads $y^2 + 3y - 1 = 0$.

   Now apply the quadratic formula to this equation.

   $$y = \frac{-3 \pm \sqrt{9 - 4(1)(-1)}}{2(1)} = \frac{-3 \pm \sqrt{13}}{2}$$

   This is what y is equal to, but you want what t is equal to. Replace the y with t^2.

   $$t^2 = \frac{-3 \pm \sqrt{13}}{2}$$

   Now, take the square root of each side:

   $$\sqrt{t^2} = \sqrt{\frac{-3 \pm \sqrt{13}}{2}}$$

   $$t = \pm\sqrt{\frac{-3 \pm \sqrt{13}}{2}}$$

   There are four different answers here.

   $$t = +\sqrt{\frac{-3 + \sqrt{13}}{2}}, \ t = +\sqrt{\frac{-3 - \sqrt{13}}{2}}, \ t = -\sqrt{\frac{-3 + \sqrt{13}}{2}}, \ t = -\sqrt{\frac{-3 - \sqrt{13}}{2}}$$

   Actually, the second and fourth solutions aren't really solutions, because there is a negative value under each radical. They'll have to be eliminated.

Work Problems

Use these problems to give yourself additional practice.

1. Solve for m in $m^8 - 3m + 2 = 0$.

2. Solve for z in $z^{4/3} - 13z^{2/3} = -36$.

3. Solve for p in $p^{-2} - 4 = 0$.

4. Solve for x in $3x^4 - 75x^2 = 0$.

5. Solve for y in $y^6 + 4y^3 = 5$.

Worked Solutions

1. $m = 1$, $m = -1$, $m = \sqrt[4]{2}$, $m = -\sqrt[4]{2}$

   Factor into the product of two binomials: $(m^4 - 1)(m^4 - 2) = 0$.

   Set the factors equal to 0.

   $m^4 - 1 = 0$ factors into $(m^2 - 1)(m^2 + 1) = 0$; $m^2 - 1 = 0$ factors into $(m - 1)(m + 1) = 0$.

   There are two solutions from these last two factors, $m = 1$ and $m = -1$.

   $m^4 - 2 = 0$ doesn't factor, but, after adding 2 to each side, the fourth root can be taken of each side.

   $\sqrt[4]{m^4} = \sqrt[4]{2}$, which gives you $m = \pm\sqrt[4]{2}$.

   There are four solutions: $m = 1$, $m = -1$, $m = \sqrt[4]{2}$, and $m = -\sqrt[4]{2}$.

2. $z \pm 27$, $z \pm 8$ Add 36 to each side. Then factor into the product of two binomials.

   $$z^{4/3} - 13z^{2/3} + 36 = (z^{2/3} - 9)(z^{2/3} - 4) = 0$$

   Set the binomials equal to 0.

   When $z^{2/3} - 9 = 0$, $z^{2/3} = 9$. Then raise each side to the $\frac{3}{2}$ power.

   $\left(z^{2/3}\right)^{3/2} = (9)^{3/2}$, which means that $z = \pm\left(\sqrt[2]{9}\right)^3 = \pm(3)^3 = \pm 27$ when $z^{2/3} - 4 = 0$, $z^{2/3} = 4$.

   Then raise each side to the $\frac{3}{2}$ power, $\left(z^{2/3}\right)^{3/2} = \pm(4)^{3/2}$, which means that

   $z = \pm\left(\sqrt[2]{4}\right)^3 = \pm(2)^3 = \pm 8$.

   The four solutions are $z = \pm 27$ and $z = \pm 8$.

3. $p = \frac{1}{2}$, $p = -\frac{1}{2}$

   Factor the binomial using the difference of two squares, $p^{-2} - 4 = (p^{-1} - 2)(p^{-1} + 2) = 0$.

   Now set each factor equal to 0.

   When $p^{-1} - 2 = 0$, $p^{-1} = 2$. Write both sides as fractions and flip them, $\frac{1}{p} = \frac{2}{1}$, which gives you $\frac{p}{1} = \frac{1}{2}$ or $p = \frac{1}{2}$.

When $p^{-1}+2=0$, $p^{-1}=-2$. Write both sides as fractions and flip them, $\frac{1}{p}=\frac{-2}{1}$, which gives you $\frac{p}{1}=\frac{1}{-2}$ or $p=-\frac{1}{2}$.

The solutions are $p=\pm\frac{1}{2}$.

4. **$x=0$, $x=5$, $x=-5$**   First factor the binomial by dividing by the greatest common factor, $3x^2$.

$$3x^4-75x^2=3x^2(x^2-25)=0$$

The binomial can be factored, and then each of the factors is set equal to 0.

$$3x^2(x-5)(x+5)=0$$

When $x^2=0$, $x=0$; when $x-5=0$, $x=5$; when $x+5=0$, $x=-5$.

The solutions are $x=0,\pm5$.

5. **$y=-\sqrt[3]{5}$, $y=1$**   Subtract 5 from each side. Then factor the quadratic-like trinomial.

$$y^6+4y^3-5=0$$
$$(y^3+5)(y^3-1)=0$$

Set each factor equal to 0, when $y^3+5=0$, $y^3=-5$. The number -5 isn't a perfect cube. That factor doesn't give a nice fraction or decimal, but you could take the cube root of each side to get $\sqrt[3]{y^3}=\sqrt[3]{-5}$, $y=-\sqrt[3]{5}$.

When $y^3-1=0$, $y^3=1$ and $y=1$.

Quadratic and Other Inequalities

A quadratic inequality is an inequality with a quadratic or second-degree term in it. In Chapter 5, there are sections on solving linear inequalities and absolute value inequalities. The inequalities in this section take some special handling. As soon as a second-degree or higher degree term gets involved, the Multiplication Property of Zero comes into play. The same is true of fractional terms in an inequality. The general plan for dealing with these problems is to get all the terms on one side of the inequality symbol and completely factor the expression. The Multiplication Property of Zero determines where each factor might change signs (negative to positive or positive to negative). A solution is found by determining whether the expression should be positive or negative, so it's important to know when which factor will be which sign. The solution of the inequality $x^2-4x>0$ is when the statement is true, and it's true when the left side is positive. If $x=10$, then the left side of the inequality is $100-40$, which is positive. That means that the number 10 is part of the solution of the inequality. If $x=1$, then the left side of the inequality is $1-4$, which is -3. This is negative, not positive, so the number 1 is not a part of the solution. If $x=-10$, then the left side of the inequality is $100+40$, which is positive, so -10 is part of the solution. These numbers that work seem to be all over the place! But, really any negative number is part of the solution, and any number bigger than 4 works, too. The examples that follow will show you how to solve the inequalities.

Example Problems

These problems show the answers and solutions.

1. Solve $y^2 + 7y - 8 > 0$.

    answer: $y < -8$ or $y > 1$

    The solution will include all values of y that make the left side positive.

    Factor the left side into $(y + 8)(y - 1) > 0$. Now find out where the factors change sign by determining where they equal 0. $y + 8$ changes sign at -8 because numbers smaller than -8 make the expression negative and numbers larger than -8 make the expression positive.

    The two places where the two factors change sign are at -8 and 1. Now graph these two values on a number line.

    Put $+$ and $-$ in parentheses above the number line to show whether the factors are $+$ or $-$ in that interval.

    For numbers less than -8, for instance choose $y = -10$, the factor $y + 8$ is negative and the factor $y - 1$ is negative. Putting this on the number line to represent $(y + 8)(y - 1)(+)(-)$.

    $$(-)(-) \quad (+)(-)$$

    Now check on those two factors for numbers between -8 and 1. A good number to choose is $y = 0$. The factor $y + 8$ is positive, and the factor $y - 1$ is negative. Put this on the number line to represent $(y + 8)(y - 1)$.

    $$(-)(-) \quad (+)(-)$$

    The last place to check is for values bigger than 1. Letting $y = 2$, both factors are positive, so

    $$(-)(-) \quad (+)(-) \quad (+)(+)$$

    Going back to the original problem, which is to determine when the expression on the left is positive, just look at the factors. Two negatives multiplied together give a positive product, and two positives give a positive product. A positive times a negative is negative, so you just want the numbers smaller than -8 and bigger than 1 in your solution. This is written $y < -8$ or $y > 1$.

2. Solve $x^2(x + 3)(x - 8) \leq 0$.

    answer: $-3 \leq x \leq 8$

    There are three factors to consider. Their product has to be either negative or equal to zero. The Multiplication Property of Zero takes care of the zeros. The factors are equal to

0 at $x = 0$, $x = -3$, and $x = 8$. Now all that has to be done is determine the signs of the factors between the numbers that make the factors 0. The number line will contain the three values already determined.

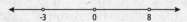

Choosing a number less than -3, I'll use $x = -5$. Putting that into the factors $x^2(x+3)(x-8)$, I find that the signs of the factors are $(+)(-)(-)$, which will go to the left of the -3.

Choose a number between -3 and 0; I'll use $x = -1$. Put that into the factors $x^2(x+3)(x-8)$; the factors are $(+)(+)(-)$, which will go between the -3 and 0.

Choose a number between 0 and 8; I'll use $x = 1$. Putting that into the factors $x^2(x+3)(x-8)$, I find that the factors are $(+)(+)(-)$, which will go between the 0 and 8.

Choose a number bigger than 8; I'll use $x = 10$. Putting that into the factors $x^2(x+3)(x-8)$, I get that the factors are $(+)(+)(+)$, which will go to the right of the 8.

$$(+)(-)(-) \quad (+)(+)(-) \quad (+)(+)(-) \quad (+)(+)(+)$$

Now to go back to what the original problem was. Where are the x's that make the expression 0 or negative? Whenever there is an odd number of negative factors, the whole product is negative, so the numbers between -3 and 8 will be in the solution, including the -3 and the 8, because they make the expression equal to 0. The answer is written $-3 \le x \le 8$.

3. Solve for z in $\frac{z+5}{3-z} \ge 0$.

answer: $-5 \le z < 3$

Even though this is a fraction, the numerator and denominator can be treated the same way as the factors in a multiplication problem. Determine where they change sign and use that to find out when the expression is positive or 0.

The numerator is 0 and changes sign when $z = -5$. The denominator is 0 and changes sign when $z = 3$. So, make a number line with those two numbers on it.

This time, stack the signs. When you choose a number smaller than -5, such as $z = -6$, the top factor $z + 5$ is negative, and the factor $3 - z$ is positive. Put $\frac{-}{+}$ on the number line.

When you choose a number between -5 and 3, such as $z = 0$, the top factor $z + 5$ is positive, and the factor $3 - z$ is positive. Put $\frac{+}{+}$ on the number line.

When you choose a number bigger than 3, such as $z = 4$, the top factor $z + 5$ is positive, and the factor $3 - z$ is negative. Put $\frac{+}{-}$ on the number line.

$$\frac{-}{+} \quad \frac{+}{+} \quad \frac{+}{-}$$

Look for where the result is positive or 0. The numbers between −5 and 3 will always give a positive result, so they're part of the solution. The number −5 makes the numerator equal to 0, so the whole fraction is equal to 0. It can be part of the solution. You can't use the 3, even though it makes a factor 0. The fact that it makes the denominator 0 is not allowed. There's no number that can be the result of dividing by 0.

So, the solution is written $-5 \leq z < 3$.

Work Problems

Use these problems to give yourself additional practice.

1. Solve for x in $x^2 + 5x - 24 > 0$.

2. Solve for z in $4z^2 + 17z - 15 \leq 0$.

3. Solve for m in $m(m - 2)(m + 4)(m - 6) \geq 0$.

4. Solve for n in $\dfrac{1}{2-n} > 0$.

5. Solve for p in $\dfrac{p-3}{p-8} \leq 0$.

Worked Solutions

1. **$x < -8$ and $x > 3$**   Factor the trinomial. $(x + 8)(x - 3) > 0$

   The values where the factors change sign are $x = -8$ and $x = 3$.

   Make a number line and indicate the signs of the factors in those intervals.

   $$(-)(-) \quad (+)(-) \quad (+)(+)$$

   The values that make the left side positive are $x < -8$ and $x > 3$.

2. **$-5 \leq z \leq \dfrac{3}{4}$**   Factor the trinomial. $(4z - 3)(z + 5) \leq 0$

   The values where the factors change sign are $z = \dfrac{3}{4}$ and $z = -5$.

   Make a number line and indicate the signs of the factors in those intervals.

   $$(-)(-) \quad (-)(+) \quad (+)(+)$$

   The values that make the left side negative or 0 are $-5 \leq z \leq \dfrac{3}{4}$.

3. **$m \leq -4$, $0 \leq m \leq 2$, and $m \geq 6$**   The values where the factors change sign are $m = 0$, 2, −4, and 6.

   Make a number line and indicate the signs of the factors in those intervals.

$$(-)(-)(-)(-) \quad (-)(-)(+)(-) \quad (+)(-)(+)(-) \quad (+)(+)(+)(-) \quad (+)(+)(+)(+)$$

The values which make the left side positive or 0 are $m \leq -4$, $0 \leq m \leq 2$, and $m \geq 6$.

4. **$n < 2$**   The only value for which a factor changes sign is 2, for the factor in the denominator.

   If n is a number bigger than 2, then the denominator is negative. If n is smaller than 2, the denominator is positive. The denominator cannot be 0, so n is never equal to 2. The solutions, which make the fraction positive, are the numbers smaller than 2, $n < 2$.

5. **$3 \leq p < 8$**   The values where the factors change sign are $p = 3$ and $p = 8$.

   Make a number line and indicate the signs of the factors in those intervals.

   $$(-)(-) \quad (+)(-) \quad (+)(+)$$

   The values that make the left side negative or 0 are $3 \leq p < 8$. You can use the 3, because that makes the numerator equal to 0. The denominator is 0 when p is 8, so that value can't be part of the solution.

Radical Equations with Quadratics

Equations including radical expressions are most easily solved by first getting rid of the radical. This process is covered in Chapter 5, where all of the situations there deal with changing from radical equations to linear equations. In this section, you'll see how to handle the special situations that can arise when you change from a radical form to a quadratic equation. There are even more opportunities for false or extraneous solutions to arise. The solution of the corresponding quadratic equation can result in two solutions to the radical equation, just one solution, or none at all. The important thing to do when solving those radical equations is to be sure to check your final answer. The procedure will be to get the radical term on one side of the equation and the rest of the terms on the other side. Then square both sides, solve the resulting equation, and then, of course, check your answers.

Example Problems

These problems show the answers and solutions.

1. $\sqrt{2x-1} = x - 2$.

   answer: $x = 5$

   Square both sides. $\left(\sqrt{2x-1}\right)^2 = (x-2)^2$

   The left side will be the binomial under the radical. The right side will be the perfect square binomial that you get from multiplying the two binomials together.

   $$2x - 1 = x^2 - 4x + 4$$

Set the quadratic equal to zero by subtracting 2x from each side and adding 1 to each side.

$$0 = x^2 - 6x + 5$$

Factor and solve for x. $0 = (x - 1)(x - 5)$, so $x = 1$ or $x = 5$

Now these both need to be checked. They're solutions of the quadratic equation but not necessarily of the radical equation.

When $x = 1$, $\sqrt{2(1) - 1} = (1) - 2$. The value of the left side is 1, and the right side is $1 - 2$ or -1, so the number 1 is not a solution.

When $x = 5$, $\sqrt{2(5) - 1} = (5) - 2$; $\sqrt{10 - 1} = 3$. This one does work. So the only solution is $x = 5$.

2. Solve $\sqrt{8x + 9} - 8 = 2x - 17$.

   answer: $x = 9$

   Before both sides can be squared, add 8 to each side to get the radical alone on the left.

   $$\sqrt{8x + 9} = 2x - 9$$

   Now square both sides.

   $$\left(\sqrt{8x + 9} \right)^2 = \left(2x - 9 \right)^2$$
   $$8x + 9 = 4x^2 - 36x + 81$$

   Subtract 8x and 9 from each side to set the quadratic equal to 0. Then factor and solve for x.

   $$0 = 4x^2 - 44x + 72$$
   $$0 = 4(x^2 - 11x + 18) = 4(x - 2)(x - 9)$$

   The two factors give solutions of $x = 2$ and $x = 9$. These both need to be checked in the original equation.

   If $x = 2$, $\sqrt{8(2) + 9} - 8 = 2(2) - 17$, $\sqrt{25} - 8 = 4 - 17$ or $5 - 8 = -13$. This is not true, so 2 is not a solution of the original equation.

   If $x = 9$, $\sqrt{8(9) + 9} - 8 = 2(9) - 17$, $\sqrt{81} - 8 = 18 - 17$ or $9 - 8 = 1$. This is true, so 9 is the only solution of the original equation.

3. Sometimes, both solutions of the quadratic work. Solve $\sqrt{5x - 1} = x + 1$.

   answer: $x = 2, x = 1$

   Squaring both sides and setting the quadratic equal to 0,

$$\left(\sqrt{5x-1}\right)^2 = (x+1)^2$$
$$5x - 1 = x^2 + 2x + 1$$
$$0 = x^2 - 3x + 2$$

Now factor and solve for the two solutions of the quadratic.

$$0 = (x - 2)(x - 1)$$
$$x = 2 \text{ or } x = 1$$

Check these in the original equation:

If $x = 2$, $\sqrt{5(2) - 1} = (2) + 1$, $\sqrt{9} = 3$. This one works.

If $x = 1$, $\sqrt{5(1) - 1} = (1) + 1$, $\sqrt{4} = 2$. This one also works.

Work Problems

Use these problems to give yourself additional practice.

1.　Solve $\sqrt{x + 6} = x$.

2.　Solve $\sqrt{z^2 + 9} = 2z - 3$.

3.　Solve $\sqrt{3n + 1} + 9 = n$.

4.　Solve $\sqrt{t^2 + 25} = t + 5$.

5.　Solve $\sqrt{8x + 1} = x + 2$.

Worked Solutions

1.　**$x = 3$**　Square both sides and set the quadratic equal to 0.

$$\left(\sqrt{x + 6}\right)^2 = x^2$$
$$x + 6 = x^2$$
$$0 = x^2 - x - 6$$

Factor and solve for x, $x = (x - 3)(x + 2)$, which means that $x = 3$ or $x = -2$.

If $x = 3$, then $\sqrt{3 + 6} = 3$ or $\sqrt{9} = 3$. The 3 works.

If $x = -2$, then $\sqrt{-2 + 6} = -2$. This says that the left side is equal to 2, and the right side is equal to -2. This is not true, so the -2 doesn't work.

2.　**$z = 4$**　Square both sides and set the quadratic equal to 0.

$$\left(\sqrt{z^2 + 9}\right)^2 = (2z - 3)^2$$
$$z^2 + 9 = 4z^2 - 12z + 9$$

$$0 = 3z^2 - 12z$$
$$0 = 3z(z - 4)$$
$$z = 0 \text{ or } z = 4$$

Check solutions: $z = 0$ doesn't work; $z = 4$ does.

3. **$n = 16$** First subtract 9 from each side. Then square both sides.

$$\sqrt{3n + 1} = n - 9$$
$$\left(\sqrt{3n + 1}\right)^2 = (n - 9)^2$$
$$3n + 1 = n^2 - 18n + 81$$

Set the quadratic equal to zero, factor it, and determine the solutions.

$$0 = n^2 - 21n + 80$$

$0 = (n - 16)(n - 5)$. The solutions are $n = 16$ and $n = 5$.

Test $n = 16$, $\sqrt{3(16) + 1} + 9 = 16$ or $\sqrt{49} + 9 = 16$ or $7 + 9 = 16$. The 16 works.

Test $n = 5$, $\sqrt{3(5) + 1} + 9 = 5$ or $\sqrt{16} + 9 = 5$ or $4 + 5 = 9$. No, the 5 doesn't work.

The solution of the radical equation is just $n = 16$.

4. **$t = 0$** Square both sides and solve the quadratic equation.

$$\left(\sqrt{t^2 + 25}\right)^2 = (t + 5)^2$$
$$t^2 + 25 = t^2 + 10t + 25$$
$$0 = 10t \text{ giving you } t = 0$$

Test this in the original equation, $\sqrt{0 + 25} = 0 + 5$ or $\sqrt{25} = 5$. The only solution is $t = 0$.

5. **$x = 1, x = 3$** Square both sides and solve the quadratic equation.

$$\left(\sqrt{8x + 1}\right)^2 = (x + 2)^2$$
$$8x + 1 = x^2 + 4x + 4$$
$$0 = x^2 - 4x + 3$$
$$0 = (x - 1)(x - 3) \text{ so } x = 1 \text{ or } x = 3.$$

Test $x = 1$, $\sqrt{8(1) + 1} = (1) + 2$ or $\sqrt{9} = 1 + 2$. The 1 works.

Test $x = 3$, $\sqrt{8(3) + 1} = (3) + 2$ or $\sqrt{25} = 3 + 2$. The 3 works, also. There are two solutions for this radical equation, $x = 1$ and $x = 3$.

Chapter Problems and Solutions

Problems

In problems 1 through 4, solve the quadratic equation by factoring.

1. $2x^2 + x - 15 = 0$

2. $16y^2 - 8y + 1 = 0$

3. $2x^2 - 6x - 8 = 0$

4. $10t^2 + 23t + 6 = 0$

In problems 5 through 8, solve the quadratic equation using the quadratic formula.

5. $x^2 - 10x + 16 = 0$

6. $20w^2 - 17w - 24 = 0$

7. $z^2 - 4z - 8 = 0$

8. $3r^2 + r - 1 = 0$

In problems 9 through 12, solve the quadratic equation using completing the square.

9. $y^2 + 10y + 24 = 0$

10. $z^2 + 4z - 21 = 0$

11. $3x^2 - 8x + 5 = 0$

12. $w^2 + 2w - 9 = 0$

13. Solve the quadratic equation using any method: $24x^2 - 26x - 63 = 0$.

In problems 14 through 17, solve the quadratic-like equation.

14. $m^{-2} + 11^{-1} + 24 = 0$

15. $p^1 - 29p^{1/2} + 100 = 0$

16. $x^{1/2} - x^{1/4} - 2 = 0$

17. $y^6 + y^3 - 2 = 0$

In problems 18 through 21, solve the inequality.

18. $(x - 3)(2x + 5) > 0$

19. $y^2 - y - 12 \leq 0$

20. $z(z - 4)(2z + 5) \geq 0$

21. $\dfrac{w+1}{w-8} \leq 0$

In problems 22 through 25, solve the radical equation for the value of the variable.

22. $\sqrt{1 - 2x} = x + 7$

23. $\sqrt{3x + 1} = 1 - x$

24. $\sqrt{5x + 11} - 3 = 3x + 4$

25. $\sqrt{x + 10} = x - 10$

Answers and Solutions

1.   **Answer:** $x = -3$, $x = \dfrac{5}{2}$ Factor the quadratic. $(x + 3)(2x - 5) = 0$. Setting $x + 3 = 0$, $x = -3$. Setting $2x - 5 = 0$, $x = \dfrac{5}{2}$.

2.   **Answer:** $y = \dfrac{1}{4}$ Factor the quadratic. $\left(4y - 1\right)^2 = 0$. There's only one solution. When $4y - 1 = 0$, $y = \dfrac{1}{4}$.

3.   **Answer:** $x = 4$, $x = -1$ Factor the quadratic. $2(x - 4)(x + 1) = 0$. Setting $x - 4 = 0$, $x = 4$. Setting $x + 1 = 0$, $x = -1$.

4.   **Answer:** $t = -\dfrac{3}{10}$, $t = -2$ Factor the quadratic. $(10t + 3)(t + 2) = 0$. Setting $10t + 3 = 0$, $t = -\dfrac{3}{10}$. Setting $t + 2 = 0$, $t = -2$.

5.   **Answer:** $x = 8$, $x = 2$

$$x = \frac{10 \pm \sqrt{(-10)^2 - 4(1)(16)}}{2(1)} = \frac{10 \pm \sqrt{100 - 64}}{2} = \frac{10 \pm \sqrt{36}}{2} = \frac{10 \pm 6}{2}$$

$$x = \frac{10 + 6}{2} = \frac{16}{2} = 8, \quad x = \frac{10 - 6}{2} = \frac{4}{2} = 2$$

6.   **Answer:** $w = -\dfrac{3}{4}$, $w = \dfrac{8}{5}$

$$w = \frac{17 \pm \sqrt{(-17)^2 - 4(20)(-24)}}{2(20)} = \frac{17 \pm \sqrt{289 + 1920}}{40} = \frac{17 \pm \sqrt{2209}}{40} = \frac{17 \pm 47}{40}$$

$$w = \frac{17 + 47}{40} = \frac{64}{40} = \frac{8}{5}, \quad w = \frac{17 - 47}{40} = \frac{-30}{40} = -\frac{3}{4}$$

7.   **Answer:** $z = 2 + 2\sqrt{3}$, $z = 2 - 2\sqrt{3}$

$$z = \frac{4 \pm \sqrt{(-4)^2 - 4(1)(-8)}}{2} = \frac{4 \pm \sqrt{16 + 32}}{2} = \frac{4 \pm \sqrt{48}}{2} = \frac{4 \pm 4\sqrt{3}}{2} = 2 \pm 2\sqrt{3}$$

8.   **Answer:** $r = \dfrac{-1 + \sqrt{13}}{6}$, $r = \dfrac{-1 - \sqrt{13}}{6}$

$$r = \frac{-1 \pm \sqrt{(1)^2 - 4(3)(-1)}}{2(3)} = \frac{-1 \pm \sqrt{1 + 12}}{6} = \frac{-1 \pm \sqrt{13}}{6}$$

9. **Answer: $y = -4$, $y = -6$** Start with the original equation.

$$y^2 + 10y + 24 = 0$$
$$y^2 + 10y = -24$$
$$y^2 + 10y + 5^2 = -24 + 5^2$$
$$y^2 + 10y + 25 = -24 + 25$$
$$(y + 5)^2 = 1$$
$$\sqrt{(y + 5)^2} = \sqrt{1}$$
$$y + 5 = \pm 1$$
$$y = -5 \pm 1$$

10. **Answer: $z = 3$, $z = -7$** Start with the original equation.

$$z^2 + 4z - 21 = 0$$
$$z^2 + 4z = 21$$
$$z^2 + 4z + (2)^2 = 21 + 2^2$$
$$z^2 + 4z + 4 = 21 + 4$$
$$(z + 2)^2 = 25$$
$$\sqrt{(z + 2)^2} = \sqrt{25}$$
$$z + 2 = \pm 5$$
$$z = -2 \pm 5$$

11. **Answer: $x = \frac{5}{3}$, $x = 1$** Start with the original equation.

$$\frac{3x^2}{3} - \frac{8x}{3} + \frac{5}{3} = 0$$
$$x^2 - \frac{8x}{3} + \frac{5}{3} = 0$$
$$x^2 - \frac{8x}{3} = -\frac{5}{3}$$
$$x^2 - \frac{8x}{3} + \left(\frac{4}{3}\right)^2 = -\frac{5}{3} + \left(\frac{4}{3}\right)^2$$
$$x^2 - \frac{8x}{3} + \frac{16}{9} = -\frac{5}{3} + \frac{16}{9}$$
$$\left(x - \frac{4}{3}\right)^2 = \frac{-15 + 16}{9}$$
$$\sqrt{\left(x - \frac{4}{3}\right)^2} = \sqrt{\frac{1}{9}}$$
$$x - \frac{4}{3} = \pm \frac{1}{3}$$
$$x = \frac{4}{3} \pm \frac{1}{3}$$

12. **Answer: $w = -1 + \sqrt{10}$, $w = -1 - \sqrt{10}$**   Start with the original equation.

$$w^2 + 2w - 9 = 0$$
$$w^2 + 2w = 9$$
$$w^2 + 2w + 1 = 9 + 1$$
$$(w + 1)^2 = 10$$
$$\sqrt{(w + 1)^2} = \sqrt{10}$$
$$w + 1 = \pm\sqrt{10}$$
$$w = -1 \pm \sqrt{10}$$

13. **Answer: $x = \dfrac{9}{4}$, $x = -\dfrac{7}{6}$**   This problem can be solved by factoring. The quadratic factors into $(4x - 9)(6x + 7) = 0$. This wouldn't be easy to factor, though, so the quadratic formula—even with rather large numbers—might have been the way to go.

14. **Answer: $m = -\dfrac{1}{3}$, $m = -\dfrac{1}{8}$**   First factor the quadratic. $\left(m^{-1} + 3 \right)\left(m^{-1} + 8 \right) = 0$. Setting $m^{-1} + 3 = 0$, $m^{-1} = -3$. This is then written $\dfrac{1}{m} = -\dfrac{3}{1}$. Flipping the proportion, $\dfrac{m}{1} = -\dfrac{1}{3}$. Setting $m^{-1} + 8 = 0$, $m^{-1} = -8$. This is then written $\dfrac{1}{m} = -\dfrac{8}{1}$. Flipping the proportion, $\dfrac{m}{1} = -\dfrac{1}{8}$.

15. **Answer: $p = 16$, $p = 625$**   First factor the quadratic. $(p^{1/2} - 4)(p^{1/2} - 25) = 0$. Setting $p^{1/2} - 4 = 0$, $p^{1/2} = 4$.

    Raising each side of the equation to the second power, $\left(p^{1/2} \right)^2 = \left(4 \right)^2$, $p = 16$. Setting $p^{1/2} - 25 = 0$, $p^{1/2} = 25$. Raising each side of the equation to the second power, $\left(p^{1/2} \right)^2 = \left(25 \right)^2$, $p = 625$.

16. **Answer: $x = 16$**   First factor the quadratic. $\left(x^{1/4} - 2 \right)\left(x^{1/4} + 1 \right) = 0$. Setting $x^{1/4} - 2 = 0$, $x^{1/4} = 2$. Raising each side to the fourth power, $\left(x^{1/4} \right)^4 = \left(2 \right)^4$, $x = 16$. Setting $x^{1/4} + 1 = 0$, $x^{1/4} = -1$.

    Here, it's tempting to raise each side to the fourth power—which would give a solution of 1. But that's an extraneous root. There's no number, multiplying itself four times, that can give a negative answer. An even number of negatives multiplied together results in a positive.

17. **Answer: $y = 1$, $y = -\sqrt[3]{2}$**   First factor the quadratic. $(y^3 - 1)(y^3 + 2) = 0$. Setting $y^3 - 1 = 0$, $y^3 = 1$. Taking the cube root of each side, $y = 1$. Setting $y^3 + 2 = 0$, $y^3 = -2$. Taking the cube root of each side, $y = \sqrt[3]{-2} = -\sqrt[3]{2}$.

18. **Answer: $x < -\dfrac{5}{2}$ or $x > 3$**   Draw a number line and place hollow circles at $x = -\dfrac{5}{2}$ and $x = 3$, where the factors are equal to 0. Determine the signs of the factors in the three sections.

$$(-)(-) \quad (-)(+) \quad (+)(+)$$

The solution contains values of x that cause the product to be positive. That occurs when both factors are negative or both factors are positive.

19. **Answer:** $-3 \leq y \leq 4$ First factor the expression to get $(y-4)(y+3) \leq 0$. Draw a number line with solid circles at -3 and 4. Those two values will be included in the solution, because the expression is to be less than 0 or equal to 0. Determine the signs of the factors in the three sections. The solution will occur where there are an odd number of negative signs.

$$(-)(-) \quad (-)(+) \quad (+)(+)$$

20. **Answer:** $-\dfrac{5}{2} \leq z \leq 0$ **or** $z \geq 4$ Draw a number line with solid circles at $-\dfrac{5}{2}$, 0, and 4. Determine the signs of the factors in the four sections.

$$(-)(-)(-) \quad (-)(-)(+) \quad (+)(-)(+) \quad (+)(+)(+)$$

The solution includes the three values that make the expression 0 and, also, wherever the product of the factors is positive. This happens when there is an even number of negative signs.

21. **Answer:** $-1 \leq w < 8$ This solution can't include the value 8, because that would result in a 0 in the denominator. Draw a number line with a solid circle at the -1 and a hollow circle at the 8.

$$\dfrac{(-)}{(-)} \quad \dfrac{(+)}{(-)} \quad \dfrac{(+)}{(+)}$$

22. **Answer:** $x = -4$ Square both sides of the radical and simplify the quadratic equation.

$$\left(\sqrt{1-2x}\right)^2 = (x+7)^2$$
$$1 - 2x = x^2 + 14x + 49$$
$$0 = x^2 + 16x + 48$$

Solve the quadratic equation by factoring.

$$x^2 + 16x + 48 = 0$$
$$(x+4)(x+12) = 0$$

Setting the factors equal to 0 gives answers of $x = -4$ and $x = -12$. The answer $x = -12$ is extraneous—it doesn't work in the original problem.

23. **Answer: $x = 0$** Square both sides and simplify the quadratic equation.

$$\left(\sqrt{3x+1}\right)^2 = (1-x)^2$$
$$3x+1 = 1-2x+x^2$$
$$0 = x^2 - 5x$$

Factor the quadratic to find the solution.

$$x^2 - 5x = x(x-5) = 0$$

Setting the factors equal to 0 gives answers of $x = 0$ and $x = 5$. The answer $x = 5$ is extraneous—it doesn't work in the original problem.

24. **Answer: $x = -2$ and $x = \dfrac{-19}{9}$** First add three to each side of the equation. Then square both sides and simplify the quadratic equation.

$$\sqrt{5x+11} - 3 = 3x + 4$$
$$\left(\sqrt{5x+11}\right)^2 = (3x+7)^2$$
$$5x + 11 = 9x^2 + 42x + 49$$
$$0 = 9x^2 + 37x + 38$$

The quadratic can be factored. Then set the factors equal to 0.

$$9x^2 + 37x + 38 = (9x+19)(x+2) = 0$$

Setting the factors equal to 0 gives answers of $x = -\dfrac{19}{9}$ and $x = -2$.

25. **Answer: $x = 15$** Square both sides and simplify the quadratic.

$$\left(\sqrt{x+10}\right)^2 = (x-10)^2$$
$$x + 10 = x^2 - 20x + 100$$
$$0 = x^2 - 21x + 90$$

The quadratic factors as follows.

$$x^2 - 21x + 90 = (x-6)(x-15) = 0$$

Setting the factors equal to 0, there are two solutions, $x = 6$ and $x = 15$. Only $x = 15$ works.

Supplemental Chapter Problems

Problems

In problems 1 through 5, solve the equation by factoring.

1. $x^2 + 5x - 84 = 0$

2. $6y^2 - 11y - 35 = 0$

3. $9z^2 + 12z + 4 = 0$

4. $4w^2 - 49 = 0$

5. $2w^2 + 15w - 8 = 0$

In problems 6 through 9, use the quadratic formula to solve the equation.

6. $36t^2 - 13t + 1 = 0$

7. $m^2 + 2m - 15 = 0$

8. $x^2 - 8x + 1 = 0$

9. $3z^2 + z - 7 = 0$

In problems 10 through 13, use completing the square to solve the equation.

10. $y^2 + 4y - 12 = 0$

11. $z^2 + 10z + 21 = 0$

12. $2x^2 + 11x - 6 = 0$

13. $w^2 - 4w - 1 = 0$

In problems 14 through 17, solve the quadratic-like equations.

14. $x^{-2} + 7x^{-1} - 30 = 0$

15. $y^6 - 26y^3 - 27 = 0$

16. $2z^{1/2} - 5z^{1/4} - 3 = 0$

17. $m^4 - m^2 - 1 = 0$

In problems 18 through 21, solve the inequality.

18. $(x + 3)(x - 1) \geq 0$

19. $y^2 - 49 < 0$

20. $(z + 4)(z + 1)(z - 2) > 0$

21. $\dfrac{x}{x - 5} \leq 0$

In problems 22 through 25, solve the radical equations.

22. $\sqrt{2x + 7} = x - 4$

23. $\sqrt{5y - 6} = 3y - 4$

24. $\sqrt{z + 3} - z = z - 4$

25. $\sqrt{2q - 3} - 4 = q - 7$

Answers

1. $x = 7, x = -12$ (Solving Quadratic Equations by Factoring, p. 224)

2. $y = \dfrac{7}{2}, y = -\dfrac{5}{3}$ (Solving Quadratic Equations by Factoring, p. 224)

3. $z = -\dfrac{2}{3}$ (Solving Quadratic Equations by Factoring, p. 224)

4. $w = \dfrac{7}{2}, w = -\dfrac{7}{2}$ (Solving Quadratic Equations by Factoring, p. 224)

5. $w = -8, w = \dfrac{1}{2}$ (Solving Quadratic Equations by Factoring, p. 224)

6. $t = \dfrac{1}{4}, t = \dfrac{1}{9}$ (Solving Quadratic Equations with the Quadratic Formula, p. 228)

7. $m = 3, m = -5$ (Solving Quadratic Equations with the Quadratic Formula, p. 228)

8. $x = 4 + \sqrt{15}, x = 4 - \sqrt{15}$ (Solving Quadratic Equations with the Quadratic Formula, p. 228)

9. $z = \dfrac{-1 + \sqrt{85}}{6}, z = \dfrac{-1 - \sqrt{85}}{6}$ (Solving Quadratic Equations with the Quadratic Formula, p. 228)

10. $y = 2, y = -6$ (Solving Quadratic Equations by Completing the Square, p. 231)

11. $z = -3, z = -7$ (Solving Quadratic Equations by Completing the Square, p. 231)

12. $x = \dfrac{1}{2}, x = -6$ (Solving Quadratic Equations by Completing the Square, p. 231)

13. $w = 2 + \sqrt{5}, w = 2 - \sqrt{5}$ (Solving Quadratic Equations by Completing the Square, p. 231)

14. $x = \dfrac{1}{3}, x = -\dfrac{1}{10}$ (Solving Quadratic-Like Equations, p. 237)

15. $y = 3, y = -1$ (Solving Quadratic-Like Equations, p. 237)

16. $z = 81$ (Solving Quadratic-Like Equations, p. 237)

17. $m = \sqrt{\dfrac{1 + \sqrt{5}}{2}}$, $m = -\sqrt{\dfrac{1 + \sqrt{5}}{2}}$ (Solving Quadratic-Like Equations, p. 237)

18. $x \leq -3$ or $x \geq 1$ (Quadratic and Other Inequalities, p. 241)

19. $-7 < y < 7$ (Quadratic and Other Inequalities, p. 241)

20. $-4 < z < -1$ or $z > 2$ (Quadratic and Other Inequalities, p. 241)

21. $0 \leq x < 5$ (Quadratic and Other Inequalities, p. 241)

22. $x = 9$ (Radical Equations with Quadratics, p. 245)

23. $y = 2$ (Radical Equations with Quadratics, p. 245)

24. $z = \dfrac{13}{4}$ (Radical Equations with Quadratics, p. 245)

25. $q = 6$ (Radical Equations with Quadratics, p. 245)

Chapter 7
Graphing and Systems of Equations

Coordinate System

Equations and inequalities can be used to express some relationship or pattern in a particular situation. In Chapters 5 and 6, you found ways of solving equations and inequalities and determining, under certain circumstances, what values of the variable make a true statement. Graphs of points on a number line are a nice visual description of solutions and/or relationships between values. By taking two number lines and setting them perpendicular to one another, you can create a way of showing the relationship between the two **different** variables. The number lines are then called **axes** (singular, **axis**), and usually the horizontal axis is the *x*-axis, and the vertical axis is the *y*-axis.

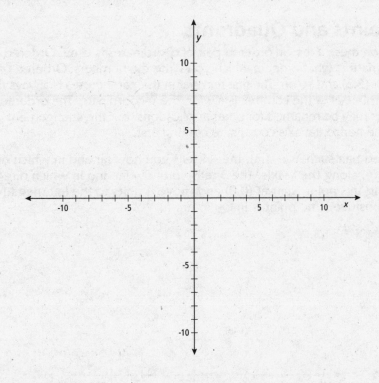

The point at which the vertical axis intersects the horizontal axis is zero (0) for both axes. The horizontal axis is labeled just like a number line, with the negative numbers to the left of zero and the positive numbers to the right of zero. On the vertical axis, the positive numbers are above the *x*-axis, and the negative numbers are below. This arrangement of the axes is called **coordinate axes** or the **Cartesian plane**. The name coordinate axes comes from the placing of points (coordinates) around the axes. The Cartesian plane is so named for Rene Descartes, a philosopher and mathematician, who made big contributions to formalizing parts of algebra.

The two intersecting axes form four **quadrants**, which are traditionally identified by Roman numerals as shown. Where they intersect is the **origin**.

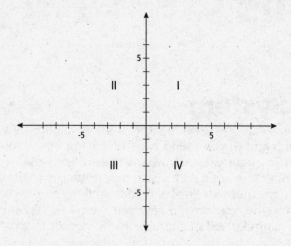

Graphing Points and Quadrants

To identify a point on these axes, an ordered pair of coordinates is used. Ordered pairs are so named because it matters what *order* in which you write the numbers. Ordered pairs look like this: (3,−2), (−4,3), (1,0), (5,5) and so on. The first number in the parentheses is always the *x*-coordinate, and the second number, after the comma, is always the *y*-coordinate. The ordered pair is of the form (*x*,*y*). The axes may be renamed for other applications, but the arrangement of the coordinates always has the horizontal axis coordinate coming first.

To graph an ordered pair such as (−4,3), the −4 tells you how far and in which direction to go from the origin (0,0), along the *x*-axis. The 3 tells you how far and in which direction to go on the *y*-axis. Graphing this point, start at (0,0) and move 4 units to the left (negative). From there, move 3 units up (positive). The point is in Quadrant II.

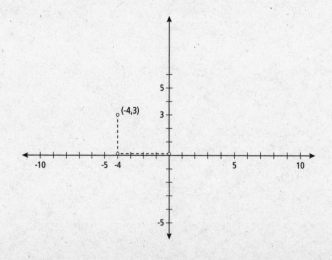

To graph the point (3,−6) start at the origin and move 3 units to the right (positive). From there, move 6 units down (negative). The point is in Quadrant IV.

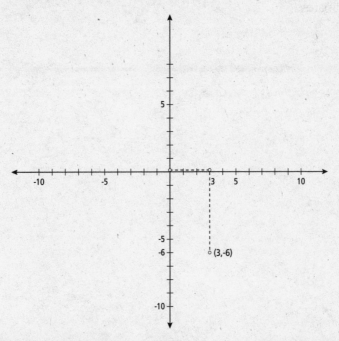

Points that lie on one axis or the other have a 0 for one of the coordinates. The point (5,0) lies on the x-axis, to the right of the origin. The 5 indicates that you move 5 units to the right of the origin. The 0 indicates that you don't go up or down from there—stay put. The point (0,−2) lies on the y-axis, below the origin. You have to be careful when graphing points that lie on one axis or the other. It's easy to confuse which axis they belong on, so count carefully and remember the arrangement in the ordered pairs.

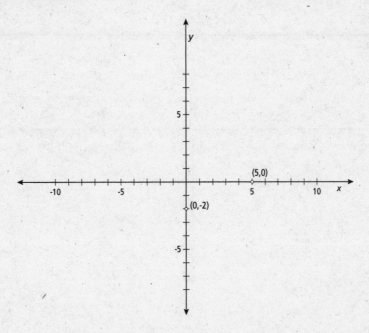

Example Problem

Graph each of the points listed here on coordinate axes. Points can be labeled with capital letters instead of their coordinates.

A. (3,4)

B. (6,0)

C. (9,−2)

D. (2,−5)

E. (−1,7)

F. (−3,−1)

G. (−10,0)

H. (−2,3)

I. (8,0)

J. (0,0)

answer:

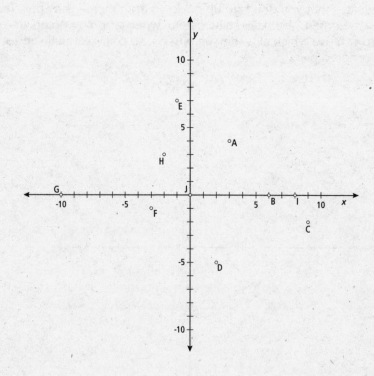

Work Problem

Graph and label the following points.

1. A: (−6,2)

2. B: (0,−4)

3. C: (−2,−2)

4. D: (3,1)

5. E: (4,−5)

Worked Solution

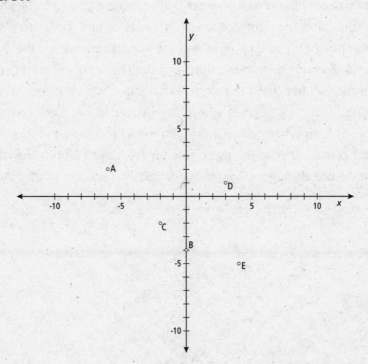

Graphing Lines

There are many ways to write relationships between variables, and one of the most common relationships that occurs naturally is a linear relationship. In business, there's "straight line depreciation." In science, the relationship between degrees Fahrenheit and degrees Celsius is linear. By linear, it means that if you graph all the ways the variables relate to one another as ordered pairs, the graph looks like a straight line.

There are two different forms of the equations of a line shown in the example problems. Look at the two different ways of writing the equation of the same line:

$$y = -3x + 8y$$
$$3x + y = 8$$

The first equation, $y = -3x + 8$, is written in the **slope-intercept** form. It's called that because the -3, the multiplier of x, represents the **slope** of the line, and the 8, the constant, is the y-intercept of the line; it's where the line crosses the y-axis. The slope is a number that tells you how steep or flat a line is and whether it rises or falls as you go from left to right. The greater the absolute value of the slope, the steeper the line. Lines with positive slopes rise from left to right, and lines with negative slopes fall from left to right.

The other form of the equation, $3x + y = 8$, is called the **standard form**. It has uses in finite mathematics and linear programming problems.

The slope-intercept form of a line is nice to work with, because you can quickly graph a line by finding two points. You graph the y-intercept first. It's easy to find, because it's the constant. Then you find another point by counting over and then up or down from the y-intercept. The slope tells you how far to count over to the right and then how far to count up or down from the y-intercept. Think of the number that's the slope as a fraction with the property: $\frac{\text{change in } y}{\text{change in } x}$. So a slope of -3 would be the fraction $\frac{-3}{1}$ where the change in x is right 1, and the change in y is down 3 (from the -3) from where you left off with the x. To graph $y = -3x + 8$, start with the y-intercept, $(0,8)$ and count 1 unit to the right. From there, count 3 units down. That's where the second point is. Draw a line through the y-intercept and that second point that you found.

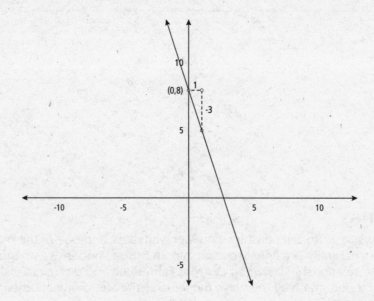

Example Problems

These problems show the answers and solutions.

1. Graph the line $y = 3x + 1$.

   answer:

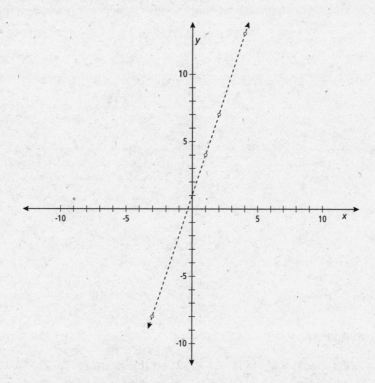

This equation says that y is a variable that is always 1 more than 3 times as big as the variable x. Some points that satisfy this equation are (1,4), (2,7), (5,16), (−3,−8), and so on. To solve for these points, first pick a number for the x-coordinate, such as $x = 4$. Then put the 4 into the equation and solve for y.

$y = 3(4) + 1 = 13$. The corresponding point on the line is (4,13). The graph of the five points mentioned here satisfies this equation.

2. Graph the line $y = -6x + 4$.

answer:

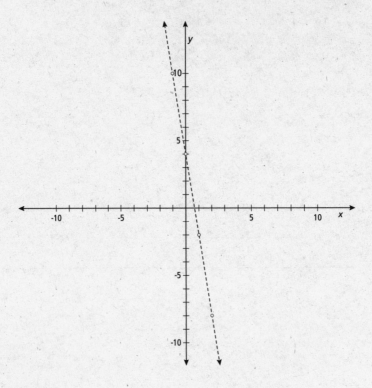

Choosing some points,

> let $x = 1$, $y = -6(1) + 4 = -2$, so the point is $(1,-2)$.
> let $x = 2$, $y = -6(2) + 4 = -8$, so the point is $(2,-8)$.
> let $x = -1$, $y = -6(-1) + 4 = 10$, so the point is $(-1,10)$.
> let $x = 0$, $y = -6(0) + 4 = 4$, so the point is $(0,4)$.

The graph is of the four points and a line drawn through them. There are an infinite number of points on that line, and only four of them are named.

3. Graph the line $4x + 3y = 12$.

answer:

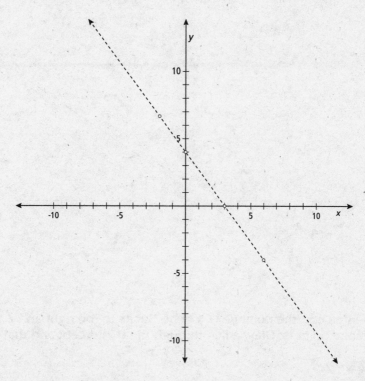

This time, the equation is written in standard form. I'll still try some values, though, and determine some points on the line.

If $x = 0$, $4(0) + 3y = 12$ or $3y = 12$ which means that $y = 4$. The point is $(0,4)$.

If $y = 0$, $4x + 3(0) = 12$ or $4x = 12$ which means that $x = 3$. The point is $(3,0)$.

These two points, with the 0s in them, are called the **intercepts** of the line. The intercept is where the line crosses the axes. These intercepts are usually the easiest points to find on a line, because of the 0s. Find two more points.

If $x = 6$, $4(6) + 3y = 12$, $24 + 3y = 12$ or $3y = -12$, which means that $y = -4$. The point is $(6,-4)$.

If $x = -2$, $4(-2) + 3y = 12$, $-8 + 3y = 12$, or $3y = 20$, which means that $y = \frac{20}{3}$. The point is $\left(-2, \frac{20}{3}\right)$.

4. Graph $y = \frac{2}{3}x + 1$ using the y-intercept and slope.

    answer:

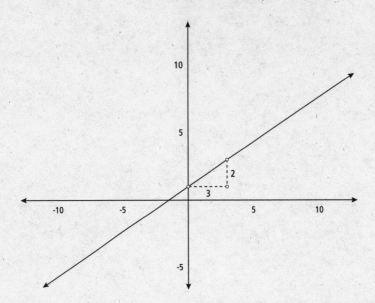

Start at the y-intercept, the point (0,1). Count 3 units to the right and 2 units up. That's where the second point is. Draw a line through the y-intercept and that second point.

5. Graph $y = -\frac{4}{3}x - 2$.

    answer:

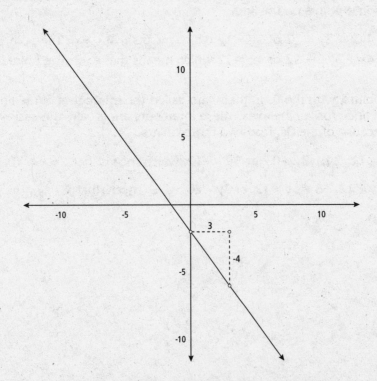

Start with the y-intercept, (0,−2). Count 3 units to the right and then 4 units down from there. Draw a line through the points.

6.   Graph the lines $y = 4$ and $x = -3$.

answer:

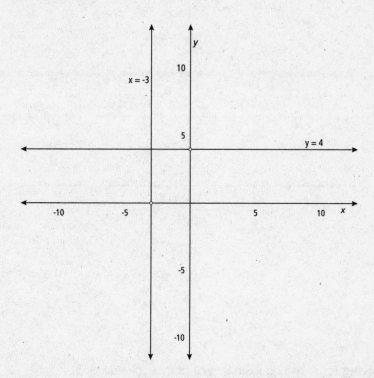

These two lines are easy to graph and, yet, hard to graph. One is horizontal and the other is vertical. Note that there's only an x or a y in each equation; the other variable is missing. The best way to graph these is to name a few points, graph them, and then draw the line. The line $y = 4$ has points where the x-coordinate can be anything and the y-coordinate always has to be 4. Some points would be $(1,4)$, $(-3,4)$, $(0,4)$, and so on. The line $x = -3$ has points where the x-coordinate is always -3, and the y-coordinate can be anything. Some points would be $(-3,4)$, $(-3,-1)$, $(-3,0)$, and so on.

Work Problems

Use these problems to give yourself additional practice.

1.   Graph by solving for some coordinates of points: $y = -2x + 3$.

2.   Graph using the y-intercept and the slope: $y = \frac{1}{3}x - 2$.

3.   Graph by solving for y to put it into the slope-intercept form $5x - 2y = 20$.

4.   Graph the line $2x + y = 14$.

5.   Graph the lines $x = 1$ and $y = -2$.

Worked Solutions

1.

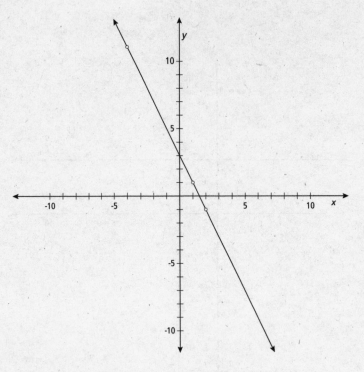

Graph by solving for some coordinates of points $y = -2x + 3$.

Some sample coordinates are $(1,1)$, $(2,-1)$, $(-4,11)$.

2.

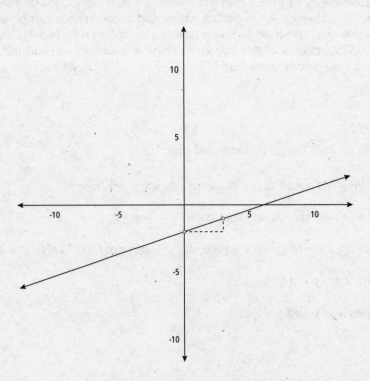

Graph using the y-intercept and the slope: $y = \frac{1}{3}x - 2$.

Graph the point $(0,-2)$. Then, from that intercept, count 3 units to the right and 1 unit up from there. Draw a line through that new point and the intercept.

3.

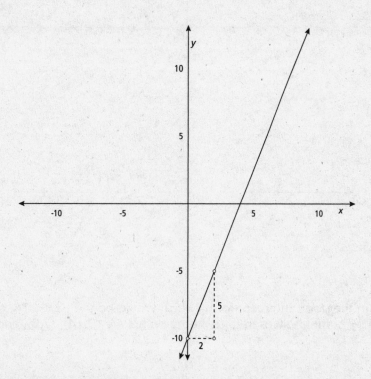

Graph by solving for y to put it into the slope-intercept form $5x - 2y = 20$.

Solve for y by subtracting $5x$ from each side and then dividing every term by -2.

$$5x - 2y = 20$$
$$-2y = -5x + 20$$
$$\frac{-2y}{-2} = \frac{-5x}{-2} + \frac{20}{-2}$$

$y = \frac{5}{2}x - 10$. The y-intercept is -10, and the slope is $\frac{5}{2}$. To graph this, first graph the point $(0,-10)$. Then, from that point, count 2 units to the right and from there 5 units up. Draw a line through that new point and the intercept.

4. Graph the line $2x + y = 14$.

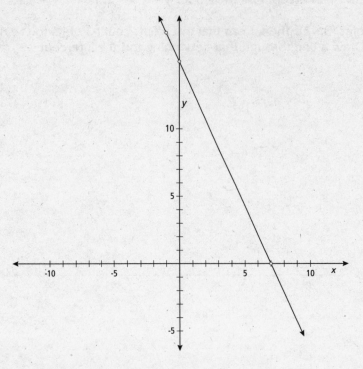

Either put it in the slope-intercept form, which would be $y = -2x + 14$, or just solve for some points by putting values in for x. Some points are $(0,14)$, $(7,0)$, and $(-1, 16)$.

5.

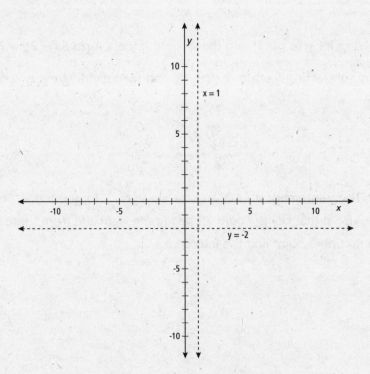

The line $x = 1$ is vertical through the point $(1,0)$, and the line $y = -2$ is horizontal through the point $(0,-2)$.

Graphing Other Curves

There are many types of relationships, besides lines, that associate variables. These relationships are found in science, business, engineering, medicine, statistics, and so on. Some of the equations of these relationships have graphs that are distinctive—they have a predictable shape. Being able to tell what the graph of some points will look like—before even graphing the points—is a big help. It cuts down on the work and possible errors. In this section, three different graphs of relations are discussed, graphs of **parabolas**, **cubic polynomials**, and **exponentials**.

Parabolas

Parabolas are the conic sections that appear to be ∪ or ∩ shaped. The parabolic curve is found in headlight reflectors, paths of objects in outer space, and so on. The bottom (or top, if it's turned over) of a parabola is called its **vertex**. That's always the highest or lowest point, depending on the direction it's going. A parabola can be graphed if you find the vertex and a couple other points.

The standard form for the equation of a parabola is $y = a(x - h)^2 + k$. The letters h and k represent the coordinates of the vertex. The letter a acts like slope—making the parabola steeper or flatter. If a is positive, then the parabola opens upward. If it's negative, the parabola opens downward. A parabola is always symmetric on either side of a vertical line that runs through the vertex, called the axis of the parabola.

Cubic Polynomials

A cubic polynomial is a smooth curve that looks like a sideways S. The cubic polynomial moves from negative infinity to positive infinity or vice versa, and it does its curving or turning or flattening in between. The graphs of cubics can look like the following:

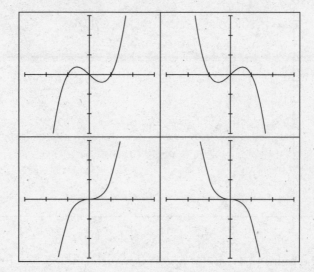

The cubic polynomial can have one, two, or three x-intercepts. These places where the curve crosses the x-axis are very helpful when sketching the graph of the curve. If the cubic polynomial can be factored into two or three different factors, then the factored form can be set equal to zero, and the x-intercepts can be solved for. Any x that makes $y = 0$ is where there's an x-intercept. If the coefficient of the term that's cubed is positive, then the curve starts down low to the left and goes up to positive infinity to the right. If the coefficient is negative, then the opposite occurs.

Exponential Curves

By using an exponent (or power) on a number or variable, you indicate how many times you want that number or variable multiplied times itself. An exponential curve is one with the variable in the exponent's place and a constant number as the base. The standard form is $y = a^x$ where a is a positive number. The graphs of these exponential curves are very simple and predictable. They're nice smooth curves that look like flattened C's. They all cross the y-axis at (0,1). The difference between one exponential curve and another is how steep they are and whether they rise or fall from left to right. The a is always positive, but if a is between 0 and 1 (it's a proper fraction or decimal), then the curve falls toward the x-axis. It never touches the axis; it just gets very close to it. If a is greater than 1, then the curve rises. It starts low on top of the x-axis but never touches it. If a is equal to 1, then the curve isn't a curve at all but the horizontal line $y = 1$.

Example Problems

These problems show the answers and solutions.

1. Sketch the graph of the parabola $y = 2(x - 3)^2 + 1$.

answer:

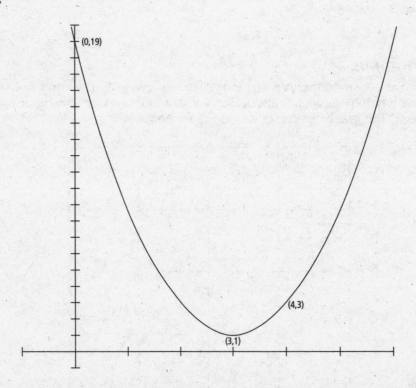

The vertex is at the point (3,1), and it opens upward. Finding two other points to help graph this parabola, choose some values for x, substitute them into the equation, and solve for y.

When $x = 4$, $y = 2(4 - 3)^2 + 1 = 2(1) + 1 = 3$. So the point (4,3) is on the parabola.

When $x = 0$, $y = 2(0 - 3)^2 + 1 = 2(9) + 1 = 19$. So the point (0,19) is on the parabola.

Graph the vertex and the two other points. Keep in mind that there's another point, on the other side of the vertex, that's the same height as the point (4,3). The same goes for the point (0,19). This is the way to use the symmetry of the parabola.

2. Sketch the graph of the parabola $y = -\frac{1}{3}(x+4)^2 + 5$.

answer:

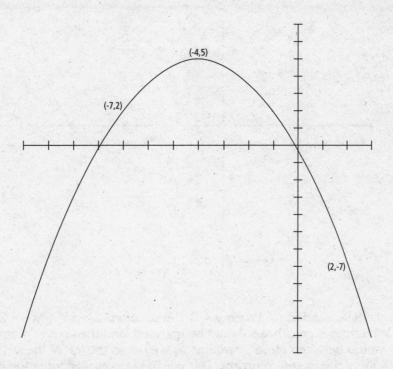

This equation isn't quite in the standard form needed to determine the vertex of the parabola. To write the terms in the parentheses with a subtraction sign, also change the sign of the 4. It now reads $y = -\frac{1}{3}(x-(-4))^2 + 5$.

The vertex is at the point (−4,5), and it opens downward. Find two other points to help graph this parabola; choose some values for x, substitute them into the equation, and solve for y.

When $x = 2$, $y = -\frac{1}{3}(2+4)^2 + 5 = -12 + 5 = -7$. So the point (2,−7) is on the parabola.

When $x = -7$, $y = -\frac{1}{3}(-7+4)^2 + 5 = -3 + 5 = 2$. So the point (−7,2) is on the parabola.

Graph the vertex and the two other points.

3. Graph $y = 2(x - 1)(x + 2)(x - 3)$.

answer:

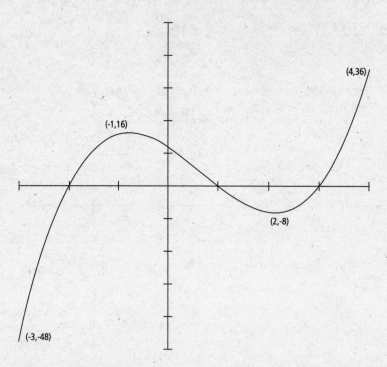

This is already in factored form. When $y = 0$, the solutions are $x = 1$, $x = -2$, $x = 3$. This is where the x-intercepts are. Those should be graphed, and then some points are picked, choosing x values between these intercepts as well as to the left of the smallest intercept and to the right of the largest intercept. This will help determine what the graph looks like.

Some points you might try are

$x = -3$ gives you $y = 2(-3 - 1)(-3 + 2)(-3 - 3) = -48$, so the point $(-3,-48)$ is on the curve.
$x = -1$ gives you $y = 2(-1 - 1)(-1 + 2)(-1 - 3) = 16$, so the point $(-1,16)$ is on the curve.
$x = 2$ gives you $y = 2(2 - 1)(2 + 2)(2 - 3) = -8$, so the point $(2,-8)$ is on the curve.
$x = 4$ gives you $y = 2(4 - 1)(4 + 2)(4 - 3) = 36$, so the point $(4,36)$ is on the curve.

The multiplier in front of the factored form is positive, so the coefficient of the cubed term will be positive. This curve starts down low at negative infinity, winds its way through the points and intercepts, and then goes on up infinitely high.

4. Graph $y = -x(x + 2)(x - 2)$.

answer:

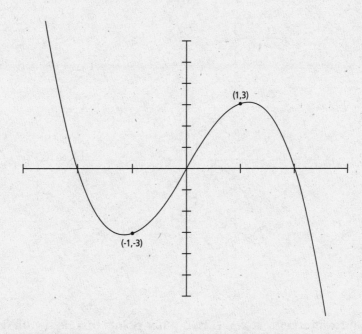

This is already in factored form. When $y = 0$, the solutions are $x = 0$, $x = -2$, $x = 2$. This is where the x-intercepts are. Those should be graphed, and then some points are picked choosing x values between these intercepts as well as to the left of the smallest intercept and to the right of the largest intercept. This will help determine what the graph looks like.

Some points you might try are

$x = -3$ gives you $y = -(-3)(-3 + 2)(-3 - 2) = 15$, so the point $(-3,15)$ is on the curve.

$x = -1$ gives you $y = -(-1)(-1 + 2)(-1 - 2) = -3$, so the point $(-1,-3)$ is on the curve.

$x = 1$ gives you $y = -(1)(1 + 2)(1 - 2) = 3$, so the point $(1,3)$ is on the curve.

$x = 3$ gives you $y = -(3)(3 + 2)(3 - 2) = -15$, so the point $(3,-15)$ is on the curve.

The multiplier in front of the factored form is negative, so the coefficient of the cubed term will be negative. This curve starts up at positive infinity, winds its way through the points and intercepts, and then goes down infinitely low.

5. Graph $y = (x-1)^3$.

answer:

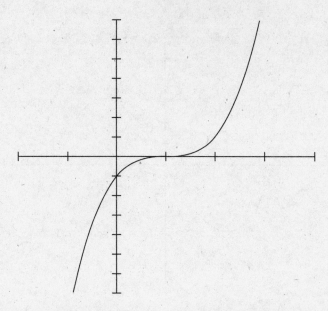

There's only one x-intercept, at $x = 1$. Two other points that help with the graphing are $(0,-1)$ and $(2,1)$.

6. Graph $y = 2^x$ and $y = 3^x$ on the same graph.

answer:

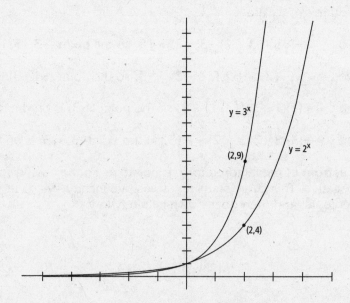

Find some points for the graph of $y = 2^x$:

$x = 0$ then $y = 2^0 = 1$, and the point $(0,1)$ is on the graph.

$x = 1$ then $y = 2^1 = 2$, and the point $(1,2)$ is on the graph.

$x = 2$ then $y = 2^2 = 4$, and the point $(2,4)$ is on the graph.

$x = -1$ then $y = 2^{-1} = \frac{1}{2}$, and the point $\left(-1, \frac{1}{2}\right)$ is on the graph.

$x = -2$ then $y = 2^{-2} = \frac{1}{4}$, and the point $\left(-2, \frac{1}{4}\right)$ is on the graph.

Go through the same process for $y = 3^x$; the following points are on the graph:

$$\left(-2, \frac{1}{9}\right), \left(-1, \frac{1}{3}\right), (0,1), (1,3), (2,9)$$

7. Graph $y = \left(\frac{1}{3}\right)^x$ and $y = \left(\frac{1}{5}\right)^x$ on the same graph.

answer:

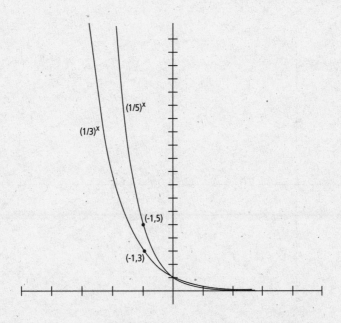

Note that the bases here are numbers between 0 and 1. These two graphs will fall as the curve moves from left to right. Finding points on the curves, first, if $y = \left(\frac{1}{3}\right)^x$,

$x = -1$ then $y = \left(\frac{1}{3}\right)^{-1} = \left(\frac{3}{1}\right)^1 = 3$, and the point $(-1,3)$ is on the graph.

$x = 0$ then $y = \left(\frac{1}{3}\right)^0 = 1$, and the point $(0,1)$ is on the graph.

$x = 1$ then $y = \left(\frac{1}{3}\right)^1 = \frac{1}{3}$, and the point $\left(1, \frac{1}{3}\right)$ is on the graph.

$x = 2$ then $y = \left(\frac{1}{3}\right)^2 = \frac{1}{9}$, and the point $\left(2, \frac{1}{9}\right)$ is on the graph.

Likewise, for $y = \left(\frac{1}{5}\right)^x$, some points are $(-1,5)$, $(0,1)$, $\left(1, \frac{1}{5}\right)$, and $\left(2, \frac{1}{25}\right)$.

Work Problems

Use these problems to give yourself additional practice.

Graph each of the following.

1. $y = (x - 4)^2 - 2$

2. $y = -2(x + 1)^2 + 5$

3. $y = -3(x + 5)(x + 1)(x - 1)$

4. $y = 2(x + 3)(x - 1)(x - 3)$

5. $y = 4^x$ and $y = \left(\frac{1}{4}\right)^x$ on the same graph.

Worked Solutions

1.

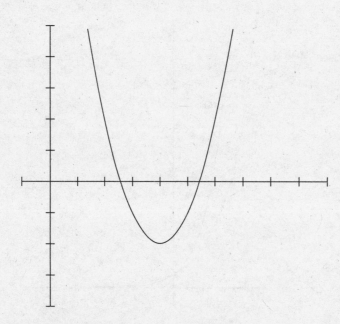

This is a parabola that opens upward. Graph the vertex, $(4, -2)$, and find a few more points to help you draw the graph.

2.

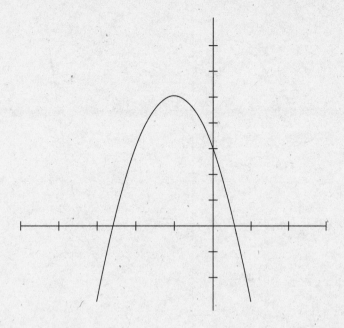

This is a parabola that opens downward. Graph the vertex, $(-1,5)$, and find a few more points to help you draw the graph.

3.

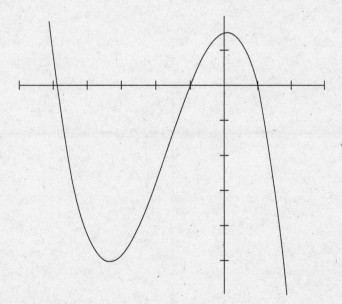

This is a cubic equation with x-intercepts intercepts at $(-5,0)$, $(-1,0)$, and $(1,0)$ and a y-intercept at $(0,15)$. Find some other points to help you draw the graph.

4.

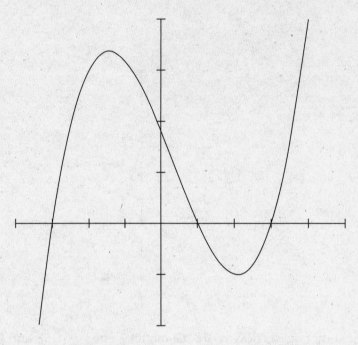

This is a cubic equation with *x*-intercepts at (−3,0), (1,0), and (3,0) and a *y*-intercept at (0,18). Find some other points to help you draw the graph.

5.

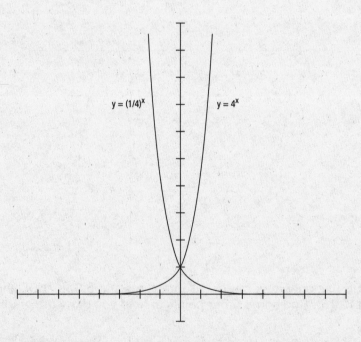

These are exponential functions, each with a *y*-intercept of (0,1). Find a few points to help you draw the graphs.

Finding the Equation of a Line

Graphing a line is rather straightforward (pardon the pun). Once you've found a couple of points, or if you have the y-intercept and slope, then you can confidently draw a line to show where all of the points that satisfy the equation lie. Going the other direction is just as easy. If you have just two points that lie on the line, you can figure out its equation. The same holds true if you know the slope and any point on the line (it doesn't have to be the y-intercept). A tool used to write the equation of a line given this information is the point-slope form.

Point-slope form: $y - y_1 = m(x - x_1)$

The x and y are going to be part of the final equation. The y_1 and x_1 are the coordinates of a point, (x_1, y_1), that lies on the line. The m is the slope of the line. This point-slope form may remind you of the slope-intercept form, $y = mx + b$. Simplifying the values in this point-slope form gives you the slope-intercept form.

Slope of a line: $m = \dfrac{y_2 - y_1}{x_2 - x_1}$

The (x_1, y_1) and (x_2, y_2) are two different points on the line.

Example Problems

These problems show the answers and solutions.

1. Write the equation of a line in slope-intercept form, $y = mx + b$, if the slope is 3 and a point on the line is $(2, -1)$.

   answer: $y = 3x - 7$

   Use the point-slope form and substitute the values given:

   $$y - y_1 = m(x - x_1)$$
   $$y - (-1) = 3(x - 2)$$
   $$y + 1 = 3x - 6$$
   $$y = 3x - 7$$

2. Write the equation of a line in **standard form**, $ax + by = c$, if the slope is $-\dfrac{2}{5}$ and a point on the line is $(1, 3)$.

   answer: $2x + 5y = 17$

   Use the point-slope form and substitute the values given:

   $$y - y_1 = m(x - x_1)$$
   $$y - 3 = -\frac{2}{5}(x - 1)$$
   $$y - 3 = -\frac{2}{5}x + \frac{2}{5}$$
   $$y = -\frac{2}{5}x + \frac{17}{5}$$

   This is the slope-intercept form. To change it to the standard form, multiply each term by 5 and add the x-term to each side to get it over to the left, $5y = -2x + 17$. Adding $2x$ to each side, $2x + 5y = 17$.

3. Write the equation of a line in slope-intercept form if two points on the line are (4,−3) and (−1,2).

answer: $y = -x + 1$

First find the slope of the line.

$$m = \frac{y_2 - y_1}{x_2 - x_1} = \frac{2 - (-3)}{-1 - 4} = \frac{5}{-5} = -1$$

Now, use the point-slope form and the point (4,−3):

$$y - (-3) = -1(x - 4)$$
$$y + 3 = x + 4$$
$$y = -x + 1$$

You could have used the other point, (−1,2), and the answer would have come out exactly the same.

Work Problems

Use these problems to give yourself additional practice.

1. Write the equation of the line, in slope-intercept form, that has a slope −2 and point (3,4).

2. Write the equation of the line, in standard form, that has a slope $\frac{4}{3}$ and point (−6,5).

3. Write the equation of the line, in slope-intercept form, that goes through both (4,1) and (6,−3).

4. Write the equation of the line, in standard form, that goes through both (2,2) and (−4,−1).

5. Write the equation of the line through (3,−2) and (−5,−2).

Worked Solutions

1. **$y = -2x + 10$**   Substitute into $y - y_1 = m(x - x_1)$,

$$y - 4 = -2(x - 3)$$
$$y - 4 = -2x + 6$$

Add 4 to each side, $y = -2x + 10$.

2. **−4x + 3y = 39**   Substitute into $y - y_1 = m(x - x_1)$,

$$y - 5 = \frac{4}{3}(x - (-6))$$

$$y - 5 = \frac{4}{3}x + 8$$

$$y = \frac{4}{3}x + 13$$

Now multiply each term by 3 and subtract the *x* term from each side.

$$3y = 4x + 39$$

Subtract 4x from each side, −4x + 3y = 39.

3. **y = −2x +9**   First find the slope of the line. Then substitute that slope and the coordinates of one of the points into the point-slope form.

$$m = \frac{y_2 - y_1}{x_2 - x_1} = \frac{(-3) - 1}{6 - 4} = \frac{-4}{2} = -2$$

$$y - 1 = -2(x - 4)$$

$$y - 1 = -2x + 8$$

Add 1 to each side, $y = -2x + 9$.

4. **−x + 2y = 2**   First find the slope. Then use the slope and one of the points in the point-slope form.

$$m = \frac{y_2 - y_1}{x_2 - x_1} = \frac{2 - (-1)}{2 - (-4)} = \frac{3}{6} = \frac{1}{2}$$

$$y - 2 = \frac{1}{2}(x - 2)$$

$$y - 2 = \frac{1}{2}x - 1$$

$$y = \frac{1}{2}x + 1$$

$$2y = x + 2$$

5. **y = −2**   First find the slope.

$$m = \frac{-2 - (-2)}{-5 - 3} = \frac{0}{-8} = 0$$

The slope is 0, so this is a horizontal line. The two *y*-values are both −2, so the equation of the line is $y = -2$.

Graphing Inequalities

The graph of an inequality on a number line, as seen in Chapter 5, involves shading portions of the number line to the left of, to the right of, or between two points. The graph of an inequality with two variables, on the coordinate plane, looks like half the page shaded in. That's because, with inequalities, you pick a starting point and shade everything above it, below it, along side it, or a mixture of those things.

The procedure used to graph these inequalities is to first graph the corresponding **line**, and then figure out which side of the line to shade. The line is solid, with no breaks, if the inequality is either $\leq$ or $\geq$. The line is dashed or dotted if the inequality is either $<$ or $>$. When the line is solid, it means that all of the points on the line are included in the solution. Otherwise, they are not included.

Example Problems

These problems show the answers and solutions.

1. Graph the inequality $x + y \geq 2$.

    answer:

    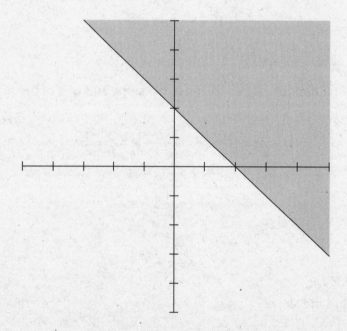

    First, graph the corresponding line, $x + y = 2$. Use a solid line, because it's greater than or equal to, so the points that lie on the line are included in the solution. Then determine which side of the line to shade in. To do this, choose some point that is clearly on one side of the line or the other. In the case of this line, the origin, (0,0), is clearly to the left of the line—the line doesn't run close to it. This is the test point. If it makes the inequality into a true statement, then it and every point on that side of the line are in the solution. If it doesn't work, then the other side gets shaded.

    Testing the point (0,0) in $x + y \geq 2$, $0 + 0 \geq 2$ is not a true statement, so (0,0) isn't in the solution.

Trying another point, on the other side, test (5,5) in the inequality. $5 + 5 \geq 2$ is a true statement, so that point and that side of the line are shaded to show all of the points in the solution.

2. Graph the inequality $3x - 2y < 6$.

    answer:

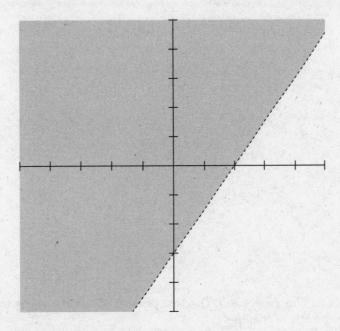

The corresponding line is $3x - 2y = 6$, and it'll be drawn as a dashed line, because the points on the line aren't included in the solution.

Testing the point (0,0) in the inequality, $3(0) - 2(0) < 6$ or $0 < 6$, which is true. So all the points on the same side of the line as the origin are in the solution. Check a point on the other side of the line, such as (3,0), $3(3) - 2(0) < 6$ or $9 < 6$, which is not true. The point (3,0) and all the points on that side of the line do not belong in the solution.

Work Problems

Use these problems to give yourself additional practice.

1. Graph $2x + y \leq 3$.

2. Graph $x - y > 1$.

3. Graph $x + 5y < 6$.

4. Graph $x + 3y \geq 3$.

5. Graph $x > -2x$.

Worked Solutions

1.

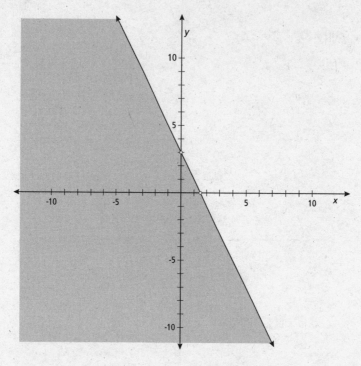

First graph the solid line $2x + y = 3$. The test point (0,0) works in the inequality, so shade in under and to the left of the line.

2.

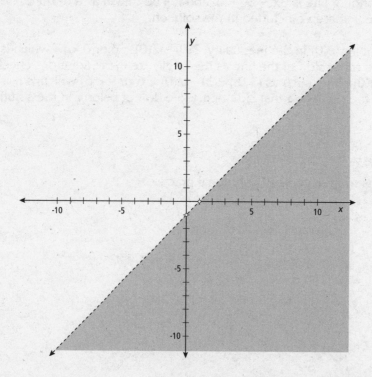

First graph the dashed line $x - y = 1$. The test point (0,0) doesn't work in the inequality, so shade in to the right and below the line.

3.

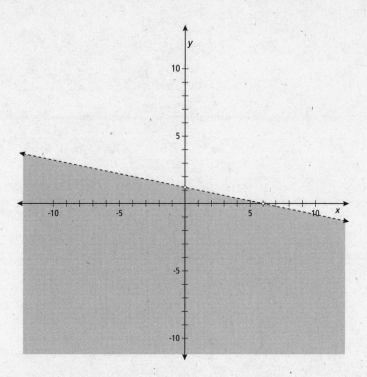

First graph the dashed line $x + 5y = 6$. The test point $(0,0)$ works in the inequality, so shade in under and to the left of the line.

4.

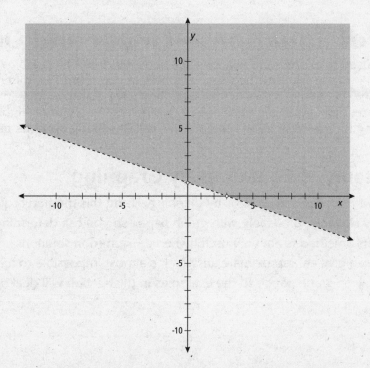

First graph the solid line $x + 3y = 3$. The test point $(0,0)$ doesn't work in the inequality, so shade in above and to the right of the line.

5.

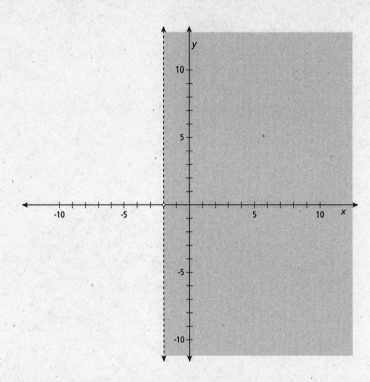

First graph the dashed line $x = -2$. This is a vertical line. All values of x greater than -2 are to the right of the line. Shade in to the right.

Systems of Equations—Linear and Quadratic

Equations that are used to express the relationships between different variables can also be compared to one another to see whether they have anything in common. Two different lines have lots of points on each line, but, if they cross or intersect, they have that one point in common—and only one. Two different parabolas can have one or two points in common—if they intersect at all. When you're looking for these common solutions, it's called **solving systems of equations**.

Solving Systems of Equations by Graphing

This section deals with solving the systems by graphing. Using this method requires that you graph the lines very carefully, preferably with graph paper, so you can determine the intersection points accurately. This method is not very useful when the common solutions involve fractions or decimals, but it can give an approximate answer. It's almost impossible to tell the difference between $2\frac{1}{3}$ and $2\frac{1}{2}$ on graph paper, so the examples in this section will deal only with answers that are integers.

Example Problems

These problems show the answers and solutions.

1. Find the common solution for the lines $y = -3x + 2$ and $y = x - 2$.

    answer: $(1, -1)$ This means that $x = 1$ and $y = -1$ are the common solution.

    Graph the two lines on the same graph. Determine where they intersect.

    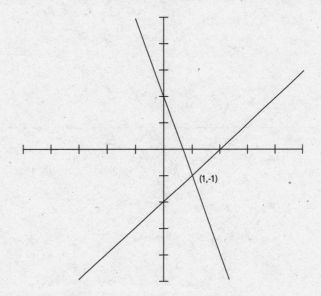

2. Find the common solution for the lines $3x + 4y = 6$ and $x + y = 1$.

    answer: $(-2, 3)$.   This means that $x = -2$ and $y = 3$ are the common solution.

    Graph the two lines and determine where they intersect.

    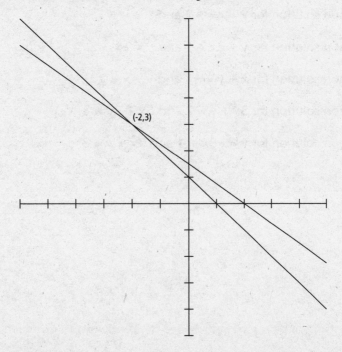

3. Find the common solution for the lines $y = 2x - 1$ and $2x - y = 3$.

answer: There is no solution. The two lines are parallel.

Graph the two lines. They appear to be parallel, which means that they have the same slope. The lines will never intersect, so there's no solution.

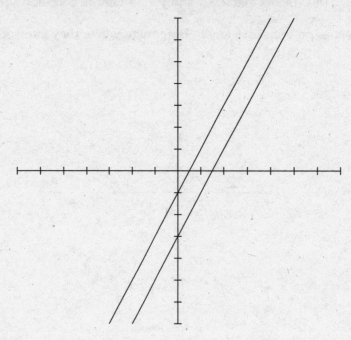

Work Problems
Use these problems to give yourself additional practice.

1. Find the common solution for $y = 4x - 2$ and $y = x + 1$.

2. Find the common solution for $y = 2x + 1$ and $y = -x - 2$.

3. Find the common solution for $x + y = -1$ and $y - x = 7$.

4. Find the common solution for $3x + y = 2$ and $5x + 2y = 3$.

5. Find the common solution for $y = -3x + 1$ and $3x + y = 3$.

Worked Solutions

1. **(1,2)** Graph the lines on the same graph.

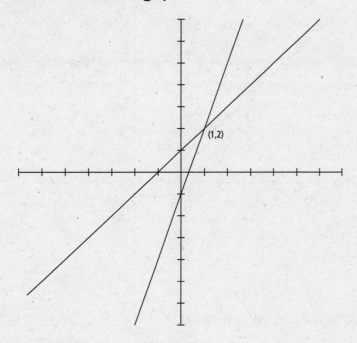

2. **(–1,–1)** Graph the lines on the same graph.

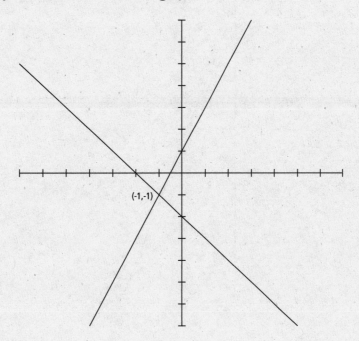

3. **(–4,3)**   Graph the lines on the same graph.

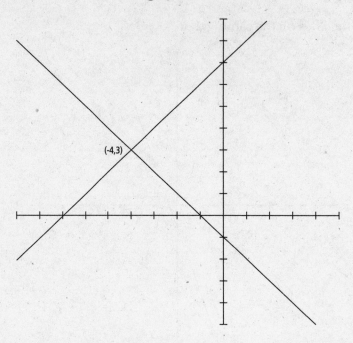

4. **(1,–1)**   Graph the lines on the same graph.

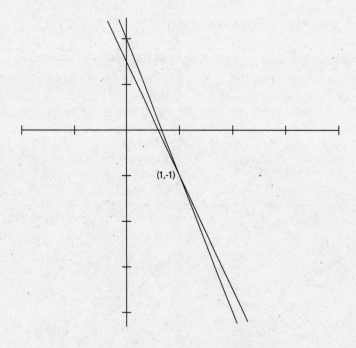

5. **No solution**   Graph the lines on the same graph. The lines are parallel, so they never intersect. There's no solution.

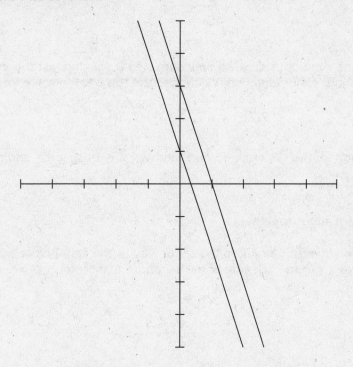

Solving Systems of Equations Using Addition

The main problem with solving systems of equations by graphing is that you can rarely find an exact solution if it involves a fraction or decimal. Also, you may not always have graph paper and a ruler handy. When the two equations you're trying to find the common solution to are lines, you can use a method called **linear combinations**, or more familiarly, **the addition method**. The goal is to add the two equations together and have the result be at least one part of the answer—the value of one of the variables. This will happen if the other pair of variables in the two equations are opposites of one another, so they add up to 0. Sometimes it takes a little manipulating of one or more of the equations, such as multiplying through by some number, to make this work successfully.

Example Problems

These problems show the answers and solutions.

1.   Find the solution for the system $2x + y = 7$ and $3x - y = 3$ using the addition method.

     answer: (2,3)

     Add the two equations together.

$$2x + y = 7$$

$$\underline{3x - y = 3}$$

$$5x \quad\;\; = 10$$

$$x = 2$$

The two y terms were opposites, so their sum was 0, leaving just the x terms. Now put the $x = 2$ back into either equation (I'll choose the top one) to solve for y.

$$2(2) + y = 7,\; 4 + y = 7,\; y = 3$$

2. Find the solution for the system $x - 3y = 8$ and $y = x - 4$ using the addition method.

   answer: $(2, -2)$

   Add the two equations together.

   The second equation isn't arranged the same way as the first. Before adding them together, subtract x from each side of the second equation to get $-x + y = -4$.

$$x - 3y = 8$$

$$\underline{-x + y = -4}$$

$$-2y = 4$$

$$y = -2$$

The two x terms were opposites, so their sum was 0, leaving just the y terms. Now put $y = -2$ back into either equation (I'll choose the bottom one) to solve for x.

$$-x + (-2) = -4$$

$$-x = -2$$

$$x = 2$$

3. Find the solution for the system $3x + 4y = 9$ and $x + y = 2$ using the addition method.

   answer: $(-1, 3)$

   Adding the equations together, the way they are, won't solve for either variable.

$$3x + 4y = 9$$

$$x + y = 2$$

   Either the x terms or the y terms need to be opposites of one another. One way to accomplish this is to multiply the bottom equation through by -3. (Multiplying by -4 would work, too. It would just result in finding a different variable first.)

$$3x + 4y = 9$$

$$\underline{-3x - 3y = -6}$$

$$y = 3$$

Put this back into the first equation, $3x + 4(3) = 9$, $3x = -3$, $x = -1$.

Work Problems

Use these problems to give yourself additional practice.

1. Find the solution to the system using the addition method: $x + 3y = 5$ and $2x - 3y = 4$.

2. Find the solution to the system using the addition method: $x - 2y = 5$ and $-x + 3y = 7$.

3. Find the solution to the system using the addition method: $3x - 4y = 9$ and $y = 4x + 1$.

4. Find the solution to the system using the addition method: $x = 7y + 2$ and $x - 8y = 1$.

5. Find the solution to the system using the addition method: $4x + y = 3$ and $8x + 2y = 1$.

Worked Solutions

1. $\left(3, \frac{2}{3}\right)$ Add the two equations together.

$$x + 3y = 5$$
$$\underline{2x - 3y = 4}$$
$$3x \qquad = 9$$
$$x = 3$$

Now put 3 in for x in the first equation.

$$3 + 3y = 5, \; 3y = 2, \; y = \frac{2}{3}$$

2. **(29,12)**   Add the two equations together.

$$x - 2y = 5$$
$$\underline{-x + 3y = 7}$$
$$y = 12$$

Put this into the first equation to get x.

$$x - 2(12) = 5, \; x - 24 = 5, \; x = 29$$

3. **(−1,−3)**   First rewrite the second equation so it'll have the same arrangement as the first, with the x and y terms on the left.

$y = 4x + 1$ becomes $-4x + y = 1$.

Now add the two equations together.

$$3x - 4y = 9$$
$$-4x + y = 1$$

Even with this arrangement, adding them together won't achieve anything. Neither of the variables has its opposite in the other equation. Multiply the terms in the bottom equation by 4 so the y terms will be opposites.

Then add the two equations together.

$$3x - 4y = 9$$
$$\underline{-16x + 4y = 4}$$
$$-13x \quad\quad = 13$$
$$x = -1$$

Now substitute the -1 for x in the first equation to solve for y.

$$3(-1) - 4y = 9,\ -3 - 4y = 9,\ -4y = 12,\ y = -3$$

4. **(9,1)**　Subtract $7y$ from each side of the top equation so they can be added together.

$x = 7y + 2$ becomes $x - 7y = 2$.

$$x - 7y = 2$$
$$x - 8y = 1$$

Multiply the terms in the bottom equation by -1 and add the two equations together.

$$x - 7y = 2$$
$$\underline{-x + 8y = -1}$$
$$y = 1$$

Substitute the 1 for y in the first equation to solve for x.

$$x = 7(1) + 2 = 9$$

5. **No solution**　Multiply the terms in the top equation by -2. Then add the equations together.

$$-8x - 2y = -6$$
$$\underline{8x + 2y = 1}$$
$$0 + 0 = -5$$

This last equation says that $0 = -5$. That's impossible, which means that a solution is impossible. These lines are parallel, so there's no common solution.

Chapter Problems and Solutions

Problems

1. Graph the points A (2,2), B (3,−5), C (−3,5), D (−4,−6), E (3,0). Label them with the corresponding letters.

In problems 2 through 4, graph the lines on the same coordinate axes.

2. $y = -2x + 3y$

3. $y = \frac{3}{4}x - 2$

4. $4x + 5y = 20$

In problems 5 and 6, graph the parabolas on the same coordinate axes.

5. $y = -2(x-3)^2 + 1$

6. $y = \frac{1}{2}(x+1)^2 - 4$

7. Graph the cubic $y = x(x-4)(x+3)$.

In problems 8 and 9, graph the exponentials on the same coordinate axes.

8. $y = 4^x$

9. $y = \left(\frac{1}{5}\right)^x$

10. Write the slope-intercept form of the equation of the line with a slope of 4 and that goes through $(-1,2)$.

11. Write the standard form of the equation of the line with a slope of $-\frac{4}{5}$ that goes through $(10,7)$.

12. Write the slope-intercept form of the equation of the line that goes through $(3,6)$ and $(-2,11)$.

13. Write the slope-intercept form of the equation of the line that goes through $(4,2)$ and $(-6,2)$.

14. Graph the inequality $2x - 3y \geq 8$.

15. Graph the inequality $x + 5y < 2$.

16. Find the common solution of $x + y = 11$ and $2x - 3y = 2$ by graphing.

17. Find the common solution of $3x - y = 5$ and $y = 4$ by graphing.

In problems 18 through 20, find the common solution using the addition method.

18. $5x - 3y = 16$ and $3x + y = 4$

19. $x - 4y = 8$ and $y = \frac{1}{4}x + 3$

20. $3x + 8y = 2$ and $y = 5x + 11$

Answers and Solutions

1. **Answer:**

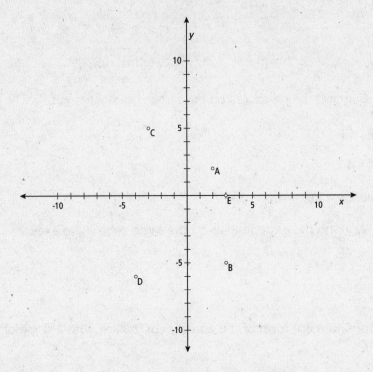

The point A lies in Quadrant I (upper right); the point B lies in Quadrant IV (lower right); the point C lies in Quadrant II (upper left); the point D lies in Quadrant III (lower left); and the point E lies on the *x*-axis, to the right of the origin.

2. **Answer:**

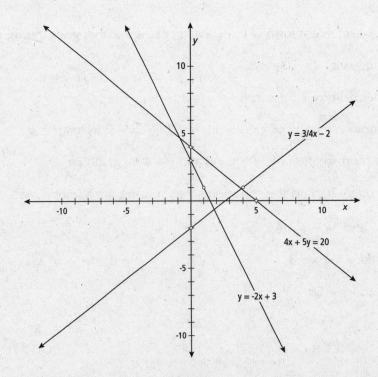

Graph the point (0,3). Count over 1 unit to the right and 2 units down from there to find a second point. Draw the line through those two points.

3. **Answer: (See preceding figure.)**   Graph the point (0,−2). Count over 4 units to the right and 3 units up from there to find a second point. Draw the line through those two points.

4. **Answer: (See preceding figure.)**   Find the two intercepts. Let $x = 0$, then $y = 4$. Let $y = 0$, then $x = 5$. Graph those two points, and draw the line through them.

5. **Answer:**

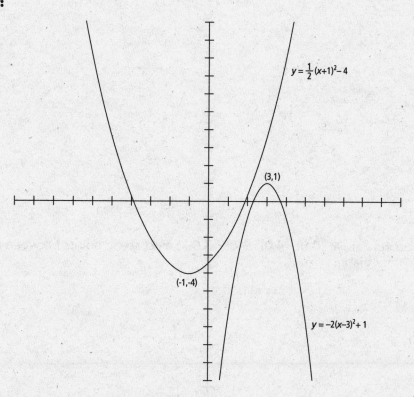

$$y = \tfrac{1}{2}(x+1)^2 - 4$$

(3,1)

(−1,−4)

$$y = -2(x-3)^2 + 1$$

Graph the vertex, (3,1). This is the high point of the parabola. Two other points that are also on the parabola are (2,−1) and (0,−17).

6. **Answer: (See preceding figure.)**   Graph the vertex, (−1,−4). This is the low point of the parabola. Two other points on the parabola are (1,−2) and (−3,−2).

7.   Answer:

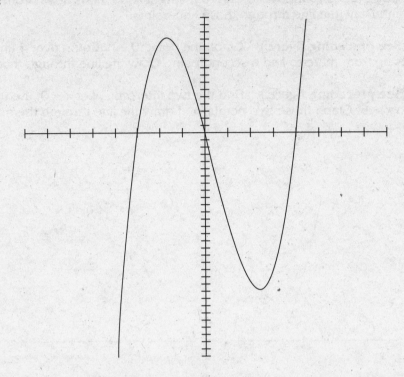

The *x*-intercepts are at (0,0), (4,0), and (−3,0). Select some points between the intercepts and sketch the graph.

8.   Answer:

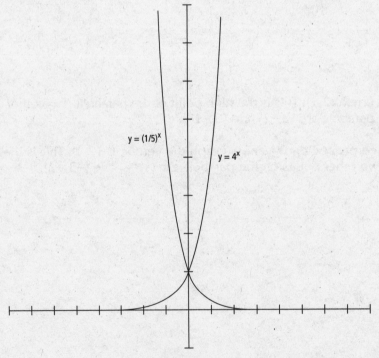

The curve crosses the *y*-axis at (0,1). Two other points on the graph are (1,4) and $\left(-2, \frac{1}{16}\right)$.

9. **Answer: (See preceding figure.)**   The curve crosses the y-axis at $(0,1)$. Two other points on the graph are $\left(1, \frac{1}{5}\right)$ and $(-2, 25)$.

10. **Answer: $y = 4x + 6$**   Using the point-slope form, $y - 2 = 4(x - (-1))$, $y - 2 = 4(x + 1)$, $y - 2 = 4x + 4$.

11. **Answer: $4x + 5y = 75$**   Using the point-slope form, $y - 7 = -\frac{4}{5}(x - 10)$, $y - 7 = -\frac{4}{5}x + 8$, $y = -\frac{4}{5}x + 15$. This is in the slope-intercept form. Multiply through by 5 and add $4x$ to each side.

12. **Answer: $y = -x + 9$**   First find the slope. $m = \frac{11 - 6}{-2 - 3} = \frac{5}{-5} = -1$. Put this and one of the points into the slope-intercept form. $y - 6 = -1(x - 3)$, $y - 6 = -x + 3$

13. **Answer: $y = 2$**   First find the slope. $m = \frac{2 - 2}{-6 - 4} = \frac{0}{-10} = 0$. This is a horizontal line with y values all equaling 2.

14. **Answer:**

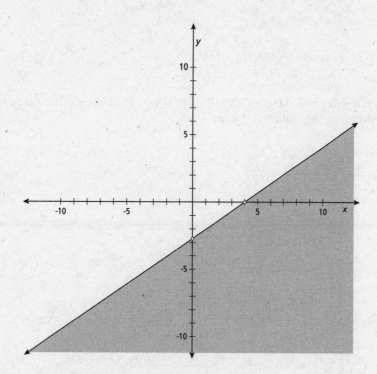

Graph the line $2x - 3y = 8$. The test point $(0,0)$ doesn't work, so shade all the points below and to the right of the line.

15.   Answer:

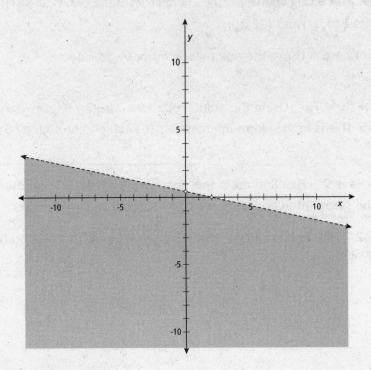

Graph the line $x + 5y = 2$. Use a dashed line. The test point $(0,0)$ works, so shade all the points below and to the left of the line.

16.   Answer: (7,4)

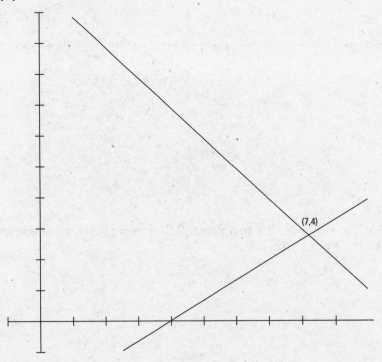

17. Answer: (3,4)

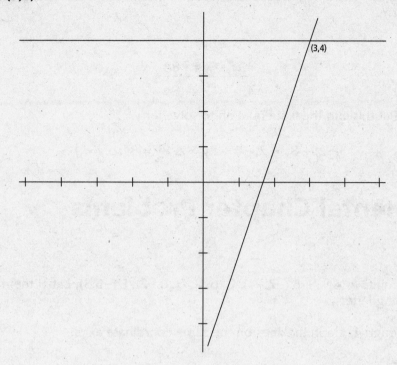

18. Answer: (2,–2)   Multiply the second equation through by 3 before adding them together.

$$5x - 3y = 16$$
$$\underline{9x + 3y = 12}$$
$$14x \quad\ = 28$$

So $x = 2$. Put this into the first equation,

$$5(2) - 3y = 16,\ 10 - 3y = 16,\ -3y = 6,\ \text{or } y = -2.$$

19. Answer: No answer   These lines are parallel. When the second equation is multiplied through by 4 and written in the standard form

$$x - 4y = 8$$
$$\underline{-x + 4y = 12}$$
$$0 + 0\ = 20$$

This is an impossible statement.

20. **Answer: (–2,1)** Multiply the second equation through by –8 and change it to the standard form.

$$3x + 8y = 2$$
$$40x - 8y = -88$$
$$\overline{43x \qquad = -86}$$

So $x = -2$. Put this into the first equation to solve for y.

$$3(-2) + 8y = 2, \; -6 + 8y = 2, \; 8y = 8, \text{ so } y = 1.$$

Supplemental Chapter Problems

Problems

1. Graph the points A (4,–7), B (–2,–4), C (6,1), D (0,–2), E (–5,3). Label them with the corresponding letters.

In problems 2 through 4, graph the lines on the same coordinate axes.

2. $y = -\dfrac{8}{5}x - 2$

3. $y = 4x + 1$

4. $x - 7y = 7$

In problems 5 and 6, graph the parabolas on the same coordinate axes.

5. $y = -\dfrac{3}{5}(x - 1)^2 - 1$

6. $y = 4(x + 2)^2 - 6$

7. Graph the cubic $y = -(x + 5)(x - 2)(x - 4)$.

In problems 8 and 9, graph the exponentials on the same coordinate axes.

8. $y = \left(\dfrac{1}{6}\right)^x$

9. $y = 3^x$

10. Write the slope-intercept form of the equation of the line with a slope of $\dfrac{3}{2}$ and that goes through (4,–7).

11. Write the standard form of the equation of the line with a slope of –5 that goes through (3,2).

12. Write the slope-intercept form of the equation of the line that goes through $(-4, 5)$ and $(6, 0)$.

13. Write the slope-intercept form of the equation of the line that goes through $(8, -3)$ and $(5, -3)$.

14. Graph the inequality $x - 4y \leq 7$.

15. Graph the inequality $3x + 8y > 4$.

16. Find the common solution for $3x + 2y = 13$ and $x - y = 6$ by graphing.

17. Find the common solution for $4x + y = 7$ and $x = 1$ by graphing.

In problems 18 through 20, find the common solution using the addition method.

18. $4x + y = 7$ and $x - y = 8$

19. $2x - 5y = 13$ and $y = x - 2$

20. $x - 3y = 8$ and $y = \frac{1}{3}x + 1$

Answers

1.

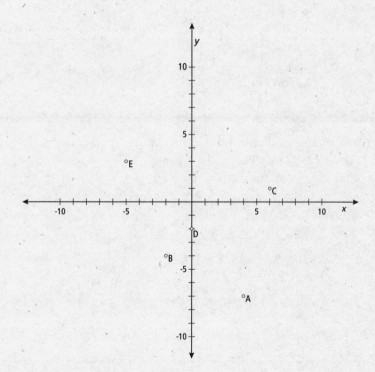

(Graphing Points and Quadrants, p. 260)

2–4.

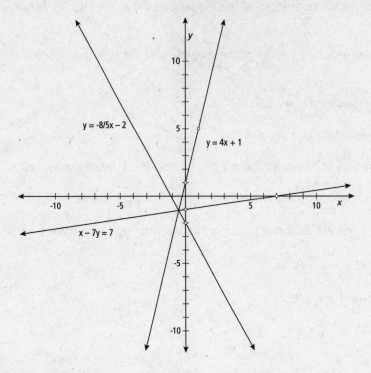

(Graphing Lines, p. 263)

5–6.

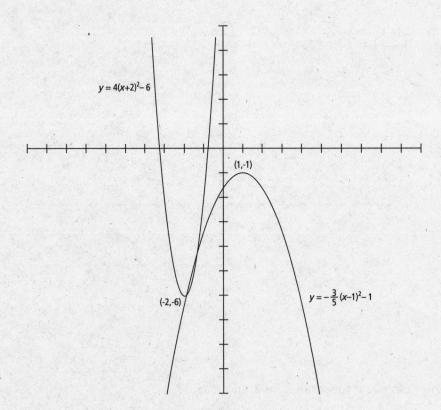

(Graphing Other Curves—Parabolas, p. 273)

7.

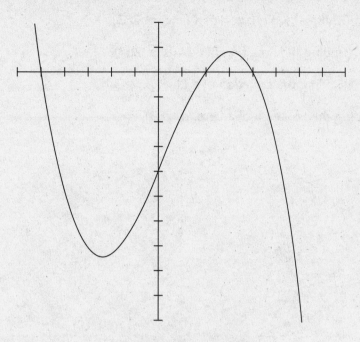

(Graphing Other Curves—Cubic Polynomials, p. 273)

8–9.

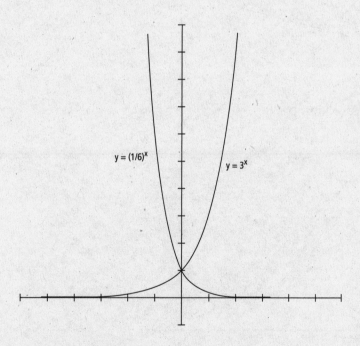

(Graphing Other Curves—Exponential Curves, p. 274)

10. $y = \frac{3}{2}x - 13$ (Finding the Equation of a Line, p. 283)

11. $5x + y = 17$ (Finding the Equation of a Line, p. 283)

12. $y = -\frac{1}{2}x + 3$ (Finding the Equation of a Line, p. 283)

13. $y = -3$ (Finding the Equation of a Line, p. 283)

14.

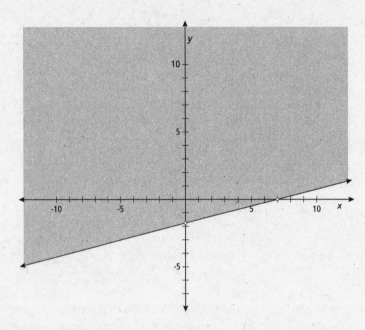

(Graphing Inequalities, p. 286)

15.

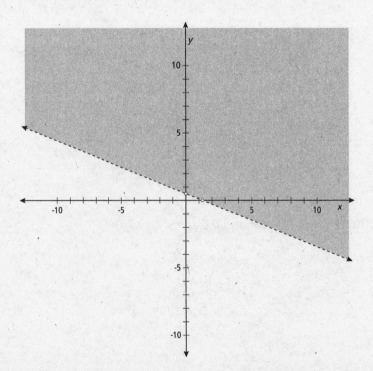

(Graphing Inequalities, p. 286)

16. (5,−1)   (Solving Systems of Equations by Graphing, p. 290)

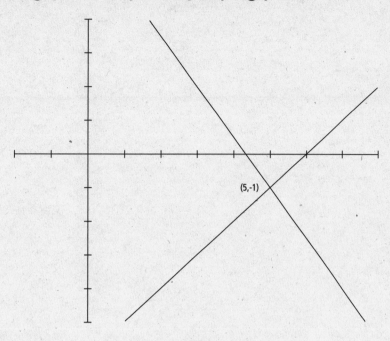

17. (1,3)   (Solving Systems of Equations by Graphing, p. 290)

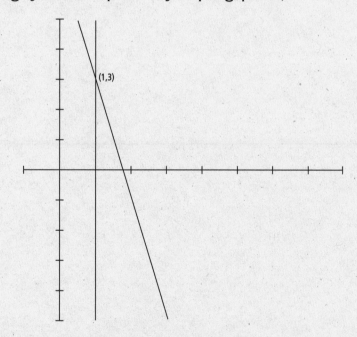

18. (3,−5)   (Solving Systems of Equations Using Addition, p. 295)

19. (−1,−3)   (Solving Systems of Equations Using Addition, p. 295)

20. No common answer   (Solving Systems of Equations Using Addition, p. 295)

Chapter 8
Functions

Functions and Relations

Y ou can express the relationship between two variables in many ways. The usual ways are pictures (graphs), tables of numbers, and formulas with equations or inequalities. A **function** is a particular type of relationship that is considered most frequently in algebra. What sets a function aside from all other relations between two variables is that one of the variables is said to be a function of the other if the operations involved in the equation always produce a **unique** result.

The equation $y = x^2 + 3$ represents a function where y is a function of x. The equation $y^2 = x^2 + 16$ is **not** a function, even though it represents a relationship between the variables y and x. What distinguishes a function from just any relationship? Look at the charts of values that follow. They represent how the variables are related. They also can represent coordinates on the graphs of the equations.

$$y = x^2 + 3 \qquad\qquad\qquad y^2 = x^2 + 16$$

x	y
−3	12
−2	7
−1	4
0	3
1	4
2	7
3	12

x	y
−3	5
−3	−5
−1	$\sqrt{17}$
−1	$-\sqrt{17}$
0	4
0	−4
3	5
3	−5

The first equation is a function, because every x value results in just one answer when that value is put into the equation. The second equation is not a function, because two answers exist for every x value shown. A function has to have one unique output for every input.

When y is a function of x, it's driven by x. The x variables are the input variables, and the y variables are the output variables. When y is a function of x, **exactly one** output variable exists for every input variable.

The main interest in whether or not an equation represents a function is its usefulness as a formula or model in a practical situation. If an equation or formula is supposed to tell a store owner how much it costs to stock x number of an item, then she wants the formula to have one answer for any x that's input. It isn't very helpful if the formula gives two different answers.

Example Problems

These problems show the answers and solutions.

1.   Is $y = \pm \sqrt{x^2 + 15}$ a function or not?

     answer: No, it isn't a function.

     The $\pm$ symbol in front of the radical is the biggest clue that it isn't a function. It says, automatically, that there are two answers. If you let $x = 7$, for example, then $y = \pm \sqrt{7^2 + 15} = \pm \sqrt{49 + 15} = \pm \sqrt{64} = \pm 8$. Two answers exist for the input value of 7.

2.   Is $|x + 1| = |y|$ a function or not?

     answer: No, it isn't a function.

     Making a table of values will show this.

x	y
−2	−1
−2	1
0	−1
0	1
1	2
1	−2

3.   Is $y = \dfrac{x^2}{x^2 - 3}$ a function or not?

     answer: Yes, it's a function.

     A table of values doesn't really prove that it's a function, but it demonstrates how there's only one y value for every x value shown.

x	y
−3	$\frac{3}{2}$
−2	4
−1	$-\frac{1}{2}$
0	0
1	$-\frac{1}{2}$
2	4
3	$\frac{3}{2}$

Work Problems

Use these problems to give yourself additional practice.

Determine whether or not y is a function of x by making a chart that finds several sets of coordinates.

1. $y = \sqrt{x^2 - 1}$

2. $x = |y + 2|$

3. $x^2 + y^2 = 25$

4. $y = |x^3 + 2|$

5. $x + y^2 = 40$

Worked Solutions

1. **Yes, it's a function.** There's no sign or double sign like ± in front of the radical, so only the positive value is considered here. There's only one y for every x. There's no y value when x is 0, because that would put a negative number under the radical. This situation is discussed more in the next section, "Domain and Range."

$$y = \sqrt{x^2 - 1}$$

x	−3	−2	−1	0	1	2	3
y	$\sqrt{8}$	$\sqrt{3}$	0		0	$\sqrt{3}$	$\sqrt{8}$

2. **No, it isn't a function.** Look at the table of values that follows. In one instance the x produces a unique value, but the others all have two y's.

$$x = |y + 2|$$

x	0	1	1	2	2	3	3
y	−2	−1	−3	0	−4	1	−5

3. **No, it isn't a function.** Look at the table of values that follows. In one instance the x produces a unique value, but the others all have two y's.

$$x^2 + y^2 = 25$$

x	−4	−4	−3	−3	−1	−1	0	0	3	3	5
y	−3	3	−4	4	$-\sqrt{24}$	$\sqrt{24}$	−5	5	−4	4	0

4. **Yes, it's a function.**

$$y = \left| x^3 + 2 \right|$$

x	−3	−2	−1	0	1	2	3
y	25	6	1	2	3	10	29

5. **No, it isn't a function.** Look at the table of values that follows. Lots of values exist between those on the table. (I just stayed away from the radicals and made the numbers come out nicely for this demonstration.)

$$x + y^2 = 40$$

x	−9	−9	4	4	15	15	39	39
y	−7	7	−6	6	−5	5	−1	1

Domain and Range

When you study functions, for instance those in which y is a function of x, some properties and characteristics are important when you are choosing and using them. The **domain** and **range** of a function are such properties.

Domain

The domain of a function contains all of the possible **input** values that you can use—every number that can be put into the formula or equation and get a real answer. The function $y = \sqrt{x - 2}$ has a domain that contains the number 2 and every number greater than 2, written $x \geq 2$. The domain can't include anything smaller than 2, because then a negative number would be under the radical, and that kind of number isn't real.

Range

The range of a function contains all of the possible **output** values—every number that is a result of putting input values into the formula or equation. The function $y = \sqrt{x - 2}$ has a range that contains the number 0 and also contains every positive number. The range doesn't include any negatives, because this radical operation results in only positive answers or 0.

When determining the domain and range of functions, just a few operations cause restrictions or special attention. Functions with radicals that have even roots will have restricted domains. You can't take the square root or fourth root of a negative number, so any x value that would create that situation has to be eliminated from the domain. Fractions also have to be considered carefully. Any x value that creates a 0 in the denominator has to be eliminated from the domain. Functions with even-powered radicals or absolute value will have restricted ranges. They'll produce just positive results. Other "special" cases will have to be determined by trying a few coordinates or by putting x values into the function equation.

Example Problems

These problems show the answers and solutions.

1. Find the domain and range of the function $y = \sqrt{8-x}$.

    answer: The domain is $x \leq 8$; the range is $y \geq 0$.

    Whenever there's an even-powered radical in the function equation, you have to determine what values of x will create a negative number under the radical—and then eliminate them from the domain. This radical is positive and will produce only positive results and 0 for the range.

2. Find the domain and range of the function $y = \frac{1}{x}$.

    answer: The domain is all real numbers except for 0, $x \neq 0$; the range is all real numbers except for 0, $y \neq 0$.

    Whenever you have a fraction, the restriction is that the denominator cannot be equal to 0. No result is possible where you divide by 0. Any other number in the denominator will work. That's why the domain contains everything except 0. As far as the range is concerned, the only result that's not possible is also 0. The only way to get a 0 when you have a fraction is to have a 0 in the numerator, but the numerator is always equal to 1.

3. Find the domain and range of the function $3x + 2y = 19$.

    answer: The domain is all real numbers, $(-\infty, \infty)$; the range is all real numbers, $(-\infty, \infty)$.

    This is the equation of a straight line. Written in slope-intercept form, the equation is $y = -\frac{3}{2}x + \frac{19}{2}$. The y-intercept is $\frac{19}{2}$, and the slope is negative, making the line go down from left to right. The graph of the function goes on forever, from negative infinity to positive infinity in the domain and from positive infinity to negative infinity in the range.

4. Find the domain and range of the function $y = x^2 + 5$.

    answer: The domain is all real numbers, $(-\infty, \infty)$; the range is all numbers 5 and greater, $y \geq 5$.

    No restrictions are on the domain. Squaring any real number results in another real number. The range is 5 and above, because all of the results of squaring numbers in the domain are either positive or, the smallest value, 0. Add a 5 onto 0 or a positive number, and the result is 5 or greater. The range doesn't go below the number 5.

5. Find the domain and range of the function $y = -\sqrt{x^2 + 1}$.

    answer: The domain is all real numbers, $(-\infty, \infty)$; the range is all numbers -1 and lower, $y \leq -1$.

    Even though there's a radical here, the operation under the radical takes all positive and negative numbers, squares them, which makes them positive, and then adds 1. There'll never be a negative under the radical, so any real number can be used in the domain. The range consists of the **opposite** of the values that are obtained by squaring the x value, adding 1, and then taking the square root of that sum. The smallest number that will ever be under the radical is a 1, when x is 0. The negative sign in front of the radical makes all results negative.

Work Problems

Use these problems to give yourself additional practice.

Find the domain and range of each of the following.

1. $y = -\sqrt{x-5}$

2. $x + y = 7$

3. $y = x^2 - 4$

4. $y = \sqrt{x^2 - 9}$

5. $y = \dfrac{1}{x+1}$

Worked Solutions

1. **Domain is $x \geq 5$; range is $y \leq 0$.**   A negative value cannot be under the radical, so only numbers 5 and larger can be in the domain. The radical itself produces results that are either positive or 0. The opposite of that is negative or 0.

2. **Domain is all real numbers, $(-\infty, \infty)$; range is all real numbers, $(-\infty, \infty)$.**   This is the equation of a line with a slope of -1 and a y-intercept of 7. It goes from negative infinity to positive infinity.

3. **Domain is all real numbers, $(-\infty, \infty)$; range is all numbers greater than or equal to -4, $y \geq -4$.**   Squaring real numbers results in real numbers, so no numbers need to be excluded from the domain. The operation in the function squares all values in the domain, which results in positive numbers and 0, at the least. Subtracting 4 from any positive result or 0 gives a low of -4 and then all numbers higher than that.

4. **Domain is $x \leq -3$ or $x \geq 3$; range is $y \geq 0$.**   A negative number cannot be under the radical. Any number between -3 and 3, when squared, results in a number smaller than 9, so they can't be used in the domain—the result under the radical would be negative. The radical is positive, so all of the values in the domain will give a positive or 0 result when put into the formula.

5. **Domain is all real numbers except -1, $x \neq -1$; range is all real numbers except 0, $y \neq 0$.**   The number -1 would make the denominator of the function equal to 0. You can't have a 0 in the denominator. Any other number works. Also, because the numerator is always equal to 1, it can never be equal to 0, and so 0 can't be in the range.

Graphs of Functions—Vertical Line Test

When you use a graph to illustrate the equation of a function, one very important characteristic of the graph distinguishes the function from other relationships and their graphs. The graph of a function may go far to the left, right, or both, or it may just be in a small area, depending on the domain, but it never doubles back on itself or occupies more than one point with a given x value. The definition of a function is that one output value exists for any input value, so there can't be more than one y for any given x. Making a chart to determine whether a particular equation represents a function is helpful, but performing the **vertical line test** on the graph of an equation is very dramatic and definitive. The vertical line test consists of drawing as many vertical lines through the graph of the function as necessary to satisfy yourself that there's no place on the graph where more than one y value exists for a given x value.

Example Problems

These problems show the answers and solutions.

1. Perform the vertical line test on the following graph, the graph of $x^2 + y^2 = 25$, to determine whether or not it represents a function of x.

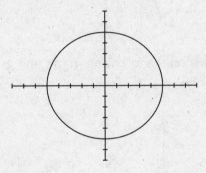

    answer: This is **not** a function of x.

    This graph failed the vertical line test, because vertical lines cross through the graph of the curve more than once.

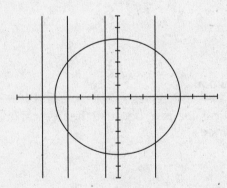

2. Perform the vertical line test on the following graph, the graph of $y = x^2 - 2$, to determine whether or not it's a function.

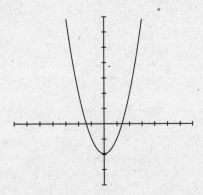

    answer: This **is** a function.

This graph passed the vertical line test, because the vertical lines cross through the graph only once.

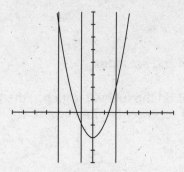

3. Perform the vertical line test on the following graph, the graph of $x = |y|$, to determine whether or not it's a function.

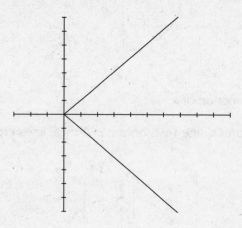

answer: This is **not** a function.

This graph failed the vertical line test, because the vertical lines cross through the graph of the function more than once.

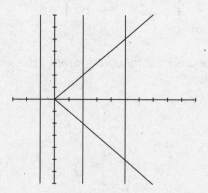

4. Perform the vertical line test on the following graph, the graph of $y = \sqrt{x^2 - 1}$, to determine whether or not it's a function.

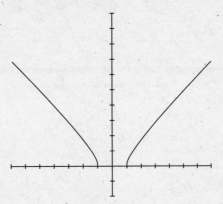

answer: This **is** a function.

This graph passed the vertical line test, because the vertical lines cross through the graph of the function only once.

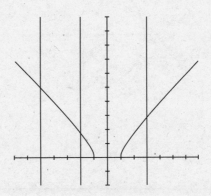

Work Problems

Use these problems to give yourself additional practice.

In each case, determine whether the graph of the relation passes the vertical line test or not.

1. The graph of $x = y^2 - 9$

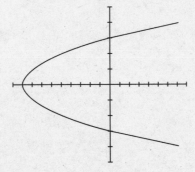

2. The graph of $y = 5x - 3$

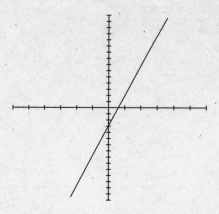

3. The graph of $x^2 + y^2 = 16$

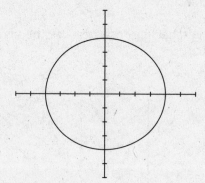

4. The graph of $y = \sqrt{5 - x}$

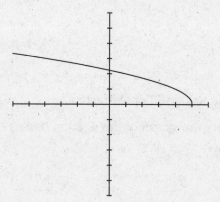

5. The graph of $y = |x + 1|$

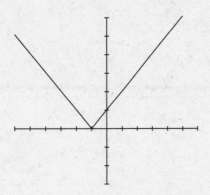

Worked Solutions

1. **This is not a function.**

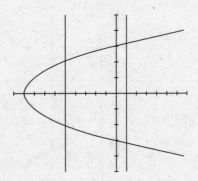

This graph failed the vertical line test.

2. **This is a function.**

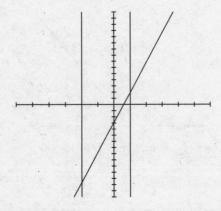

This graph passed the vertical line test.

3. This is not a function.

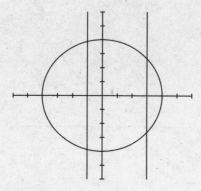

This graph failed the vertical line test.

4. This is a function.

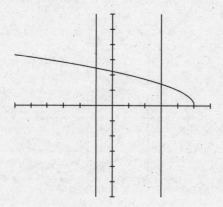

This graph passed the vertical line test.

5. This is a function.

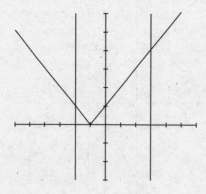

This graph passed the vertical line test.

Inverse Functions and Function Notation

A function is a relationship between two variables such that there's exactly one output value for every input value—only one value in the range for every value in the domain. Some special functions also have exactly one input value for every output value. For instance, the function $y = 3x + 2$ has this property. Look at a chart of some of the coordinates that satisfy the equation.

x	−3	−2	−1	0	1	2	3
y	−7	−4	−1	2	5	8	11

None of the y values repeat. Now look at the function $y = 3x^2 + 2$. This function does **not** have the property in which there's only one input value for every output value. The output value 14 has two input values. The same is true of the output values 5, 29, and many more.

x	−3	−2	−1	0	1	2	3
y	29	14	5	2	5	14	29

Functions that have the property in which there's only one x value for every y value, as well as just one y value for every x value, which is what makes it a function, are called **one-to-one** functions. What's special about one-to-one functions is that they have **inverses**.

See Chapter 1 for a discussion on the inverses of numbers. A number's additive inverse is the number that, when added to it, results in 0. The numbers 3 and −3 are additive inverses. A number's multiplicative inverse is the number that, when multiplied by it, gives you 1. The numbers $\frac{2}{3}$ and $\frac{3}{2}$ are multiplicative inverses. Some functions have inverses, too. An inverse of a function undoes what the function did. If you perform a function on a number and then perform the inverse function on the result, you're back to the original number. Using the inverse function on an output variable tells you what the input variable was.

First, look at some new notation for functions. The notation used so far has been in terms of x's and y's, and, usually, y is a function of x. The functions look like $y = x + 8$ or $y = \sqrt{4x - 1}$, and so on. A common notation, called **function notation**, would write these last two functions as $f(x) = x + 8$ or $g(x) = \sqrt{4x - 1}$. This is read, "f of x equals x plus 8." or "g of x equals the square root of $4x$ minus 1." This function notation allows you to name functions with the letters f, g, h, and so on. It also is handy when you want to write down coordinates of points or numbers from a chart. If you're considering the function $f(x) = x + 8$ and want to know what the output is if $x = 2$, write $f(2) = 2 + 8 = 10$. The input is 2; the output is 10; the coordinates of the point are (2, 10).

Only one-to-one functions can have inverses. An example of a function and its inverse is $f(x) = x^3 - 7$ and $f^{-1}(x) = \sqrt[3]{x + 7}$. The notation $f^{-1}(x)$ doesn't mean to write the reciprocal of the function f; it's the standard notation for a function's inverse. Because an inverse of a function takes the function's output and tells you what the original input was, consider the output 1 for the function $f(x) = x^3 - 7$. What value of x resulted in that output of 1? Put the 1 into the inverse function, $f^{-1}(1) = \sqrt[3]{1 + 7} = \sqrt[3]{8} = 2$. This tells you that the original input was a 2. Try it here, $f(2) = 2^3 - 7 = 8 - 7 = 1$. It worked!

Example Problems

These problems show the answers and solutions.

1. Test to see whether $f^{-1}(x) = \frac{x+1}{2}$ is the inverse of $f(x) = 2x - 1$.

   answer: Yes, this is the inverse.

   To test the inverse, choose a number, such as $x = 7$. Put it into the function equation, put the result into the inverse equation, and see whether the original number, the 7, is the result. $f(7) = 2(7) - 1 = 14 - 1 = 13$. Now put the 13 into the inverse, $f^{-1}(13) = \frac{13+1}{2} = \frac{14}{2} = 7$. This isn't a formal proof that this is the inverse—it's more of a demonstration, but it works for our purposes here.

2. Test to see whether $g^{-1}(x) = \sqrt{x-1}$ is the inverse of $g(x) = \sqrt{x+1}$.

   answer: No, this is **not** the inverse.

   To test the inverse, choose a number, such as $x = 8$. Put it into the function equation, put the result into the inverse equation, and see whether the original number, the 8, is the result: $g(8) = \sqrt{8+1} = \sqrt{9} = 3$. Now put the 3 into the inverse, $g^{-1}(3) = \sqrt{3-1} = \sqrt{2}$. This is **not** the inverse function.

3. Test to see whether $h^{-1}(x) = \frac{3x}{1-x}$ is the inverse of $h(x) = \frac{x}{x+3}$.

   answer: Yes, this is the inverse.

   Choose a number, such as $x = -4$. Put it into the function equation,

   $h(-4) = \frac{-4}{-4+3} = \frac{-4}{-1} = 4$. Now, put this result into the inverse function equation,

   $h^{-1}(4) = \frac{3(4)}{1-4} = \frac{12}{-3} = -4$. This is the inverse function.

Work Problems

Use these problems to give yourself additional practice.

Determine, in each case, if the two functions are inverses of one another.

1. $f(x) = \frac{1-x}{4}$ and $f^{-1}(x) = 1 - 4x$

2. $g(x) = 5x - 3$ and $g^{-1}(x) = 3x + 5$

3. $h(x) = \frac{1+4x}{x}$ and $h^{-1}(x) = \frac{1}{x-4}$

4. $k(x) = \sqrt{x+2}$ and $k^{-1}(x) = \sqrt{2-x}$

5. $p(x) = \frac{4}{x}$ and $p^{-1}(x) = \frac{4}{x}$

Worked Solutions

1. **Yes, they are inverses.**   Choose $x = -3$. (This isn't the only possible choice. It's just convenient not to have the result be a fraction.) $f(-3) = \dfrac{1-(-3)}{4} = \dfrac{4}{4} = 1$. Now use that result in $f^{-1}(x)$: $f^{-1}(1) = 1 - 4(1) = 1 - 4 = -3$. This is the value with which you started. This is evidence that the functions are inverses, not a formal proof.

2. **No, they are not inverses.**   Choose $x = 2$: $g(2) = 5(2) - 3 = 10 - 3 = 7$. Now use that result in $g^{-1}(x)$: $g^{-1}(7) = 3(7) + 5 = 21 + 5 = 26$. This is not the same value with which you started.

3. **Yes, they are inverses.**   Choose $x = 2$: $h(2) = \dfrac{1+4(2)}{2} = \dfrac{9}{2}$. Now use this result in $h^{-1}(x)$:
    $h^{-1}\left(\dfrac{9}{2}\right) = \dfrac{1}{\dfrac{9}{2}-4} = \dfrac{1}{\dfrac{1}{2}} = 2$. It's the value with which you started. This is evidence that they are inverses, not a formal proof.

4. **No, they are not inverses.**   Choose $x = 7$: $k(7) = \sqrt{7+2} = \sqrt{9} = 3$. Now use this result in $k^{-1}(x)$: $k^{-1}(3) = \sqrt{2-3} = \sqrt{-5}$. This is not the number with which you started. There is a negative number under the radical, which means that it is not even a real number.

5. **Yes, they are inverses.**   This function is its own inverse. Choose $x = 4$: $p(4) = \dfrac{4}{4} = 1$. Now use this result in $p^{-1}(x)$: $p^{-1}(1) = \dfrac{4}{1} = 4$. It's the same thing with which you started. This is evidence that they are inverses, not a formal proof.

Chapter Problems and Solutions

Problems

In problems 1 through 5, determine whether or not the relation y is a function of x.

1. $y = \sqrt[3]{x^2 + 1}$

2. $x = \sqrt{y^2 + 7}$

3. $x + |y| = 4$

4. $y + |x| = 3$

5. $x^2 + y^2 = 100$

For problems 6 through 10, find the domain of the function.

6. $y = 3x + 4$

7. $y = \sqrt{x+5}$

8. $y = x^2 + 3$

9. $y = |x + 1|$

10. $y = \dfrac{3}{x + 1}$

In problems 11 through 13, find the range of the function.

11. $y = \sqrt{x + 5}$

12. $y = x^2 + 3$

13. $y = |x - 2|$

For problems 14 and 15, use the vertical line test to determine whether the function is one-to-one.

14. The graph of $y = \sqrt{8 - x^2}$

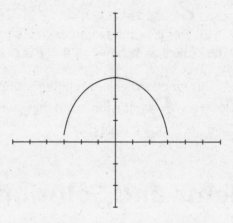

15. The graph of $x = 4 - y^2$

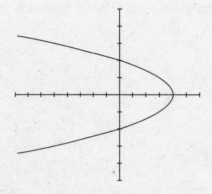

In problems 16 through 20, determine whether or not $f^{-1}(x)$ is an inverse of $f(x)$.

16. $f(x) = 3x + 2$, $f^{-1}(x) = \dfrac{x - 2}{3}$

17. $f(x) = \sqrt[3]{x + 1}$, $f^{-1}(x) = x^3 - 1$

18. $f(x) = \dfrac{1}{x-2}$, $f^{-1}(x) = \dfrac{1+2x}{x}$

19. $f(x) = \sqrt{x^2 - 1}$, $f^{-1}(x) = x + 1$

20. $f(x) = \dfrac{6x-3}{5}$, $f^{-1}(x) = \dfrac{5x+3}{6}$

Solutions

1. **Answer: Yes, it's a function.**   In a table of values, there's only one y value for every x value.

2. **Answer: No, it isn't a function.**   In a table of values, there is more than one y value for many x values.

x	4	4	$\sqrt{11}$	$\sqrt{11}$
y	3	−3	2	−2

3. **Answer: No, it isn't a function.**   In a table of values, there is more than one y value for many x values.

x	1	1	2	2
y	3	−3	2	−2

4. **Answer: Yes, it's a function.**   In a table of values, there's only one y value for every x value.

5. **Answer: No, it isn't a function.**   In a table of values, there is more than one y value for many x values.

x	6	6	8	8
y	8	−8	6	−6

6. **Answer: All real numbers, $(-\infty, \infty)$.**   This is a line with a slope of 3. It goes from negative infinity to positive infinity.

7. **Answer: All numbers −5 and larger, $x \geq -5$.**   A negative value cannot be under the radical; all numbers smaller than −5 would create a negative under the radical and must be excluded.

8. **Answer: All real numbers, $(-\infty, \infty)$.**   This is a parabola. The x values can go from negative infinity to positive infinity.

9. **Answer: All real numbers, $(-\infty, \infty)$.**   The x values can go from negative infinity to positive infinity.

10. **Answer: All numbers except −1, $x \neq -1$.** The denominator of a fraction can't equal 0, so −1 has to be excluded.

11. **Answer: All positive numbers and 0, $y \geq 0$.** The result of taking the square root of the numbers under the radical always will be positive or 0.

12. **Answer: All numbers 3 and greater, $y \geq 3$.** The square of any real number is either positive or 0, and when that's added onto 3, the result can't be less than 3.

13. **Answer: All positive numbers and 0, $y \geq 0$.** The result of taking the absolute value of numbers will always be positive or 0.

14. **Answer: It passes the vertical line test and is one-to-one.**

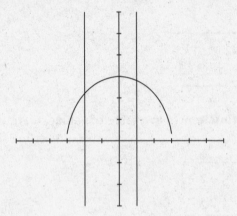

15. **Answer: It fails the vertical line test and is not one-to-one.**

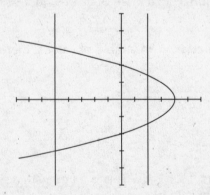

16. **Answer: It is the inverse.** **Evidence:** Use $x = 4$, then $f(4) = 3(4) = 3(4) + 2 = 14$ and $f^{-1}(14) = \dfrac{14 - 2}{3} = \dfrac{12}{3} = 4$.

17. **Answer: It is the inverse.** **Evidence:** Use $x = 7$, then $f(7) = \sqrt[3]{7 + 1} = \sqrt[3]{8} = 2$ and $f^{-1}(2) = 2^3 - 1 = 8 - 1 = 7$.

18. **Answer: It is the inverse.** **Evidence:** Use $x = 3$, then $f(3) = \dfrac{1}{3 - 2} = \dfrac{1}{1} = 1$ and $f^{-1}(1) = \dfrac{1 + 2(1)}{1} = \dfrac{3}{1} = 3$.

19. **Answer: It is <u>not</u> the inverse.**   **Proof:** Use $x = 4$, then $f(4) = \sqrt{4^2 - 1} = \sqrt{16 - 1} = \sqrt{15}$ and $f^{-1}(\sqrt{15}) = \sqrt{15} + 1$. This is not the value that you started with, $x = 4$.

20. **Answer: It is the inverse.**   **Evidence:** Use $x = 4$, then $f(4) = \frac{6(4) - 3}{5} = \frac{21}{5}$ and $f^{-1}\left(\frac{21}{5}\right) = \frac{5\left(\frac{21}{5}\right) + 3}{6} = \frac{21 + 3}{6} = 4.$

Supplemental Chapter Problems

Problems

In problems 1 through 5, determine whether or not the relation is a function y in terms of x.

1. $y = \sqrt{x^3 + 3}$

2. $y = \left|x^2 - 2\right|$

3. $x + y^2 = 11$

4. $x^2 - y = 2$

5. $x^2 - y^2 = 5$

For problems 6 through 10, find the domain of the function.

6. $y = -\sqrt{x - 7}$

7. $y = \sqrt[3]{x + 13}$

8. $y = 6$

9. $y = \frac{1}{x + 6}$

10. $y = x$

In problems 11 through 15, find the range of the function.

11. $y = -\sqrt{x - 7}$

12. $y = \frac{1}{x + 6}$

13. $y = 6$

14. $y = 4x + 2$

15. $y = 4 - x^2$

For problems 16 through 20, determine whether or not $f^{-1}(x)$ is an inverse of $f(x)$.

16. $f(x) = \sqrt{5-x}$, $f^{-1}(x) = \sqrt{5+x}$

17. $f(x) = \dfrac{3+x}{x}$, $f^{-1}(x) = \dfrac{3}{x-1}$

18. $f(x) = 4x-3$, $f^{-1}(x) = \dfrac{x+4}{3}$

19. $f(x) = \dfrac{2-x}{x}$, $f^{-1}(x) = \dfrac{x+1}{2x}$

20. $f(x) = 9x^2$, $f^{-1}(x) = \dfrac{1}{9}\sqrt{x}$

Answers

1. Yes, it is a function of x.   (Functions and Relations, p. 313)

2. Yes, it is a function of x.   (Functions and Relations, p. 313)

3. No, it isn't a function of x.   (Functions and Relations, p. 313)

4. Yes, it is a function of x.   (Functions and Relations, p. 313)

5. No, it isn't a function of x.   (Functions and Relations, p. 313)

6. All numbers 7 and greater, $x \geq 7$.   (Domain and Range, p. 316)

7. All real numbers, $(-\infty, \infty)$.   (Domain and Range, p. 316)

8. All real numbers, $(-\infty, \infty)$.   (This is a horizontal line.)   (Domain and Range, p. 316)

9. All real numbers except for -6, $x \neq -6$.   (Domain and Range, p. 316)

10. All real numbers, $(-\infty, \infty)$.   (Domain and Range, p. 316)

11. All negative numbers and 0, $y \leq 0$.   (Domain and Range, p. 316)

12. All numbers except 0, $y \neq 0$.   (Domain and Range, p. 316)

13. Just the number 6, $y = 6$.   (This is a horizontal line.)   (Domain and Range, p. 316)

14. All real numbers, $(-\infty, \infty)$.   (Domain and Range, p. 316)

15. All numbers 4 and lower, $y \leq 4$.   (Domain and Range, p. 316)

16. No, it isn't the inverse.   (Inverse Functions and Function Notation, p. 325)

17. Yes, it is the inverse.   (Inverse Functions and Function Notation, p. 325)

18. No, it isn't the inverse.   (Inverse Functions and Function Notation, p. 325)

19. No, it isn't the inverse.   (Inverse Functions and Function Notation, p. 325)

20. No, it isn't the inverse.   (Inverse Functions and Function Notation, p. 325)

Chapter 9
Story Problems

Story problems or "practical applications" are one of the best uses for the techniques and procedures learned and used in algebra. Not many places of business will require you to solve a quadratic equation as a daily chore. But problem solving—determining solutions to practical, everyday challenges—is something that everyone faces, no matter what the occupation. Granted, you probably won't haul out the quadratic formula to solve many problems, but the mental exercises and organization practiced in algebra will serve you well in future situations.

There are some basic steps to follow that should help you when faced with a story problem.

1. Determine what is to be solved for; assign that as the variable. (Remember that the variable represents a number, not a person or thing.)

2. Draw a picture, if called for. Make a chart, if helpful.

3. Write an equation representing the relationships between the different parts of the problem.

4. Solve the equation for the variable.

5. Answer the question completely. Often, the value of the variable is just one part of the answer.

6. Check your answer to be sure it makes sense.

The sections of this chapter are organized according to the types of story problems. There are number problems, consecutive integer problems, age problems, distance problems, and so on. Most story problems found in algebra courses are just variations of these basic types.

Number Problems

Number problems are so designated because there isn't usually any practical problem involved—just vocabulary and operations and relationships between numbers. They're often called brain teasers. This is a good place to start in algebraic story problems because you can practice writing a statement using symbols and then solve it. The statements written for these problems pretty much follow, word for word, the verbal version—with the verb being replaced by the equal sign. Remember that "sum" is the result of addition, "difference" is the result of subtraction, "product" is the result of multiplication, and "quotient" is the result of division.

Example Problems

These problems show the answers and solutions.

1. If I double a number, the result is the same as subtracting 4 from that original number. What is the number?

answer: The number is −4.

First choose a variable to represent the number. Let x represent the number. Double means to multiply by 2, so that will be $2x$. The words "is the same as" will be replaced by "=." The sentence, written in "algebra" reads $2x = x - 4$. Now, solving for x, $x = -4$. Does this check? Doubling −4 gives you −8, and subtracting 4 from −4 is $-4 - 4 = -8$.

2. If a number is added to half that number, then the result is equal to 8 less than twice the number. What is the number?

   answer: The number is 16.

   Choose a variable to represent the number. In this case, let n be the number. Half the number can be written as either $\frac{1}{2}n$ or $\frac{n}{2}$. The operation indicated by "less than" is subtraction. The statement, written in algebraic language, reads: $n + \frac{n}{2} = 2n - 8$. Notice that even though the wording "8 less than" puts the 8 first, the subtraction has the 8 follow the subtraction operation. To solve the equation, first multiply every term by 2 to get rid of the fraction. $2n + n = 4n - 16$. Combining the terms on the left, you get $3n = 4n - 16$. Subtracting $4n$ from each side and then multiplying each side by −1, the answer is $n = 16$. To check this, 16 added to half of 16 (which is 8) is 24. Is this the same as 8 less than twice 16? Twice 16 is 32, and 8 less than that is 24.

Work Problems

Use these problems to give yourself additional practice.

1. Add 3 to half a number, and it's the same as subtracting 1 from it. What is the number?

2. Three times the sum of 2 and a number is 4 less than 5 times the number. What is the number?

3. Divide the difference of a number and 4 by 3, and it's 6 less than the number. What is the number?

4. Twice a number less 1 is equal to three times the sum of 2 and that number. What is the number?

5. Karen said, "I'm thinking of a number that, when you multiply it by 9, it's the same as adding 8 to 7 times the number. What is it?"

Worked Solutions

1. **The number is 8.**   Choose x to be the number. The equation is written $\frac{x}{2} + 3 = x - 1$. Multiply all the terms by 2 to get $x + 6 = 2x - 2$. Subtract x from each side and add 2 to each side to get $x = 8$. Checking the answer, half of 8 is 4, and adding 3 to that gives you 7. That's the same as subtracting 1 from 8.

2. **The number is 5.**   Choose y to be the number. Multiplying 3 times the sum of 2 and a number means to put the 2 and the number in a parenthesis and distribute. Writing the equation, $3(y + 2) = 5y - 4$. First, distributing the 3, the equation becomes $3y + 6 = 5y - 4$. Subtracting $5y$ from each side and subtracting 6 from each side results in $-2y = -10$. Dividing each side by -2, the answer is $y = 5$. Checking this, the sum of 2 and 5 is 7, and 3 times that is 21. Is it the same as multiplying 5 by 5 and subtracting 4? Yes, it is.

3. **The number is 7.**   Choose z to be the number. Dividing the difference of the number and 4 by 3 means to divide both terms—do the subtraction first. The equation is written $\frac{z - 4}{3} = z - 6$. Multiply each side of the equation by 3 to get $z - 4 = 3z - 18$. Subtract z from each side and add 18 to each side to get $14 = 2z$. Dividing by 2, $z = 7$. Does this work? The difference between 7 and 4 is 3. Dividing that by 3 gives you 1. That's the same as 6 less than 7.

4. **The number is −7.**   Choose w to be the number. The equation is written $2w - 1 = 3(w + 2)$. First distribute the 3 on the right to get $2w - 1 = 3w + 6$. Subtract $2w$ from each side and subtract 6 from each side to get $-7 = w$. Checking that answer, twice -7 is -14. And, -14 less 1 is -15. The sum of 2 and -7 is -5, and 3 times that is also -15.

5. **The number is 4.**   Choose n to be the number. The equation to use is $9n = 8 + 7n$. Subtract $7n$ from each side to get $2n = 8$. Dividing each side by 2 gives you $n = 4$. Checking this, 9 times 4 is 36, and 8 added to 7 times 4 (or 28) is also 36.

Consecutive Integer Problems

Consecutive integers are lists or sequences of numbers with a common difference of 1. Some examples are

3, 4, 5, 6, ...
−7, −6, −5, −4, ...
100, 101, 102, ...

Then, there are consecutive *even* integers or consecutive *odd* integers:

6, 8, 10, 12, ...
−12, −10, −8, −6, ...
1297, 1299, 1301, 1303, ...
13, 15, 17, 19, ...

And, lastly, there are consecutive multiples of 3 or multiples of 4 or multiples of 11 or other multiples:

12, 15, 18, 21, ...
−8, −4, 0, 4, ...
0, 11, 22, 33, ...

The key to finding a particular list of consecutive integers in a story problem is to find just one of the integers—usually the first or smallest—and its position in the list. The rest are then easy to find. Consecutive integers are written as $n, n + 1, n + 2, n + 3, ...$ for as many as are needed. Consecutive even and odd integers are written as $n, n + 2, n + 4, n + 6, ...$ Consecutive multiples are written with n as the first and the amount of the multiple added on to each subsequent term. For instance, multiples of 5 are written as follows: $n, n + 5, n + 10, n + 15, ...$

Example Problems

These problems show the answers and solutions.

1. The sum of three consecutive integers is 30. What are they?

    answer: The numbers are 9, 10, and 11.

    Three consecutive numbers are written n, $n + 1$, $n + 2$, so their sum is $n + n + 1 + n + 2$. This sum is equal to 30, so solve $n + n + 1 + n + 2 = 30$. Simplifying the left side of the equation, $3n + 3 = 30$. Subtract 3 from each side to get $3n = 27$. Dividing each side by 3, $n = 9$. That's the first number, and the other two are 10 and 11.

2. The sum of four consecutive multiples of 3 is 90. What are they?

    answer: The numbers are 18, 21, 24, and 27.

    The four consecutive multiples of 3 are written n, $n + 3$, $n + 6$, $n + 9$. Adding them up and setting the sum equal to 90 gives you $n + n + 3 + n + 6 + n + 9 = 90$. Simplify the left side, $4n + 18 = 90$.

    Subtract 18 from each side to get $4n = 72$. Dividing each side by 4, $n = 18$, the smallest of the 4 multiples of 3.

3. Two times the smallest of three consecutive integers is 10 more than the largest of the three. What are they?

    answer: The numbers are 12, 13, and 14.

    The three consecutive integers are n, $n + 1$, $n + 2$. Two times the smallest is $2n$. Writing 10 more than the largest, you get $10 + n + 2$. The equation, setting them equal, is $2n = 10 + n + 2$. Notice that the "is" in the original statement is replaced by the = sign, and the rest are written in the same order as the wording. Simplifying the right side of the equation, you get $2n = n + 12$. Subtract n from each side to get $n = 12$. The three consecutive integers are 12, 13, 14. Two times the smallest is 24; 10 more than the largest is 10 plus 14, or 24.

Work Problems

Use these problems to give yourself additional practice.

1. The sum of three consecutive even integers is 48. What are they?

2. The sum of four consecutive odd integers is 64. What are they?

3. The sum of five consecutive integers is −5. What are they?

4. The sum of three consecutive multiples of 5 is 90. What are they?

5. The square of the middle of three consecutive odd integers is 40 more than the largest of the three integers. What are they?

Worked Solutions

1. **The numbers are 14, 16, 18.**   The three integers are written n, $n + 2$, $n + 4$. Add them and setting the sum equal to 48, $n + n + 2 + n + 4 = 48$. Simplify the terms on the left to get $3n + 6 = 48$. Subtract 6 from each side to get $3n = 42$. Dividing each side by 3, you get $n = 14$, which is the smallest of the three integers.

2. **The numbers are 13, 15, 17, 19.**   The four integers are n, $n + 2$, $n + 4$, $n + 6$. Adding them together and setting the sum equal to 64, you get $n + n + 2 + n + 4 + n + 6 = 64$. Simplifying the right side, you get $4n + 12 = 64$. Subtract 12 from each side. $4n = 52$. Then, divide each side by 4, $n = 13$, which is the smallest of the four odd integers.

3. **The numbers are –3, –2, –1, 0, 1.**   The five integers are written n, $n + 1$, $n + 2$, $n + 3$, $n + 4$. Adding them and setting the sum equal to –5, $n + n + 1 + n + 2 + n + 3 + n + 4 = -5$. Simplifying the left side, $5n + 10 = -5$. Subtract 10 from each side. $5n = -15$. Dividing each side, by 5, $n = -3$, which is the smallest of the five integers.

4. **The numbers are 25, 30, 35.**   The three integers are written n, $n + 5$, $n + 10$. Setting their sum equal to 90, $n + n + 5 + n + 10 = 90$. Simplify the left to get $3n + 15 = 90$. Subtract 15 from each side, $3n = 75$ and $n = 25$, the smallest of the three multiples of 5.

5. **The numbers are 5, 7, 9.**   The three odd integers are written n, $n + 2$, $n + 4$. The middle one is $n + 2$, and its square is $(n + 2)^2$. The largest of the three integers is $n + 4$, and 40 more than that is $n + 4 + 40$ or $n + 44$. Set up the equation, $(n + 2)^2 = n + 44$. The left will have to be squared, and then all the terms will be moved to the left by subtracting n and 44 from each side to form a quadratic equation: $n^2 + 4n + 4 = n + 44$ and then $n^2 + 3n - 40 = 0$. Factor this equation, $(n + 8)(n - 5) = 0$. And, by the multiplication property of zero, $n = -8$ or $n = 5$. Since the problem calls for *odd* integers, ignore the –8, and use the answer $n = 5$. The number 5 is the smallest of the odd integers. The middle one is 7, and its square is 49. The number 49 is 40 bigger than the largest of the consecutive odd integers, the 9.

Age Problems

A traditional type of story problem has to do with comparative ages of people. Your father may have been 25 years old when you were born, but at different stages in your lives, his age could be described differently. He's 25 years older than you; he was 6 times your age (when you were 5 and he was 30); he was twice your age (when you were 25 and he was 50). These types of situations make for some interesting and creative story problems. The key here is to be sure to "age" all participants, when the problems calls for things like, "In 5 years, …" or "Four years ago, …". Also, remember that variables represent numbers, not people. The variable x can represent Jon's age, but it cannot represent Jon.

Example Problems

These problems show the answers and solutions.

1. Annie is 4 years older than Ben. In 7 years, Annie will be $\frac{4}{3}$ Ben's age. How old is Annie now?

    answer: Annie is 9 years old.

Let b represent Ben's age. Then Annie is $b + 4$ years old. In 7 years, Ben will be $b + 7$ years old, and Annie will be $b + 4 + 7$ years old. The equation, which will use "Annie's age in 7 years equals four-thirds Ben's age," looks like: $b + 4 + 7 = \frac{4}{3}(b + 7)$. Simplifying on the left and distributing the fraction on the right, you get $b + 11 = \frac{4}{3}b + \frac{28}{3}$. Multiply each term by 3 to get rid of the fractions: $3b + 33 = 4b + 28$. Now subtract $3b$ from each side and 28 from each side to get $b = 5$. The variable b represents Ben's age. He's 5 years old. Because Annie is 4 years older than that, she's 9. Another approach to this problem would be to let a represent Annie's age and then $a - 4$ represent Ben's age. This has the advantage of having the variable be what you're looking for (and not having to do another operation to get the answer). The disadvantage is the translation of "older than" into a negative expression. Either works, so it's really up to you.

2. Grace is twice as old as Hank, and Keith is 4 years younger than Grace. In 3 years, the sum of their ages will be 35. How old are they now?

    answer: Grace is 12; Hank is 6; Keith is 8.

    Choose the variable to be h for Hank's age. Then Grace's age is $2h$, and Keith's age is $2h - 4$. In 3 years, they'll all be 3 years older, so you add 3 years onto each age before adding them all up and setting them equal to 35. $h + 3 + 2h + 3 + 2h - 4 + 3 = 35$. Simplifying on the left, $5h + 5 = 35$. Subtract 5 from each side and then divide each side by 5 to get that $h = 6$. That's Hank's age. Grace is twice as old, so she's 12. And Keith is 4 years younger than Grace, so he's 8. In three years, they'll be 9 and 15 and 11. Sum those numbers, $9 + 15 + 11 = 35$.

Work Problems

Use these problems to give yourself additional practice.

1. Cassidy is 3 times as old as Drew. In 8 years, Cassidy will only be twice as old as Drew. How old is Cassidy right now?

2. Steven's grandfather is 4 times as old as Steven. But 16 years ago, Steven's grandfather was 4 more than 15 times Steven's age. How old is Steven now?

3. Alex is 11 years younger than Bart, and Charlie is 6 years more than twice Alex's age. Nine years ago, the sum of their ages was 30. How old are they now?

4. The twins, Mike and Ike, have a sister who is 4 years younger than they are. Six years ago, the sum of the ages of all three was twice the sister's age right now. How old are the twins now?

5. Sam is 3 years older than Tim, and Rick is 14 years less than twice Sam's age. Eight years ago, the sum of their ages was 1 more than twice Sam's age right now. How old is Sam right now?

Worked Solutions

1. **Cassidy is 24 years old.** Let d represent Drew's age. Then Cassidy's age is $3d$. In eight years, Drew will be $d + 8$ and Cassidy will be $3d + 8$. The equation will express, "Cassidy will be twice as old as Drew." So it will read, $3d + 8 = 2(d + 8)$. First distribute the 2 on the right to get, $3d + 8 = 2d + 16$. Subtract $2d$ from each side and 8 from each side to get $d = 8$. That's Drew's age now. If Cassidy is 3 times as old as Drew right now, she's 24 years old. In eight years, she'll be 32 and Drew will be 16, so she'll be twice his age then.

2. **Steven is 20 years old.** Let Steven's age be t, and his grandfather's age be $4t$. Sixteen years ago, Steven was $t - 16$ and his grandfather was $4t - 16$. At that time, Steven's grandfather was 4 more than 15 times Steven's age. Writing that as an equation, $4t - 16 = 4 + 15(t - 16)$. Note that the "was" is replaced by the = sign. Distributing the 15 on the left and simplifying, you get $4t - 16 = 4 + 15t - 240$ or $4t - 16 = 15t - 236$. Now, subtracting $4t$ from each side and adding 236 to each side, you get $220 = 11t$ or $t = 20$. So Steven is 20, his grandfather is 80, and 16 years ago Steven was 4 and his grandfather was 64.

3. **Alex is 10, Bart is 21, and Charlie is 26.** Choose a to represent Alex's age. Then Bart is $a + 11$ and Charlie is $6 + 2a$. This can also be done using b as Bart's age and then $b - 11$ as Alex's, but Charlie's age involves a distribution if you do it that way. It's not impossible—just a bit more complicated. Anyway, using a, $a + 11$ and $6 + 2a$ for the ages, 9 years ago their ages were $a - 9$, $a + 11 - 9$ and $6 + 2a - 9$ or, simplified, $a - 9$, $a + 2$, and $2a - 3$. The sum of these ages, 9 years ago, was 30: $a - 9 + a + 2 + 2a - 3 = 30$. Simplifying the left side, $4a - 10 = 30$. Adding 10 to each side gives $4a = 40$, so $a = 10$. Alex is 10 years old, Bart is 21, and Charlie is 26. Nine years ago, they were 1, 12, and 17, and the sum of those numbers is 30.

4. **The twins are 14 years old.** Let w represent the twins' ages, so their sister is $w - 4$. Six years ago, they were $w - 6$, $w - 6$, and $w - 4 - 6$. The sum of all three kids ages six years ago was $w - 6 + w - 6 + w - 10 = 3w - 22$. This sum was twice the sister's current age, so $3w - 22 = 2(w - 4)$. Distribute the 2 on the right to get $3w - 22 = 2w - 8$. Subtracting $2w$ from each side and adding 22 to each side, you get $w = 14$. The twins are 14; their sister is 10. Six years ago, the twins were 8 and their sister was 4. The sum of the three ages was 20, which is twice the sister's age right now.

5. **Sam is 21 years old.** Let m represent Sam's age. Then Tim is $m - 3$, and Rick is $2m - 14$. Eight years ago, they were $m - 8$, $m - 3 - 8$, and $2m - 14 - 8$ or, simplified, $m - 8$, $m - 11$, and $2m - 22$. The sum of those ages is $m - 8 + m - 11 + 2m - 22 = 4m - 41$. This is 1 more than twice Sam's age now, so $4m - 41 = 2m + 1$. Subtracting $2m$ from each side and adding 41 to each side, $2m = 42$ and $m = 21$. Sam is 21, Tim is 18, and Rick is 28. Eight years ago, they were 13, 10, and 20. The sum of those numbers is 43, which is 1 more than twice Sam's age of 21.

Geometric Problems

Probably some of the more practical types of story problems are those involving geometric figures. Standard room shapes are rectangular. Many home pools are circular. Some building lots are triangular. There are many opportunities for problems involving the perimeter (distance around the outside) and area (number of squares on the surface) of geometric figures. Some of the perimeter and area formulas are listed here.

Figure	Perimeter	Area	
Rectangle	$P = 2l + 2w$	$A = lw$	l: length, w: width
Square	$P = 4s$	$A = s^2$	s: side
Triangle	$P = s_1 + s_2 + s_3$	$A = \frac{1}{2}bh$	s_i: sides, b: base, h: height (drawn perpendicular to base)
Circle	$C = 2\pi r$	$A = \pi r^2$	C: circumference (perimeter), r: radius, use 3.14 for π

Example Problems

These problems show the answers and solutions.

1. The length of a rectangle is 3 more than twice its width. The perimeter is 66 feet. What are the dimensions of the rectangle?

    answer: The length is 23 feet; the width is 10 feet.

    Let w represent the width of the rectangle. Then the length is $3 + 2w$.

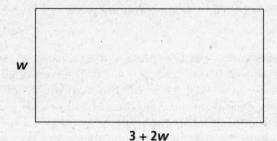

w

$3 + 2w$

    Use the perimeter formula, filling in the values given. $P = 2l + 2w$ or, in this case, $66 = 2(3 + 2w) + 2w$. Distribute the 2 on the right, $66 = 6 + 4w + 2w$. Then, simplify, $66 = 6 + 6w$. Subtract 6 from each side to get $60 = 6w$. Divide by 6, $w = 10$. The length is 3 more than twice that, or 23. The perimeter is twice the length and twice the width, so it's $2(23) + 2(10) = 46 + 20 = 66$.

2. If the sides of a square are increased by 9 inches, then the area increases by 171 square inches. What was the length of a side of the square before it was increased?

    answer: 5 inches

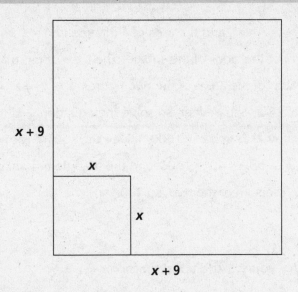

The area of a square can be found with $A = s^2$; area equals length of side squared. Let x be the length of the side of the original square. Then the square that has sides 9 inches longer has sides $x + 9$, and the area is $A = (x + 9)^2$. If this new area is 171 square inches greater than the old area, the equation is $(x + 9)^2 = x^2 + 171$. Squaring the binomial on the left (for information on squaring binomials, see page 165), you get $x^2 + 18x + 81 = x^2 + 171$. Subtract x^2 from each side and subtract 81 from each side, $18x = 90$. Divide each side by 18, $x = 5$. The original square was 5 by 5 with an area of 25 square units. The larger square is 14 by 14 with an area of 196 square units. That's 171 square units larger than the original.

3. George has a garden that's in the shape of a square. He's going to increase the size of the garden by adding a triangle onto one side. The triangle's base is a side of the square, and the height of the triangle is the same as the length of a side of the square. If the total area of his new garden is 2400 square feet, then what is the length of the side of the square?

answer: The square is 40 feet on each side.

The shape of the triangle doesn't matter. To find the area, you use the base and the height. In the figure, all of the bases are the top of the square, and all of the heights are the same. All of the triangles shown, and any others that can be drawn, will have the same area.

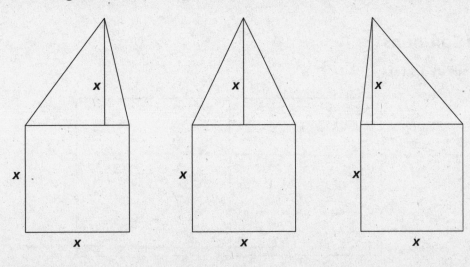

The area of a square is $A = s^2$, and the area of a triangle is $A = \frac{1}{2} bh$. In this case, let x represent the length of the sides of the square. The base of the triangle is also x, and so is the height. So the sum of the areas of the two figures is $x^2 + \frac{1}{2} x \cdot x = x^2 + \frac{1}{2} x^2 = \frac{3}{2} x^2$. This is equal to a total of 2400 square feet, so solve the equation $\frac{3}{2} x^2 = 2400$. Multiply each side of the equation by $\frac{2}{3}$ to get $x^2 = 1600$. Taking the square root of each side, $x = 40$. The square has an area of $A = 40^2 = 1600$, and the triangle has an area of $A = \frac{1}{2} 40 \cdot 40 = 800$. The sum of the two is 2400.

Work Problems

Use these problems to give yourself additional practice.

1. A square room is enlarged by adding another square to it that's the same size. If the new room has an area of 338 square feet, then what were the dimensions of the original room?

2. Sally has 600 feet of fencing to use in making a yard for her new puppy. The yard she has planned has a length that's 75 feet less than twice the width. What are the dimensions of the yard?

3. Kirsten is working on a large triangular piece of fabric that will be used as a sail. She was originally told to make it into a right triangle (a triangle with a right angle) that was twice as high as it was wide (the base). Her new directions are to increase the height and base by 5 feet each, which adds 80 square feet to the amount of material. What are these new dimensions?

4. Larry has a circular swimming pool and has a cover for it to keep the leaves out of the water. If he increases the radius of his pool by 6 feet, he'll need a new cover that has 252π square feet (about 790) more than the old cover. What is the radius of his new pool?

5. Helene's garage is 10 feet longer than it is wide. If she doubles the width and adds 10 feet onto the length, she'll have a square garage. What are the dimensions of the current garage?

Worked Solutions

1. **13 feet by 13 feet**

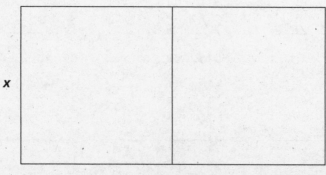

Choose x for the length of the sides of the original room. Then the area of that room was $A = x^2$. By doubling that area, the total is 338 square feet, so $2x^2 = 338$. Divide each side by 2 to get $x^2 = 169$. Take the square root of each side, $x = 13$. So the original room was 13 by 13.

2. 175 feet by 125 feet

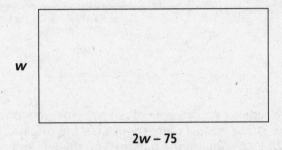

w

$2w - 75$

Choose the width of the yard to be w. Then the length is $2w - 75$. The perimeter, which is two lengths plus two widths, is 600 feet. So the equation is $2(2w - 75) + 2w = 600$. Distribute the 2 on the left to get $4w - 150 + 2w = 600$. Simplify on the left, $6w - 150 = 600$. Add 150 to each side to get $6w = 750$. Now, divide each side by 6, $w = 125$. The width is 125 feet. The length is 75 less than twice this, so it's 175 feet.

3. 23 feet high, 14 feet wide

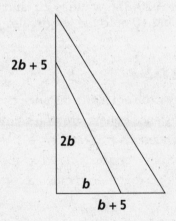

$2b + 5$

$2b$

b

$b + 5$

A right triangle has an area of $A = \frac{1}{2} bh$ in which the base and height are the two perpendicular sides. In this problem, the original height was twice the base, so the formula would read $A = \frac{1}{2} b(2b) = b^2$. When the height and base were each increased by 5 feet, the area of that new triangle became $A = \frac{1}{2}(b+5)(2b+5) = \frac{1}{2}(2b^2 + 15b + 25)$. This new area is 80 square feet more than the original, so the equation to use is $\frac{1}{2}(2b^2 + 15b + 25) = 80 + b^2$. Distribute the $\frac{1}{2}$ on the left, $b^2 + \frac{15}{2} b + \frac{25}{2} = 80 + b^2$. Now, subtract b^2 from each side and subtract $\frac{25}{2}$ from each side, $\frac{15}{2} b = \frac{135}{2}$ and $b = 9$. The original triangle was 9 feet by 18 feet. The larger triangle had the sides increased by 5 feet, so it's 14 feet by 23 feet.

4. **The radius is 24 feet.**

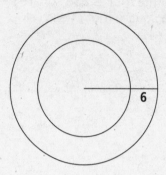

The original pool has radius r and an area of $A = \pi r^2$. The new pool has a radius of $A = \pi(r+6)^2$, and this is 252π square feet more than the original. The equation to use is $\pi(r+6)^2 = 252\pi + \pi r^2$. Square the binomial on the left and distribute the π, $\pi r^2 + 12\pi r + 36\pi = 252\pi + \pi r^2$. Subtract πr^2 from each side and subtract 36π from each side to get $12\pi r = 216\pi$. Divide each side by 12π, $r = 18$. The original pool had a radius of 18 feet. The new pool's radius is 6 feet greater, so it's 24 feet.

5. **The current garage is 30 by 20.** The current garage has a width of w. It's 10 feet longer than it is wide, so the length is $w + 10$. Doubling the width makes the new width $2w$. Adding 10 feet onto the length makes the new length $w + 20$. These new dimensions form a square, so the new width is equal to the new length. The equation is $2w = w + 20$. Subtract w from each side, $w = 20$. The width is 20, and the length is 30. Doubling the width gives you 40, and adding 10 onto the length also gives you 40.

Distance Problems

Another really practical type of story problem is the distance problem. These problems have their basis in the relationship $d = rt$, or distance equals rate times time. This equation can also be solved for either r or t to give the equations $r = \frac{d}{t}$ and $t = \frac{d}{r}$. It's a very versatile formula. Distance problems seem to be broken into two major categories: those where one distance equals another, and those in which distances are added together to equal a certain total distance.

Example Problems

These problems show the answers and solutions.

1. Cal and Carl are college roommates and are heading to their own homes for spring break. Their homes are 435 miles apart on a straight east-west road with the college in between. Cal leaves the college at 10 AM and travels west at an average speed of 50 mph. Carl leaves at 11 AM and travels east at an average speed of 60 mph. They arrive at their respective homes at exactly the same time. What time is that?

answer: 2:30 PM

This is an example of adding two distances together to get a certain total, 435 miles.

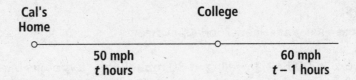

Cal's Distance + Carl's Distance = 435. Using the distance formula, and replacing the distances with rate and time, Cal's rate · Cal's time + Carl's rate · Carl's time = 435. The time is unknown, so the time that Cal traveled will be designated as t. Cal left at 10, so he traveled t hours. Carl left at 11, an hour later, so he traveled $t - 1$ hours. Now, filling in the rates and times, the equation reads $50 \cdot t + 60(t - 1) = 435$. Distribute the 60 on the left, $50t + 60t - 60 = 435$. Combine the two like terms on the left and add 60 to each side, $110t = 495$. Divide each side by 110, $t = 4.5$. Cal traveled for 4 hours and 30 minutes. Add that onto his starting time of 10 AM, and you get 2:30 PM. At 50 mph, he traveled $50(4.5) = 225$ miles. Carl traveled one hour less, so he traveled 3.5 hours, also getting him home at 2:30 PM. He traveled $60(3.5) = 210$ miles.

2. The tortoise and the hare decided to have a race. The tortoise left the starting line at 6 AM, moving at his top speed. The hare, deciding to show off a bit, took a nap on the starting line and didn't even begin to race until 10 AM. When he started, he ran 4 mph faster than the tortoise and caught up to him at 10:30 AM. How fast was each going?

    answer: The tortoise was going $\frac{1}{2}$ mph, and the hare was going $4\frac{1}{2}$ mph.

    This is an example of setting two distances equal to one another.

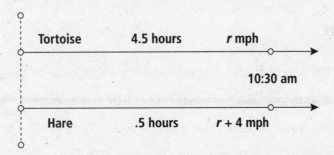

Distance traveled by tortoise = Distance traveled by hare. Replacing their distances with their respective rates and times, Rate of tortoise × time traveled by tortoise = Rate of hare × time traveled by hare. Letting the rate traveled by the tortoise be r and the time traveled be t, the rate and time for the hare can be written in relation to these. The rate of the hare, 4 mph faster than the tortoise, is written $r + 4$. Since the hare didn't leave until 4 hours after the tortoise, the time traveled is $t - 4$. The equation reads $r \cdot t = (r + 4)(t - 4)$. The time traveled by the tortoise was 4.5 hours, so the hare traveled for .5 hour. $r \cdot 4.5 = (r + 4)(4.5 - 4)$ or $4.5r = (r + 4) \cdot .5$. Distribute on the right, $4.5r = .5r + 2$. Subtract $.5r$ from each side, $4r = 2$. Divide each side by 4, $r = .5$. The tortoise is moving at a rate of .5 or $\frac{1}{2}$ mph, and the hare is moving at 4.5 or $4\frac{1}{2}$ mph.

Work Problems

Use these problems to give yourself additional practice.

1. Jake left for work at 7:00 AM, traveling at 30 mph. But he forgot his briefcase. His wife noticed the briefcase at 7:15 AM and followed after him, traveling at 45 mph. How far had he traveled when she caught up to him?

2. Dan and Stan wanted to see how far apart they could get and still talk on their car radios. Dan left the garage traveling north at 55 mph, and Stan left $\frac{1}{2}$ hour later traveling south at 60 mph. When they were 108 miles apart, the radios finally couldn't contact one another. How long after Dan left did this happen?

3. A train left the station at noon. Another train, on a parallel track, left the station at 1:30 PM and caught up to the first train at 7:00 PM. If the second train was traveling 12 mph faster than the first, then how fast was each traveling?

4. A marathon runner ran the first half hour of her race at 1 mph faster than the second half hour and the third half hour at .5 mph slower than the first half hour. The event was a 12-mile race and took her an hour and a half to complete. How fast did she run during the last part?

5. Claire left home traveling north at 45 mph. At the same time, Charlie left the same place traveling east at 60 mph. When were they 750 miles apart? (Hint: Their paths are at right angles, and the distance is the hypotenuse.)

Worked Solutions

1. **22.5 miles**   Set the two distances that Jake and his wife traveled equal to one another. Let t be the amount of time that Jake traveled and $t - \frac{1}{4}$ be the amount of time that his wife traveled. The equation will be Jake's rate × Jake's time = Wife's rate × Wife's time. Using the symbols, $30 \cdot t = 45\left(t - \frac{1}{4}\right)$. Distribute on the right, $30t = 45t - \frac{45}{4}$. Subtract $45t$ from each side to get $-15t = -\frac{45}{4}$. Now multiply each side by $-\frac{1}{15}$ to get $t = \frac{3}{4}$. Jake traveled for $\frac{3}{4}$ of an hour, or 45 minutes. His wife traveled $\frac{1}{4}$ hour less than this, or 30 minutes. If she was traveling at 45 mph, then she traveled half that or 22.5 miles in half an hour.

2. **1 hour, 12 minutes**   Add the two distances traveled together to get a sum of 108. Dan's time is t hours, and Stan's time is $t - \frac{1}{2}$ hours. Dan's rate × Dan's time + Stan's rate × Stan's time = 108. Write the equation, $55 \cdot t + 60\left(t - \frac{1}{2}\right) = 108$. Distribute the 60 to get $55t + 60t - 30 = 108$. Simplify the left and add 30 to each side. $115t = 138$. Divide each side by 115 and reduce the fraction by dividing by 23 to get $t = \frac{6}{5}$ or $1\frac{1}{5}$. The time was 1 hour plus 12 minutes.

3. **The first train was going 44 mph; the second train was going 56 mph.** Set the two distances equal to one another. First train's rate × First train's time = Second train's rate × Second train's time. The first train traveled for 7 hours before the second caught up to it. The second train traveled for 5.5 hours. The first train was going at r miles per hour. The second train was going at $r + 12$ miles per hour: $r \cdot 7 = (r + 12) \cdot 5.5$. Distribute on the right, $7r = 5.5r + 66$. Subtract $5.5r$ from each side to get $1.5r = 66$. Divide each side by 1.5, $r = 44$. That's the rate of the first train. The second train is going 12 miles per hour faster.

4. **8 mph** The sum of all the distances equals the total distance run, 12 miles. Let h be the rate at which she ran the second half hour. Then she ran the first half hour at $h + 1$ and the third half hour at $h + 1 - .5 = h + .5$ miles per hour. Add the rates and times together, $(h + 1)\frac{1}{2} + h \cdot \frac{1}{2} + (h + .5)\frac{1}{2} = 12$. Get rid of the fractions by multiplying each term by 2. $(h + 1) + h + (h + .5) = 24$. Simplify the left to get $3h + 1.5 = 24$. Subtract 1.5 from each side: $3h = 22.5$ or $h = 7.5$. She ran the second half hour at 7.5 mph, the first half hour at 8.5 mph, and the last at 8 mph.

5. **10 hours after leaving**

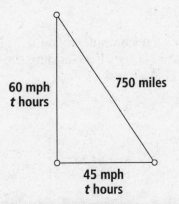

60 mph
t hours

750 miles

45 mph
t hours

Pythagorean Theorem
$$a^2 + b^2 = c^2$$

The Pythagorean Theorem says that if you square the lengths of the two shorter sides of a right triangle and add them together, that's the same as the longest side (hypotenuse) squared. So this would be Claire's distance squared + Charlie's distance squared = 750^2. $(45 \cdot t)^2 + (60 \cdot t)^2 = 750^2$. Square the three terms to get $2025t^2 + 3600t^2 = 562,500$. Add the two terms on the left, $5625t^2 = 562,500$. Divide each side by 5625, $t^2 = 100$ or $t = 10$.

Mixture Problems

Mixture problems deal with having different amounts of solutions at different concentrations. They can deal with antifreeze in a radiator, different types of nuts in a can, gasoline and ethanol, and so on. There are two considerations to deal with when doing these problems: "How much?" and "What concentration or value?" A good way to handle these problems is to consider containers whose contents are to be combined and have their combination be equal to the resulting container.

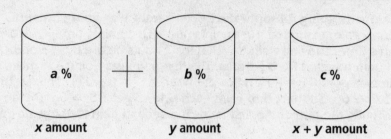

The concentrations or values and the amounts are multiplied together to form an equation to be solved.

Example Problems

These problems show the answers and solutions.

1. Six pounds of cashews that cost $4.75 per pound are added to 8 pounds of peanuts that cost $1.25 per pound. What should be charged, per pound, for the mixture?

   answer: $2.75 per pound

   Place the amounts and values (concentrations) in the figure to help write the equation. The 14 pounds for the last container comes from adding the 6 pounds and 8 pounds of the ingredients.

   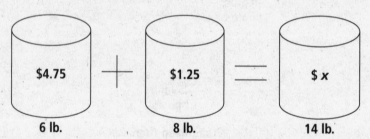

   Multiply each value times the quantity, $4.75(6) + 1.25(8) = x(14)$. Multiply and combine the terms on the left, $28.50 + 10.00 = 14x$ or $38.50 = 14x$. Divide each side by 14 to get $x = 2.75$.

2. How much 40% antifreeze must be added to 70% antifreeze to have a resulting 8 quarts of 65% antifreeze?

   answer: $1\frac{1}{3}$ quarts 40%; $6\frac{2}{3}$ quarts 70%

   Place the amounts and concentrations in the figure. Since neither amount is known—of the two solutions being added together—the first amount is designated x, and the second is $8 - x$. The two of these have to add up to 8 quarts.

   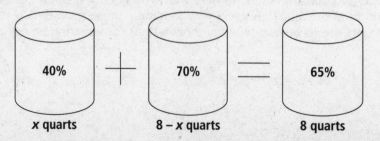

The equation can be written 40%(x) + 70%(8 − x) = 65%(8). Percentages are combined with other values by changing them to decimals. See Chapter 2 in the section "Percents" for more on this. Change the equation to .40(x) + .70(8 − x) = .65(8). Distribute on the left and simplify the other terms, .40x + 5.6 − .70x = 5.2. Combine the terms on the left to get 5.6 − .30x = 5.2. Subtract 5.6 from each side: −.30x = −.4. Divide each side by −.30 and $x = \frac{4}{3} = 1\frac{1}{3}$. This is the number of quarts of the 40% solution. Subtract that from 8 to get the $6\frac{2}{3}$ of the 70% solution.

Work Problems

Use these problems to give yourself additional practice.

1. How many quarts of 90% solution have to be added to 12 quarts of 50% solution to create 75% solution?

2. How many pounds of chocolate-covered raisins costing $3.40 per pound and how many pounds of malted milk balls costing $3.00 per pound must be combined to create 10 pounds of a mixture costing $3.25 per pound?

3. How much water needs to be added to 3 quarts of 96% alcohol to produce a mixture that's 75% alcohol?

4. Carla is making a snack mix that's to contain twice as much cereal as pretzels and also contain 6 pounds of peanuts. If the cereal costs $3.00 per pound, the pretzels cost $2.00 per pound, and the peanuts cost $1.25 per pound, then how many pounds of cereal and pretzels should she use to make a mixture costing $2.50 per pound?

5. How many ounces of chocolate syrup should be added to 8 ounces of milk to produce a mixture that's 20% chocolate syrup?

Worked Solutions

1. **20 quarts** The amount of the 75% solution is the sum of the amounts of the two being mixed together.

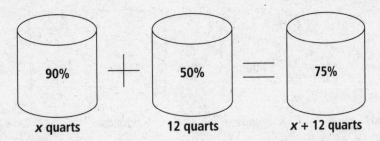

The equation reads .90(x) + .50(12) = .75(x + 12). Simplify the terms on the left and distribute on the right to get .90x + 6 = .75x + 9. Subtract .75x from each side and subtract 6 from each side to get .15x = 3. Divide each side by .15, x = 20. Adding 20 quarts of the 90% solution will result in 32 quarts of 75% solution.

2. $6\frac{1}{4}$ **pounds of chocolate raisins;** $3\frac{3}{4}$ **pounds of malted milk balls**

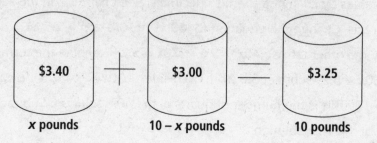

The equation is $3.40(x) + 3.00(10 - x) = 3.25(10)$. Simplify and distribute: $3.40x + 30 - 3.00x = 32.5$. Combine the terms on the left to get $.40x + 30 = 32.5$. Subtract 30 from each side: $.40x = 2.5$. Now divide each side by .40, and $x = \frac{2.5}{.40} = \frac{25}{4} = 6\frac{1}{4}$. It takes $6\frac{1}{4}$ pounds of chocolate raisins and $10 - 6\frac{1}{4} = 3\frac{3}{4}$ pounds of malted milk balls.

3. **.84 quarts of water** The water has no alcohol, so its concentration is 0%. Even though it'll make the term equal to 0, I'm including it in the process for consistency.

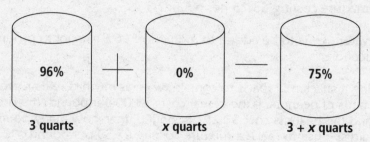

The equation is $.96(3) + 0(x) = .75(3 + x)$. Simplify on the left and distribute on the right, $2.88 = 2.25 + .75x$. Subtract 2.25 from each side to get $.63 = .75x$. Now divide each side by .75, and $x = \frac{.63}{.75} = \frac{21}{25} = .84$. Not quite a quart of water is needed.

4. **30 pounds of cereal and 15 pounds of pretzels** This problem will take 3 "containers" added together to result in the fourth. Let x represent the amount of pretzels needed; then $2x$ is the amount of cereal.

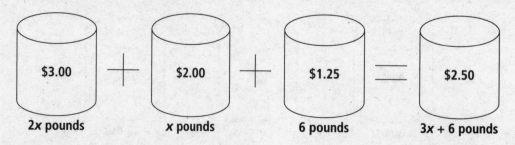

The equation needed is $3.00(2x) + 2.00(x) + 1.25(6) = 2.50(3x + 6)$. Simplify on the left and distribute on the right, $6x + 2x + 7.5 = 7.5x + 15$. Combine the terms on the left and subtract 7.5 from each side, $8x = 7.5x + 7.5$. Now subtract $7.5x$ from each side to get $.5x = 7.5$. Dividing each side by .5, $x = 15$.

5. **2 ounces**   This time, the chocolate syrup has a 100% concentration. That's written as 1.00, in decimal form. The milk has 0% concentration of chocolate syrup, in the beginning.

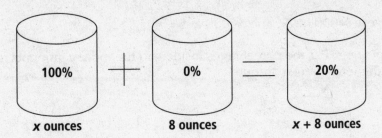

The equation needed is $1.00(x) + 0(8) = .20(x + 8)$. Simplify on the left and distribute on the right to get $x = .20x + 1.6$. Subtract $.20x$ from each side to get $.80x = 1.6$. Divide each side by $.80$: $x = 2$.

Coin and Interest Problems

These problems are similar to mixture problems, because there's both the quantity (how many coins or how much money) and the quality or concentration (denomination of the coin or percent interest rate) to consider when working with them. You can use the same "container" idea from the section on solutions—it helps keep things straight. In coin problems, multiply each denomination amount by how many coins there are. If you have 6 quarters, multiply $.25$ times 6. In interest problems, multiply each amount of money by the percent interest. Interest is computed using the formula $I = prt$, where I is the amount of interest, p is the principal or amount of money invested, i is the rate of interest, and t is the amount of time. In these problems, t is equal to one year, and simple interest is being used, not compound interest. If multiplying $6000 by 2% rate of interest, multiply 6000 times $.02$ to get $120 interest.

Example Problems

These problems show the answers and solutions.

1. Clarissa has 100 coins in dimes and quarters. She has a total of $17.05. How many of each does she have?

   answer: 47 quarters, 53 dimes

   Use the containers, putting the coin value inside and the number of coins outside. Have this equal the total amount of money. The total number of coins is 100, so one amount is x and the other is $100 - x$.

   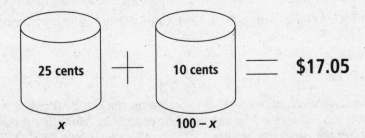

   The equation needed is $.25(x) + .10(100 - x) = 17.05$. Distribute and simplify on the left to get $.25x + 10 - .10x = 17.05$ or $.15x + 10 = 17.05$. Now subtract 10 from each side. $.15x = 7.05$. Dividing each side by $.15$, $x = 47$, which is the number of quarters. $100 - 47 = 53$, the number of dimes.

2. Jack has $10,000 to invest. He wants to earn a total of $410 in interest by investing some at 2% and the rest at a riskier $5\frac{1}{2}$%. How much should he invest at each amount?

   answer: Invest $4000 at 2% and $6000 at $5\frac{1}{2}$%

   Use the containers. Put the percentages inside and the money amounts outside. Have the sum equal the total interest amount.

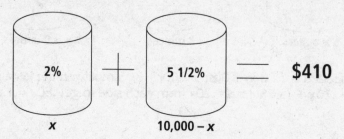

   The equation is written $.02(x) + .055(10,000 - x) = 410$. Simplify and distribute on the left: $.02x + 550 - .055x = 410$. Combine the like terms on the left and subtract 550 from each side, $-.035x = -140$. Divide each side by $-.035$, $x = 4,000$. That leaves $6000 to invest at the other amount.

Work Problems

Use these problems to give yourself additional practice.

1. Kathy has 6 more dimes than quarters and 2 fewer nickels than dimes. She has $4.00. How many of each coin does she have?

2. Jeff has 18 coins in quarters and nickels. If he reversed the number of quarters and nickels, he'd have $1.20 less than he does now. How many quarters and nickels does he have now?

3. Carlie has 175 coins in nickels and dimes. She has a total of $13.50. How many of each does she have?

4. Beth is planning on investing her $18,000 bonus, putting some in an account earning $2\frac{1}{4}$% and the rest in an account earning $6\frac{1}{2}$%. How much should she put in each to have a total of $660 in interest?

5. Patrick contributes $2500 per year into a savings plan that earns $4\frac{1}{4}$% interest. He would like to invest another $1500 per year into a different account so that his total interest per year will be $200. At what rate will he have to invest this additional money?

Worked Solutions

1. **14 dimes, 8 quarters, 12 nickels**   Let q be the number of quarters, $q + 6$ be the number of dimes, and $q + 6 - 2 = q + 4$ be the number of nickels. Multiplying each coin type by its value and adding the products, you get $4.00. $.25(q) + .10(q + 6) + .05(q + 4) = 4.00$. Distribute and simplify on the left: $.25q + .10q + .6 + .05q + .2 = 4.00$. Combine the like terms on the left to get $.40q + .8 = 4.00$. Subtract .8 from each side of the equation, $.40q = 3.20$. Divide each side by .40, $q = 8$. So there are 8 quarters, 6 more dimes, which is 14 dimes, and 2 fewer nickels than that, or 12 nickels.

2. **12 quarters, 6 nickels** Let q be the number of quarters and $18 - q$ be the number of nickels. Right now he has $.25q + .05(18 - q)$ in money. If the number of quarters and nickels is reversed, he would have $.25(18 - q) + .05(q)$ in money. The amount of money in the reversal is $1.20 less than he has now, so, to make the equation work, that much will be added onto the reversed side: $.25q + .05(18 - q) = .25(18 - q) + .05(q) + 1.20$. Distribute and simplify on both sides, $.25q + 9 - .05q = 4.5 - .25q + .05q + 1.20$. Combine like terms on each side, $.20q + .9 = 5.70 - .20q$. Now add $.20q$ to each side and subtract .9 from each side. $.40q = 4.80$. Divide each side by .40, $q = 12$. He has 12 quarters and 6 nickels for a total of $3.00 + .30 or $3.30. If he reverses the number of each coin, he'd have 12 nickels and 6 quarters for a total of $.60 + 1.50 or $2.10. That's $1.20 less than he has now.

3. **80 nickels, 95 dimes** Let n be the number of nickels and $175 - n$ be the number of dimes. Multiply each amount by the coin value and add them together, $.05(n) + .10(175 - n) = 13.50$. Simplify and distribute, $.05n + 17.5 - .10n = 13.50$. Combine the like terms on the left and subtract 17.5 from each side, $-.05n = -4$. Divide each side by $-.05$, $n = 80$.

4. **$12,000 at $2\frac{1}{4}$% and $6,000 at $6\frac{1}{2}$%** Let x be the amount invested at $2\frac{1}{4}$% and $18,000 - x$ be the amount invested at $6\frac{1}{2}$%. The equation is $.0225x + .065(18,000 - x) = 660$. Distribute on the left, $.0225x + 1170 - .065x = 660$. Combine like terms on the left and subtract 1170 from each side, $-.0425x = -510$. Divide each side by $-.0425$, $x = 12,000$. That leaves 6,000 to be invested at the other rate.

5. **$6\frac{1}{4}$%** Let x be the rate at which he'll invest the $1500. The equation should be written $.0425(2500) + x(1500) = 200$. Simplify on the left, $106.25 + 1500x = 200$. Subtract 106.25 from each side to get $1500x = 93.75$. Divide each side by 1500, and $x = .0625$ or a rate of $6\frac{1}{4}$%.

Working Together Problems

The saying, "Two hands are better than one," applies here. Working together, two people can make a job go more quickly. The challenge is to determine just how much more quickly that is—or how much work a person will have to do to get the job done. These work problems involve using fractions and the properties of proportions. Go back to Chapter 3 under "Ratios and Proportions" if you need to review these properties.

Example Problems

These problems show the answers and solutions.

1. If Stan can paint a house in 4 days, and Dan can paint that same house in 6 days, then how long will it take them if they work together to paint it?

   answer: $2\frac{2}{5}$ days

   Let x represent the number of days that it'll take for the two of them to paint the house.

   Then, in one day, they can paint $\frac{1}{x}$ of the house. If Stan was working alone, he could paint $\frac{1}{4}$ of the house in one day, and if Dan was working alone, he could paint $\frac{1}{6}$ of the house in one day. Putting Stan and Dan together, they can paint $\frac{1}{4}+\frac{1}{6}$ of the house in one day. Set this equal to $\frac{1}{x}$ to show what they can accomplish in one day by working together. Adding the two fractions on the left requires finding the common denominator, 24. $\frac{6}{24}+\frac{4}{24}=\frac{10}{24}=\frac{1}{x}$. Take the proportion $\frac{10}{24}=\frac{1}{x}$ and flip it to get $\frac{24}{10}=\frac{x}{1}$ or $x=\frac{24}{10}=\frac{12}{5}=2\frac{2}{5}$.

2. Tina can polish all the crystals on the candelabra in 9 hours. When her sister, Gina, helps her, it only takes the two of them $3\frac{3}{5}$ hours. How long would it take Gina to do the polishing job by herself?

answer: 6 hours

Tina can do $\frac{1}{9}$ of the job in one hour. The two of them working together take $3\frac{3}{5}$ hours, which is $\frac{18}{5}$ hours. In one hour they do $\frac{5}{18}$ of the job. This is found by writing the reciprocal of the amount of time taken (the same as the $\frac{1}{x}$ in the previous example). Let g represent the amount of time it would take Gina to do the job alone. Then she can do $\frac{1}{g}$ of the job in one hour. Add together how much Tina and Gina can do in an hour and set it equal to the $\frac{5}{18}$. $\frac{1}{9}+\frac{1}{g}=\frac{5}{18}$. The common denominator for the two fractions on the left is $9g$. $\frac{g}{9g}+\frac{9}{9g}=\frac{g+9}{9g}=\frac{5}{18}$. Reduce the proportion before cross multiplying. $\frac{g+9}{9g}=\frac{5}{18_2}$. Cross multiplying, $(g+9)\cdot 2=5\cdot g$ or $2g+18=5g$. Subtract $2g$ from each side to get $18=3g$ or $g=6$. Gina can do it in 6 hours.

Work Problems

Use these problems to give yourself additional practice.

1. Greg can seal coat a driveway in 4 hours, and Jon can do the same job in 2 hours. How long will it take for them to do the job if they work together?

2. Irene can check the inventory in 2 hours, but it takes Stephanie 6 hours to do the same procedure. How long will it take if they do the inventory together?

3. Ken can peel a bag of potatoes in 10 minutes, and it takes Sherrill 15 minutes to do the same job. If they work together, they can peel that bag of potatoes in how many minutes?

4. Ted can wash the van in 6 minutes. If Jim helps, it takes the two of them 4 minutes. How long would it take Jim to wash the van if he does it alone?

5. Clara and Sarah can clear all the snow from the parking lot in 7 hours and 12 minutes, each with her own snow blower. Working alone, it would take Clara 12 hours to clear the lot. How long would it take Sarah if she had to do it alone?

Worked Solutions

1. **$1\frac{1}{3}$ hours** Greg can do $\frac{1}{4}$ of the job in an hour. Jon can do $\frac{1}{2}$ of the job in an hour. Together, they can do $\frac{1}{x}$ of the job in an hour. The variable x represents how long it will take them to do the job together. $\frac{1}{4} + \frac{1}{2} = \frac{1}{x}$. Add the two fractions on the left together, $\frac{1+2}{4} = \frac{3}{4} = \frac{1}{x}$. Flip the proportion to get $\frac{4}{3} = \frac{x}{1}$. So, $x = \frac{4}{3}$ or $1\frac{1}{3}$ hours.

2. **$1\frac{1}{2}$ hours** Irene can do $\frac{1}{2}$ of the job in an hour. Stephanie can do $\frac{1}{6}$ of the job in an hour. Together, they can do $\frac{1}{x}$ of the job in an hour. The variable x represents how long it will take them to do the job together. $\frac{1}{2} + \frac{1}{6} = \frac{1}{x}$. Add the two fractions on the left together, $\frac{3+1}{6} = \frac{4}{6} = \frac{2}{3} = \frac{1}{x}$. Flip the proportion to get $\frac{3}{2} = \frac{x}{1}$. So $x = \frac{3}{2}$ or $1\frac{1}{2}$ hours.

3. **6 minutes** Ken can do $\frac{1}{10}$ of the job in a minute. Sherrill can do $\frac{1}{15}$ of the job in a minute. Together, they can do $\frac{1}{x}$ of the job in a minute. The variable x represents how long it will take them to do the job together. $\frac{1}{10} + \frac{1}{15} = \frac{1}{x}$. Add the two fractions on the left together, $\frac{3+2}{30} = \frac{5}{30} = \frac{1}{6} = \frac{1}{x}$. Flip the proportion to get $\frac{6}{1} = \frac{x}{1}$. So $x = 6$ minutes.

4. **12 minutes** Ted can do $\frac{1}{6}$ of the job in one minute. Together they can do $\frac{1}{4}$ of the job in one minute. Jim, working alone, can do $\frac{1}{y}$ of the job in one minute, where y represents how long it would take Jim to do the job alone. The equation needed is $\frac{1}{6} + \frac{1}{y} = \frac{1}{4}$. Subtract $\frac{1}{6}$ from each side and then subtract the two fractions on the right. $\frac{1}{y} = \frac{1}{4} - \frac{1}{6} = \frac{3-2}{12} = \frac{1}{12}$. The proportion now reads $\frac{1}{y} = \frac{1}{12}$. Flip the proportion, $y = 12$.

5. **18 hours** The 7 hours and 12 minutes is equal to $7\frac{12}{60} = 7\frac{1}{5} = \frac{36}{5}$ hours. Clara can clear $\frac{1}{12}$ of the lot in one hour. Sarah can clear $\frac{1}{y}$ of the lot in one hour. Together they can clear $\frac{1}{\frac{36}{5}} = \frac{5}{36}$ of the lot in one hour. So $\frac{1}{12} + \frac{1}{y} = \frac{5}{36}$. Subtract $\frac{1}{12}$ from each side and subtract the fractions on the right. $\frac{1}{y} = \frac{5}{36} - \frac{1}{12} = \frac{5-3}{36} = \frac{2}{36} = \frac{1}{18}$. Take the proportion $\frac{1}{y} = \frac{1}{18}$ and flip it to get $y = 18$.

Chapter Problems and Solutions

Problems

1. One less than 4 times a number is equal to two more than three times that same number. What is the number?

2. Doubling a number and then multiplying it by 4 is the same as tripling it and dividing by 7. What is the number?

3. Multiplying the sum of a number and 1 by 6 is the same as subtracting 9 from the number. What is the number?

4. Dividing the difference between a number and 2 by 4 is equal to dividing the sum of the number and 6 by 8. What is this number?

5. The sum of three consecutive integers is 75. What is the largest of these three integers?

6. The sum of four consecutive odd integers is 96. What are they?

7. The sum of four consecutive multiples of 3 is 78. What is the smallest of these integers?

8. Mitch is 4 years older than Rich. Six years ago, Mitch was twice as old as Rich. How old are they now?

9. Dave's grandfather is 30 times as old as Dave is. In 16 years, his grandfather will be 4 more than 4 times Dave's age. How old is Dave now?

10. The length of a rectangle is 2 more than twice its width. The area of the rectangle is 24 square feet. What are the dimensions of the rectangle?

11. A square backyard is doubled in length and halved in width, resulting in a length that's 24 feet longer than the width. What were the dimensions of the square?

12. Josh has 480 feet of fencing and wants to build a rectangular yard that's 3 times as long as it is wide. One of the long sides will be along a river, and he doesn't want to put fencing along that side. What are the dimensions of his yard?

13. Mike and Bob live 27 miles apart. They start walking toward each other's home along the same straight road. If Mike walks at a rate 1 mph faster than Bob, and it took them 3 hours until they met, how fast was each walking?

14. Ann got a 2-hour head start on the delivery route. Frank caught up to her just $\frac{1}{2}$ an hour after he left the same place that she did, following the same route. If Ann was averaging 10 mph, how fast did Frank travel?

15. Clem swims at 2 mph and bicycles at 25 mph. If he swam and bicycled for a total of 42 minutes during the 6 mile race, how long did he swim?

16. How many quarts of 80% solution do you mix with 4 quarts of 50% solution to get 60% solution?

17. How many gallons of water and how many gallons of pure lemon juice do you mix together to get 15 gallons of lemonade that's 30% lemon juice?

18. How many pounds of raisins at $2.00 per pound do you mix with granola at $3.25 per pound to get 10 pounds of a mixture that costs $2.50 per pound?

19. Tiffany has 100 coins in dimes and nickels. She has a total of $5.80. How many of each coin does she have?

20. Holly has 8 more nickels than quarters. If she has $3.40, how many of each does she have?

21. Janet has \$40,000 invested in two accounts—one earning $1\frac{3}{4}$% and the other earning 5%. If her total interest for the year is \$1025, then how much does she have invested at each rate?

22. Wanda has some money invested at 4% and twice as much invested at 8%. Her total interest earned is \$100. How much money does she have invested at each rate?

23. Joey can shovel the walk in 16 minutes, and Lou can shovel it in 12 minutes. How long will it take to shovel the walk if they work together?

24. Steffi can wax the gym floor in 3 hours, and Wendi can do it in 6 hours. How long will it take if they work together on the job?

25. Working together, Bill and Hillary can clean the house in $2\frac{2}{5}$ hours. Working alone, Hillary can do the job in 4 hours. How long would it take Bill to clean the house if he had to do it alone?

Answers and Solutions

1. **Answer: 3**   The equation needed is $4n - 1 = 3n + 2$. Add 1 to each side, and subtract $3n$ from each side to get $n = 3$. One less than 4 times 3 is 11. Three times 3 is 9; add 2, and you get 11.

2. **Answer: 0**   The equation used is $(2n) \cdot 4 = \frac{3n}{7}$. Multiply each side by 7 and simplify the left to get $56n = 3n$. This should look awfully suspicious. How can you multiply the same number by both 56 and 3 and get the same answer? Subtract $3n$ from each side to get $53n = 0$. Dividing each side by 53 gives $n = 0$. It might have been tempting to divide each side of $56n = 3n$ by n. As a rule, *never* divide or multiply each side by the variable. You can get extraneous solutions or, as in this case, not even the actual answer.

3. **Answer: –3**   The equation to use is $6(n + 1) = n - 9$. Distribute on the left to get $6n + 6 = n - 9$. Subtract 6 from each side and subtract n from each side, $5n = -15$. Divide each side by 5, $n = -3$.

4. **Answer: 10**   The equation needed is $\frac{n-2}{4} = \frac{n+6}{8}$. Multiply each side by 8 to get $2(n - 2) = n + 6$. Distribute on the left to get $2n - 4 = n + 6$. Add 4 to each side and subtract n from each side to get $n = 10$.

5. **Answer: 26**   The equation is $n + n + 1 + n + 2 = 75$. Simplifying on the left leaves $3n + 3 = 75$. Subtract 3 from each side, $3n = 72$. Divide by 3, $n = 24$, so the other two numbers are 25 and 26. The largest of the three numbers is the 26.

6. **Answer: 21, 23, 25, 27**   The equation is $n + n + 2 + n + 4 + n + 6 = 96$. Simplify on the left, $4n + 12 = 96$. Subtract 12 from each side, $4n = 84$. Divide each side by 4, $n = 21$, so the next three odd numbers are 23, 25, and 27.

7. **Answer: 15**   The equation to use is $n + n + 3 + n + 6 + n + 9 = 78$. Simplify on the left, $4n + 18 = 78$. Subtract 18 from each side, $4n = 60$. Divide each side by 4, $n = 15$, and the next three multiples of 3 are 18, 21, and 24. The smallest of these is the 15.

8. **Answer: Mitch is 14; Rich is 10.**   Let Mitch's age be m. Then Rich is $m - 4$. Six years ago, their ages were $m - 6$ and $m - 4 - 6 = m - 10$. If Mitch was twice as old as Rich at that time, 6 years ago, then $m - 6 = 2(m - 10)$. Distributing on the right, $m - 6 = 2m - 20$.

Subtract m from each side and add 20 to each side, $14 = m$. If Mitch is 14 and Rich is 4 years younger, then Rich is 10. Six years ago, their ages were 8 and 4, with Mitch's age twice Rich's.

9. **Answer: 2 years old**   Let d be Dave's age. Then his grandfather is $30d$. In 16 years, Dave will be $d + 16$ and his grandfather will be $30d + 16$. If his grandfather will be 4 more than 4 times Dave's age at that time, then the equation needed is $30d + 16 = 4(d + 16) + 4$. Distribute and simplify on the right, $30d + 16 = 4d + 68$. Subtract $4d$ from each side and subtract 16 from each side to get $26d = 52$. Divide each side by 26, $d = 2$. So Dave is 2 years old; his grandfather is 30 times that or 60. In 16 years, Dave will be 18 and his grandfather will be 76. Four times Dave's age plus 4 is $4(18) + 4 = 72 + 4 = 76$.

10. **Answer: 8 by 3**   Let w be the width of the rectangle. Then the length is $2w + 2$. The area is found by multiplying the length by the width. If the area is 24 square feet, then $w(2w + 2) = 24$. Distribute on the left, $2w^2 + 2w = 24$. This is a quadratic equation. Subtract 24 from each side to get $2w^2 + 2w - 24 = 0$. Factor out the 2 from each term and then factor the trinomial, you get $2(w + 4)(w - 3) = 0$. Using the Multiplication Property of Zero, $w = -4$ or $w = 3$. (For more on solving quadratic equations, refer to Chapter 6.) The answer $w = -4$ doesn't make sense here. A width can't be a negative number. So $w = 3$, which means that the length is 2 more than twice that or 8.

11. **Answer: 16 by 16**   Choose x to be the length of the sides of the original square. Doubling that to get the new length is $2x$. The new, halved width is $\frac{x}{2}$. If the new length is 24 more than the new width, $2x = \frac{x}{2} + 24$.

   Subtract $\frac{x}{2}$ from each side, $\frac{3}{2}x = 24$. Multiply each side by $\frac{2}{3}$, $x = 16$. The original square was 16 by 16. Twice 16 is 32 and half of 16 is 8. The 32 by 8 rectangle has a length that's 24 more than the width.

12. **Answer: 288 by 96**   The 480 feet of fencing will only be used on the three sides not along the river.

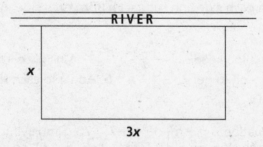

Let x represent the width of the yard and $3x$ be the length. The sum of two of the widths and the one length is the amount of fencing he has, or 480 feet: $x + x + 3x = 480$. Simplify on the left, $5x = 480$. Divide each side by 5, $x = 96$. Three times 96 is 288, and the sum of $96 + 96 + 288 = 480$.

13. **Answer: Mike walked 5 mph; Bob walked 4 mph.**   The sum of the distances that they walk has to equal 27 miles. Let Bob's rate of walking be x mph and Mike's then be $x + 1$. Use rate times time for their respective distances in the sum, so $x \cdot 3 + (x + 1) \cdot 3 = 27$. Distribute on the left, $3x + 3x + 3 = 27$. Simplify on the left, $6x + 3 = 27$. Subtract 3 from each side and divide each side by 6, $x = 4$.

14. Answer: 50 mph The distances traveled are the same, so set Ann's rate times her time = Frank's rate times his time. Since Frank caught up to Ann in half an hour, and she had a 2-hour head start, then Ann traveled for $2\frac{1}{2}$ hours, and Frank traveled for $\frac{1}{2}$ hour. The equation reads $10\left(2\frac{1}{2}\right) = r\left(\frac{1}{2}\right)$. Simplify the equation, $25 = \frac{1}{2}r$. Multiply each side by 2, $r = 50$.

15. Answer: 30 minutes The sum of Clem's swimming distance and bicycling distance equals 6 miles. The rate times the time of each added together will equal 6. Let t represent the amount of time spent swimming, then $42 - t$ minutes or $\frac{7}{10} - t$ will represent the amount of time bicycling. The minutes had to be changed to hours, $\frac{42}{60} = \frac{7}{10}$ because the speeds are in miles per hour. So the equation is $2 \cdot t + 25 \cdot \left(\frac{7}{10} - t\right) = 6$. Distribute on the left to get $2t + \frac{175}{10} - 25t = 6$. Simplify on the left, $\frac{175}{10} - 23t = 6$. Subtract $\frac{175}{10}$ from each side to get $-23t = \frac{60}{10} - \frac{175}{10} = \frac{-115}{10}$. Divide each side by -23, $t = \frac{1}{2}$. The amount of time spent swimming is half an hour or 30 minutes. The other 12 minutes was spent bicycling.

16. Answer: 2 quarts The equation is $.80(x) + .50(4) = .60(x + 4)$. Simplify on the left and distribute on the right: $.80x + 2 = .60x + 2.4$. Subtract $.60x$ from each side, and subtract 2 from each side to get $.20x = .4$. Divide each side by $.20$, $x = 2$.

17. Answer: 10.5 gallons of water and 4.5 gallons of pure lemon juice The water has 0% lemon juice, and the lemon juice is 100%. Let x be the amount of water. Then $15 - x$ would be the amount of lemon juice so that the total will be 15 gallons of lemonade. The equation reads $0(x) + 1(15 - x) = .30(15)$. Simplifying, you get $15 - x = 4.5$. Subtract 15 from each side and multiply each side by -1, $x = 10.5$.

18. Answer: 6 pounds Let x be the number of pounds of raisins. Then $10 - x$ is the number of pounds of granola. The equation is $2.00(x) + 3.25(10 - x) = 2.50(10)$. Simplify and distribute: $2x + 32.5 - 3.25x = 25$.

Simplify on the left, $32.5 - 1.25x = 25$. Subtract 32.5 from each side, $-1.25x = -7.5$. Divide each side by -1.25, $x = 6$.

19. Answer: 16 dimes, 84 nickels Let d be the number of dimes and $100 - d$ be the number of nickels. The equation reads $.10d + .05(100 - d) = 5.80$. Distribute on the left to get $.10d + 5 - .05d = 5.80$. Simplifying on the left, $5 + .05d = 5.80$. Subtract 5 from each side to get $.05d = .80$ Dividing each side by $.05$, $d = 16$. Subtract 16 from 100 to get 84—the number of nickels. 16 dimes is $1.60, and 84 nickels is $4.20.

20. Answer: 10 quarters, 18 nickels Let q be the number of quarters. Then $q + 8$ is the number of nickels. The equation is $.25q + .05(q + 8) = 3.40$. Distribute and simplify on the left: $.30q + .4 = 3.40$. Subtract $.4$ from each side to get $.30q = 3.00$. Dividing each side by $.30$, $q = 10$.

21. Answer: $30,000 at $1\frac{3}{4}$%, $10,000 at 5% Let x be the amount invested at $1\frac{3}{4}$%. Then $40,000 - x$ is the amount invested at 5%. Interest is determined by multiplying the amount times the interest rate, so the two amounts times their interest rates added together gives the total interest. $.0175(x) + .05(40,000 - x) = 1025$. Distribute and simplify on the left, $-.032x + 2000 = 1025$. Subtract 2000 from each side, $-.0325x = -975$. Divide each side by $-.0325$, $x = 30,000$. That leaves 10,000 invested at the other rate.

22. **Answer: $500 at 4% and $1000 at 8%**   Let x be the amount invested at 4%. Then $2x$ is the amount invested at 8%. The sum of the interest earned on each is $100. $.04x + .08(2x) = 100$. Simplify on the left, $.04x + .16x = .20x = 100$. Divide each side by $.20$, $x = 500$.

23. **Answer: $6\frac{6}{7}$ minutes**   Let x denote the time it will take them working together. Joey can do $\frac{1}{16}$ of the job in one minute, and Lou can do $\frac{1}{12}$ of the job in one minute. Together, they can do $\frac{1}{x}$ of the job in one minute. $\frac{1}{16} + \frac{1}{12} = \frac{1}{x}$. Add the two fractions on the left, $\frac{3+4}{48} = \frac{1}{x}$. Flip the proportion to get $\frac{48}{7} = \frac{x}{1}$ or $x = 6\frac{6}{7}$ minutes.

24. **Answer: 2 hours**   Steffi can do $\frac{1}{3}$ of the job in one hour, and Wendi can do $\frac{1}{6}$ of the job in one hour. If they can do the job in x hours working together, $\frac{1}{3} + \frac{1}{6} = \frac{1}{x}$. Add the fractions together, $\frac{2}{6} + \frac{1}{6} = \frac{3}{6} = \frac{1}{2} = \frac{1}{x}$.
Flip the proportion, $\frac{2}{1} = \frac{x}{1}$, or $x = 2$.

25. **Answer: 6 hours**   In one hour, Hillary can do $\frac{1}{4}$ of the job. Working together, in one hour they can do $\frac{1}{\frac{12}{5}} = \frac{5}{12}$ of the job. Let y be the amount of time it would take Bill to do the job working alone. Then he can do $\frac{1}{y}$ of the job in one hour, $\frac{1}{4} + \frac{1}{y} = \frac{5}{12}$. Subtract $\frac{1}{4}$ from each side to get $\frac{1}{y} = \frac{5}{12} - \frac{3}{12} = \frac{2}{12} = \frac{1}{6}$. Flip to get $y = 6$.

Supplemental Chapter Problems

Problems

1. Sixteen is equal to 1 more than 5 times a number. What is that number?

2. Subtracting 2 times the sum of a number and 1 from 4 is the same as subtracting that number from −4. What is the number?

3. Five times a number is equal to ten times that number less 2. What is the number?

4. Add 1 to the reciprocal of a number and you get the same thing as multiplying the number by 6. What is the number?

5. The sum of three consecutive even integers is 72. What are they?

6. The sum of five consecutive integers is 165. What is the middle integer in the list?

7. The sum of three consecutive odd integers is 33. What are they?

8. Tony is two years older than Tom. Thirty-nine years ago, Tony was three times as old as Tom. How old is Tom now?

9. Moe is 1 year older than Larry, and Larry is 2 years older than Curly. In 3 years, the sum of their ages will be 125. How old are they?

10. A rectangle has a length that is 2 feet longer than 3 times the width. If the perimeter is 84 feet, then what are the dimensions?

11. The height of a triangle is half its base. If the height is cut in half, and the base stays the same, the area decreases by 18 square inches. What is the original height?

12. One square has sides twice the length of a second square, and the difference in their areas is 75 square meters. What are the sizes of the squares?

13. Ed left home at 1:00 PM, traveling at 50 mph. Fred left the same place at 3:00 PM, following Ed's route and catching up to him at 8:00 PM. How fast was Fred traveling?

14. A truck left Des Moines traveling north at 45 mph. One hour later, a second truck left Des Moines traveling south at 60 mph. When are they 570 miles apart?

15. Marty drove his golf cart around the track 3 times in the same amount of time that Cheryl walked around the track twice. If Marty was driving at 12 mph, how fast was Cheryl walking?

16. How many ounces of red paint need to be mixed with 64 ounces of white paint to get a pink mixture that's 20% red?

17. How many quarts of 90% solution do you mix with 50% solution to get 20 quarts of 60% solution?

18. How many pounds of peas do you mix with a batch of 18 pounds of peas and carrots that's already half-and-half in order to make the mixture 60% peas?

19. Mindy has twice as many nickels as dimes and 3 times as many dimes as quarters. If she has $10.20, then how many of each coin does she have?

20. Sandy had 4 times as many quarters as dimes. When she added 5 of each coin to what she had, she had 35 coins total. How much money did she have at the beginning?

21. Fred had $\frac{1}{3}$ of his money invested at 6% and the rest at 2%. If he made a total of $1500 in interest this year, then how much money does he have invested at each rate?

22. Sadie wants to invest $\frac{3}{4}$ of her money at $3\frac{1}{2}$% and the rest at $5\frac{1}{4}$% and get a yearly return of $1575. How much should she invest at each rate?

23. Claudia likes to take her time when wrapping Christmas gifts and takes 12 hours to do it. When Jerry does the same job, he can do it in 4 hours. How long will it take if they work together?

24. When Dave and Greg paint a room together, it takes $3\frac{3}{5}$ hours. Dave working alone can paint the room in 6 hours. How long would it take Greg to paint it alone?

25. Charles can collect eggs in the hen house in 8 minutes. Cindy needs 12 minutes. Working together, how long will it take to collect the eggs?

Answers

1. 3   (Number Problems, p. 333)

2. 6   (Number Problems, p. 333)

3. $\frac{2}{5}$ (Number Problems, p. 333)

4. $\frac{1}{2}$ or $\frac{1}{3}$ (Number Problems, p. 333)

5. 22, 24, 26   (Consecutive Integer Problems, p. 335)

6. 33   (Consecutive Integer Problems, p. 335)

7. 9, 11, 13   (Consecutive Integer Problems, p. 335)

8. Tom is 40 years old.   (Age Problems, p. 337)

9. Moe is 40, Larry is 39, and Curly is 37.   (Age Problems, p. 337)

10. 32 by 10   (Geometric Problems, p. 339)

11. 6 inches   (Geometric Problems, p. 339)

12. 5 by 5 and 10 by 10   (Geometric Problems, p. 339)

13. 70 mph   (Distance Problems, p. 344)

14. 6 hours after the first left   (Distance Problems, p. 344)

15. 8 mph   (Distance Problems, p. 344)

16. 1.6 ounces   (Mixture Problems, p. 347)

17. 5 quarts   (Mixture Problems, p. 347)

18. $4\frac{1}{2}$ pounds   (Mixture Problems, p. 347)

19. 12 quarters, 36 dimes, 72 nickels   (Coin and Interest Problems, p. 351)

20. 20 quarters, 5 dimes   (Coin and Interest Problems, p. 351)

21. $15,000 at 6% and $30,000 at 2%   (Coin and Interest Problems, p. 351)

22. $30,000 at $3\frac{1}{2}$% and $10,000 at $5\frac{1}{4}$%   (Coin and Interest Problems, p. 351)

23. 3 hours   (Working Together Problems, p. 353)

24. 9 hours   (Working Together Problems, p. 353)

25. $4\frac{4}{5}$ minutes   (Working Together Problems, p. 353)

Customized Full-Length Exam

1. Simplify $4(6) - 3 \cdot 2^2 + 7$.

Answer: 19

If you answered **correctly**, go to problem 3.
If you answered **incorrectly**, go to problem 2.

2. Simplify $\sqrt{5^2 - 9} + 3(8) - 2$.

Answer: 26

If you answered **correctly**, go to problem 3.
If you answered **incorrectly**, review "Order of Operations" on page 31.

3. Simplify $12x - |-3x|$. Assume that x is positive.

Answer: $9x$

If you answered **correctly**, go to problem 5.
If you answered **incorrectly**, go to problem 4.

4. Simplify $x^2 \sqrt{49}$.

Answer: $7x^2$

If you answered **correctly**, go to problem 5.
If you answered **incorrectly**, review "Basic Math Operations" on page 35.

5. Simplify $4(3x + 2x) + 2(3y - y)$.

Answer: $20x + 4y$

If you answered **correctly**, go to problem 7.
If you answered **incorrectly**, go to problem 6.

6. Simplify $3(5z + 2z) - 2(3z + z)$.

Answer: $13z$

If you answered **correctly**, go to problem 7.
If you answered **incorrectly**, review "Combining 'Like' Terms" on page 38.

7. Simplify $7m^3 + 3m^2 + 2m + 1 + 8m^3 + 5$.

Answer: $15m^3 + 3m^2 + 2m + 6$

If you answered **correctly**, go to problem 9.
If you answered **incorrectly**, go to problem 8.

8.   Simplify $8x^3 + 5x - 3x^3 + 4x^2 + x - 5$.

Answer: $5x^3 + 4x^2 + 6x - 5$

If you answered **correctly**, go to problem 9.
If you answered **incorrectly**, review "Combining 'Like' Terms" on page 38.

9.   Complete the statement to make it true: $(4 \cdot 3) \cdot 2 = 4 \cdot (\underline{\quad})$.

Answer: $3 \cdot 2$

If you answered **correctly**, go to problem 11.
If you answered **incorrectly**, go to problem 10.

10.   Complete the statement to make it true: $-\dfrac{3}{2} \cdot \underline{\quad} = 1$.

Answer: $-\dfrac{2}{3}$

If you answered **correctly**, go to problem 11.
If you answered **incorrectly**, review "Properties of Algebraic Expressions" on page 41.

11.   Simplify $\left(4x^3 yz^6\right)^2$.

Answer: $16x^6 y^2 z^{12}$

If you answered **correctly**, go to problem 13.
If you answered **incorrectly**, go to problem 12.

12.   Simplify $\dfrac{15a^4 b^2}{(ab^3)^3}$.

Answer: $\dfrac{15a}{b^7}$

If you answered **correctly**, go to problem 13.
If you answered incorrectly, review "Integer Powers (Exponents)" on page 44.

13.   Solve $\sqrt{256} = .$

Answer: 16

If you answered **correctly**, go to problem 15.
If you answered **incorrectly**, go to problem 14.

14.   Solve $\sqrt[3]{8} = .$

Answer: 2

If you answered **correctly**, go to problem 15.
If you answered **incorrectly**, review "Square Roots and Cube Roots" on page 48.

15.   Estimate the square root to the nearest tenth: $\sqrt{5}$.

Answer: 2.2

If you answered **correctly**, go to problem 17.
If you answered **incorrectly**, go to problem 16.

16. Estimate the square root to the nearest tenth: $\sqrt{40}$.

Anwer: 6.3

If you answered **correctly**, go to problem 17.
If you answered **incorrectly**, review "Approximating Square Roots" on page 50.

17. Combine the terms after simplifying: $\sqrt{12} + \sqrt{48}$.

Answer: $6\sqrt{3}$

If you answered **correctly**, go to problem 19.
If you answered **incorrectly**, go to problem 18.

18. Combine the terms after simplifying: $\sqrt{90} - \sqrt{40}$.

Answer: $\sqrt{10}$

If you answered **correctly**, go to problem 19.
If you answered **incorrectly**, review "Simplifying Square Roots" on page 52.

19. Rationalize $\dfrac{2}{\sqrt{14}}$.

Answer: $\dfrac{\sqrt{14}}{7}$

If you answered **correctly**, go to problem 21.
If you answered **incorrectly**, go to problem 20.

20. Rationalize $\dfrac{10}{\sqrt{6}}$.

Answer: $\dfrac{5\sqrt{6}}{3}$

If you answered **correctly**, go to problem 21.
If you answered **incorrectly**, review "Simplifying Square Roots" on page 52.

21. Simplify $\sqrt[3]{a^4}\sqrt[3]{a^7}$.

Answer: $a^3\sqrt[3]{a^2}$

If you answered **correctly**, go to problem 23.
If you answered **incorrectly**, go to problem 22.

22. Simplify $\dfrac{9^{3/2}}{9^{1/2}}$.

Answer: 9

If you answered **correctly**, go to problem 23.
If you answered **incorrectly**, review "Exponents for Roots" on page 56.

23. Which of the following is divisible by 6?

    A. 23,456

    B. 106,006

    C. 9,800,004

Answer: C

If you answered **correctly**, go to problem 25.
If you answered **incorrectly**, go to problem 24.

24. Which of the following is divisible by 9?

    A. 32,400

    B. 59,200

    C. 11,119

Answer: A

If you answered **correctly**, go to problem 25.
If you answered **incorrectly**, review "Divisibility Rules" on page 58.

25. Which of the following is prime?

    A. 41

    B. 87

    C. 1001

Answer: A

If you answered **correctly**, go to problem 27.
If you answered **incorrectly**, go to problem 26.

26. Which of the following is prime?

    A. 49

    B. 63

    C. 89

Answer: C

If you answered **correctly**, go to problem 27.
If you answered **incorrectly**, review "Prime Numbers" on page 61.

27. Find the prime factorization for 320.

Answer: $2^6 \cdot 5$

If you answered **correctly**, go to problem 29.
If you answered **incorrectly**, go to problem 28.

28. Find the prime factorization for 300.

Answer: $2^2 \cdot 3 \cdot 5^2$

If you answered **correctly**, go to problem 29.
If you answered **incorrectly**, review "Prime Factorization" on page 63.

29. What is the interval notation for $x < 5$?

Answer: $(-\infty, 5)$

If you answered **correctly**, go to problem 31.
If you answered **incorrectly**, go to problem 30.

30. What is the inequality notation for $\longleftarrow\!\!\bullet\!\!\rule[0.5ex]{2em}{1pt}\!\!\bullet\!\!\longrightarrow$?
$\quad\quad\quad\quad\quad\quad\quad\quad\quad\quad\quad\quad {}_{-6}\quad\quad {}_{-1}$

Answer: $-6 \le x \le -1$

If you answered **correctly**, go to problem 31.
If you answered **incorrectly**, review "Number Lines" on page 73.

31. Add $-9 + (-3)$.

Answer: -12

If you answered **correctly**, go to problem 33.
If you answered **incorrectly**, go to problem 32.

32. Add $-8 + 2$.

Answer: -6

If you answered **correctly**, go to problem 33.
If you answered **incorrectly**, review "Addition of Signed Numbers" on page 75.

33. Subtract $5 - (-4)$.

Answer: 9

If you answered **correctly**, go to problem 35.
If you answered **incorrectly**, go to problem 34.

34. Subtract $-6 - 3$.

Answer: -9

If you answered **correctly**, go to problem 35.
If you answered **incorrectly**, review "Subtraction of Signed Numbers" on page 78.

35. Multiply $(-8)(-2)$.

Answer: 16

If you answered **correctly**, go to problem 37.
If you answered **incorrectly**, go to problem 36.

36. Multiply $(-6)(11)$.

Answer: -66

If you answered **correctly**, go to problem 37.
If you answered **incorrectly**, review "Multiplication and Division of Signed Numbers" on page 80.

37. Divide $\dfrac{-18}{3}$.

Answer: -6

If you answered **correctly**, go to problem 39.
If you answered **incorrectly**, go to problem 38.

38. Simplify $\dfrac{6(-8)}{-16}$.

Answer: 3

If you answered **correctly**, go to problem 39.
If you answered **incorrectly**, review "Multiplication and Division of Signed Numbers" on page 80.

39. Simplify $\dfrac{8y^{-4}}{2y^{-2}}$.

Answer: $\dfrac{4}{y^2}$

If you answered **correctly**, go to problem 41.
If you answered **incorrectly**, go to problem 40.

40. Simplify $\left(\dfrac{y^{-4}}{y}\right)^{-5}$.

Answer: y^{25}

If you answered **correctly**, go to problem 41.
If you answered **incorrectly**, review "Exponents That Are Signed Numbers" on page 82.

41. Simplify $\left(\dfrac{3x^{-2}y^2}{9xy^{-1}}\right)^{-1}$.

Answer: $\dfrac{3x^3}{y^3}$

If you answered **correctly**, go to problem 43.
If you answered **incorrectly**, go to problem 42.

42. Simplify $\left(\dfrac{3x^2y^{-2}}{9x^{-2}y^{-2}}\right)^{-3}$.

Answer: $\dfrac{27}{x^{12}}$

If you answered **correctly**, go to problem 43.
If you answered **incorrectly**, review "Exponents That Are Signed Numbers" on page 82.

43. Simplify $\dfrac{xyz - xy^2}{2xy - xyz^2}$.

Answer: $\dfrac{z-y}{2-z^2}$

If you answered **correctly**, go to problem 45.
If you answered **incorrectly**, go to problem 44.

44. Simplify $\dfrac{6x^2 - 2x^2}{4xy}$.

Answer: $\dfrac{x}{y}$

If you answered **correctly**, go to problem 45.
If you answered **incorrectly**, review "Lowest Terms and Equivalent Fractions" on page 86.

45. Add $\dfrac{3}{a^2} + \dfrac{4}{ab}$.

Answer: $\dfrac{3b + 4a}{a^2 b}$

If you answered **correctly**, go to problem 47.
If you answered **incorrectly**, go to problem 46.

46. Add $\dfrac{x}{6} + \dfrac{3}{y}$.

Answer: $\dfrac{xy + 18}{6y}$

If you answered **correctly**, go to problem 47.
If you answered **incorrectly**, review "Lowest Terms and Equivalent Fractions" on page 86.

47. Multiply $\dfrac{8xy^5}{25(x+y)^3} \cdot \dfrac{15(x+y)}{16xy^4}$.

Answer: $\dfrac{3y}{10(x+y)^2}$

If you answered **correctly**, go to problem 49.
If you answered **incorrectly**, go to problem 48.

48. Multiply $\dfrac{20(x-y)^4}{21x} \cdot \dfrac{14x^6}{25(x-y)}$.

Answer: $\dfrac{8x^5(x-y)^3}{15}$

If you answered **correctly**, go to problem 49.
If you answered **incorrectly**, review "Multiplying and Dividing Fractions" on page 89.

49. Divide $\dfrac{24r}{35rt^2} \div \dfrac{36rt^3}{7r^7t^5}$.

Answer: $\dfrac{2r^6}{15}$

If you answered **correctly**, go to problem 51.
If you answered **incorrectly**, go to problem 50.

50. Divide $\dfrac{16abc}{99xyz} \div \dfrac{8ac^2}{55z}$.

Answer: $\dfrac{10b}{9xyc}$

If you answered **correctly**, go to problem 51.
If you answered **incorrectly**, review "Multiplying and Dividing Fractions" on page 89.

51. Simplify $\dfrac{\frac{5x}{21}}{\frac{10x^2}{49}}$.

Answer: $\dfrac{7}{6x}$

If you answered **correctly**, go to problem 53.
If you answered **incorrectly**, go to problem 52.

52. Simplify $\dfrac{\frac{9}{6a}}{\frac{5}{b}}$.

Answer: $\dfrac{15b}{2a}$

If you answered **correctly**, go to problem 53.
If you answered **incorrectly**, review "Mixed Numbers and Complex Fractions" on page 91.

53. Change the decimal to a fraction: .025.

Answer: $\dfrac{1}{40}$

If you answered **correctly**, go to problem 55.
If you answered **incorrectly**, go to problem 54.

54. Change the decimal to a fraction: 0.48.

Answer: $\dfrac{12}{25}$

If you answered **correctly**, go to problem 55.
If you answered **incorrectly**, review "Decimals" on page 94.

55. Change the percent to a fraction: 5%.

Answer: $\dfrac{1}{20}$

If you answered **correctly**, go to problem 57.
If you answered **incorrectly**, go to problem 56.

56. Change the percent to a fraction: 80%.

Answer: $\dfrac{4}{5}$

If you answered **correctly**, go to problem 57.
If you answered **incorrectly**, review "Percents" on page 97.

57. Change the number to its equivalent in scientific notation: 41,700,000,000.

Answer: 4.17×10^{10}

If you answered **correctly**, go to problem 59.
If you answered **incorrectly**, go to problem 58.

58. Change the scientific notation to its equivalent number: 5.316×10^{-8}.

Answer: 0.00000005316

If you answered **correctly**, go to problem 59.
If you answered **incorrectly**, review "Scientific Notation" on page 99.

59. Simplify and write the answer in scientific notation: $(6 \times 10^{8})(7 \times 10^{-6})$.

Answer: 4.2×10^{3}

If you answered **correctly**, go to problem 61.
If you answered **incorrectly**, go to problem 60.

60. Simplify and write the answer in scientific notation: $(4 \times 10^{-8})(9 \times 10^{-2})$.

Answer: 3.6×10^{-9}

If you answered **correctly**, go to problem 61.
If you answered **incorrectly**, review "Scientific Notation" on page 99.

For problems 61 through 64, solve for the value of the variable.

61. $x + 10 = -8$

Answer: -18

If you answered **correctly**, go to problem 63.
If you answered **incorrectly**, go to problem 62.

62. $\dfrac{y}{2} = -5$

Answer: -10

If you answered **correctly**, go to problem 63.
If you answered **incorrectly**, review "Solving Linear Equations" on page 107.

63. $5x - 1 = 7x + 11$

Answer: -6

If you answered **correctly**, go to problem 65.
If you answered **incorrectly**, go to problem 64.

64. $\dfrac{3z + 6}{5} = z - 4$

Answer: 13

If you answered **correctly**, go to problem 65.
If you answered **incorrectly**, review "Solving Linear Equations with More Than One Operation" on page 111.

65. Solve for t: $d = rt$.

Answer: $t = \dfrac{d}{r}$

If you answered **correctly**, go to problem 67.
If you answered **incorrectly**, go to problem 66.

66. Solve for h: $A = 2\pi r^2 + 2\pi rh$.

Answer: $h = \dfrac{A - 2\pi r^2}{2\pi r}$

If you answered **correctly**, go to problem 67.
If you answered **incorrectly**, review "Solving Linear Formulas" on page 116.

67. Solve for x: $\dfrac{x}{14} = \dfrac{9}{21}$.

Answer: 6

If you answered **correctly**, go to problem 69.
If you answered **incorrectly**, go to problem 68.

68. Solve for x: $\dfrac{40}{52} = \dfrac{20}{x}$.

Answer: 26

If you answered **correctly**, go to problem 69.
If you answered **incorrectly**, review "Ratios and Proportions" on page 120.

69. Multiply $\dfrac{x^2(x+3)(x+4)^5}{x^3(x-4)^4(x+3)} \cdot \dfrac{x(x+3)(x-4)}{(x+4)}$.

Answer: $\dfrac{(x+3)(x+4)^4}{(x-4)^3}$

If you answered **correctly**, go to problem 71.
If you answered **incorrectly**, go to problem 70.

70. Divide $\dfrac{33y(y+2)}{y^5(y-2)^4} \div \dfrac{55(y+2)(y-2)}{10}$.

Answer: $\dfrac{6}{y^4(y-2)^5}$

If you answered **correctly**, go to problem 71.
If you answered **incorrectly**, review "Multiplying and Dividing Algebraic Fractions" on page 125.

71. Add $\dfrac{3}{4mn} + \dfrac{2}{m^2 n}$.

Answer: $\dfrac{3m+8}{4m^2 n}$

If you answered **correctly**, go to problem 73.
If you answered **incorrectly**, go to problem 72.

72. Subtract $\dfrac{5}{2z} - \dfrac{y}{8z^2}$.

Answer: $\dfrac{20z-y}{8z^2}$

If you answered **correctly**, go to problem 73.
If you answered **incorrectly**, review "Adding and Subtracting Algebraic Fractions" on page 128.

73. Solve for y: $\dfrac{3-y}{10} + \dfrac{y}{5} = \dfrac{y}{2} + \dfrac{8+y}{10}$.

Answer: -1

If you answered **correctly**, go to problem 75.
If you answered **incorrectly**, go to problem 74.

74. Solve for x: $\dfrac{x}{10} + \dfrac{1}{15} = \dfrac{x}{6} - \dfrac{1}{3}$.

Answer: 6

If you answered **correctly**, go to problem 75.
If you answered **incorrectly**, review "Equations with Fractions" on page 131.

75. Distribute: $-4x^2(5x^2 - 3x + 1)$.

Answer: $-20x^4 + 12x^3 - 4x^2$

If you answered **correctly**, go to problem 77.
If you answered **incorrectly**, go to problem 76.

76. Distribute: $2y(3y^2 - 2y + y^{-1} - 5y^{-3})$.

Answer: $6y^3 - 4y^2 + 2 - 10y^{-2}$

If you answered **correctly**, go to problem 77.
If you answered **incorrectly**, review "Multiplying Monomials" on page 145.

77. Multiply: $(3x + 7)(2x - 5)$.

Answer: $6x^2 - x - 35$

If you answered **correctly**, go to problem 79.
If you answered **incorrectly**, go to problem 78.

78. Multiply: $(9x - 1)(3x + 4)$.

Answer: $27x^2 + 33x - 4$

If you answered **correctly**, go to problem 79.
If you answered **incorrectly**, review "Multiplying Polynomials" on page 148.

79. Multiply: $(x - 1)(3x^2 + 2x - 4)$.

Answer: $3x^3 - x^2 - 6x + 4$

If you answered **correctly**, go to problem 81.
If you answered **incorrectly**, go to problem 80.

80. Multiply: $(x + 2)(x^2 - 5x - 2)$.

Answer: $x^3 - 3x^2 - 12x - 4$

If you answered **correctly**, go to problem 81.
If you answered **incorrectly**, review "Multiplying Polynomials" on page 148.

81. Multiply: $(x - 4y)(x + 4y)$.

Answer: $x^2 - 16y^2$

If you answered **correctly**, go to problem 83.
If you answered **incorrectly**, go to problem 82.

82. Multiply: $(a - 11)(a + 11)$.

Answer: $a^2 - 121$

If you answered **correctly**, go to problem 83.
If you answered **incorrectly**, review "Special Products" on page 151.

83. Multiply: $(m - 3)^2$.

Answer: $m^2 - 6m + 9$

If you answered **correctly**, go to problem 85.
If you answered **incorrectly**, go to problem 84.

84.   Multiply: $(2y + 5)^2$.

Answer: $4y^2 + 20y + 25$

If you answered **correctly**, go to problem 85.
If you answered **incorrectly**, review "Special Products" on page 151.

85.   Divide: $(5x^4 - 2x^3 - 4x^2 + 1) \div (x - 1)$.

Answer: $5x^3 + 3x^2 - x - 1$

If you answered **correctly**, go to problem 87.
If you answered **incorrectly**, go to problem 86.

86.   Divide: $(4z^4 - 8z^3 + 3z^2 - 2) \div (2z + 1)$.

Answer: $2z^3 - 5z^2 + 4z - 2$

If you answered **correctly**, go to problem 87.
If you answered **incorrectly**, review "Dividing Polynomials" on page 156.

87.   Factor out the Greatest Common Factor: $24a^2b^3c - 30ab^4c^2 + 36ab^2c$.

Answer: $6ab^2c(4ab - 5b^2c + 6)$

If you answered **correctly**, go to problem 89.
If you answered **incorrectly**, go to problem 88.

88.   Factor out the Greatest Common Factor: $8a(b - c)^2 + 6a^2(b - c)^3$.

Answer: $2a(b - c)^2[4 + 3a(b - c)]$

If you answered **correctly**, go to problem 89.
If you answered **incorrectly**, review "Greatest Common Factor" on page 163.

For problems 89 through 98, factor completely.

89. $x^2 - 100 =$

Answer: $(x - 10)(x + 10)$

If you answered **correctly**, go to problem 91.
If you answered **incorrectly**, go to problem 90.

90. $9y^2 - 25 =$

Answer: $(3y - 5)(3y + 5)$

If you answered **correctly**, go to problem 91.
If you answered **incorrectly**, review "Factoring Binomials" on page 165.

91. $y^3 - 64 =$

Answer: $(y - 4)(y^2 + 4y + 16)$

If you answered **correctly**, go to problem 93.
If you answered **incorrectly**, go to problem 92.

92. $8x^3 + 27 =$

Answer: $(2x + 3)(4x^2 - 6x + 9)$

If you answered **correctly**, go to problem 93.
If you answered **incorrectly**, review "Factoring Binomials" on page 165.

93. $x^2 + x - 72 =$

Answer: $(x + 9)(x - 8)$

If you answered **correctly**, go to problem 95.
If you answered **incorrectly**, go to problem 94.

94. $20x^2 + 7x - 6 =$

Answer: $(5x - 2)(4x + 3)$

If you answered **correctly**, go to problem 95.
If you answered **incorrectly**, review "Factoring Trinomials" on page 168.

95. $m^6 - 12m^3 + 27 =$

Answer: $(m^3 - 9)(m^3 - 3)$

If you answered **correctly**, go to problem 97.
If you answered **incorrectly**, go to problem 96.

96. $y^8 + 2y^4 - 35 =$

Answer: $(y^4 - 5)(y^4 + 7)$

If you answered **correctly**, go to problem 97.
If you answered **incorrectly**, review "Factoring Other Polynomials" on page 172.

97. $ac + ad^2 - bc - bd^2 =$

Answer: $(a - b)(c + d^2)$

If you answered **correctly**, go to problem 99.
If you answered **incorrectly**, go to problem 98.

98. $ax - xz - 3a + 3z =$

Answer: $(x - 3)(a - z)$

If you answered **correctly**, go to problem 99.
If you answered **incorrectly**, review "Factoring Other Polynomials" on page 172.

99. Which of the choices is in the solution of $-9 < 2x - 3 \leq 3$?

    A. 0

    B. 4

    C. 6

Answer: A

If you answered **correctly**, go to problem 101.
If you answered **incorrectly**, go to problem 100.

100. Which of the choices is in the solution of $4y + 1 < 3y - 1$?

    A. -3

    B. 0

    C. -1

Answer: A

If you answered **correctly**, go to problem 101.
If you answered **incorrectly**, review "Inequalities" on page 183.

101. Find the common solution of $-1 \leq y < 8$ and $y + 2 > 4$.

Answer: $2 < y < 8$

If you answered **correctly**, go to problem 103.
If you answered **incorrectly**, go to problem 102.

102. Find the common solution of $x \geq -2$ and $x < 6$.

Answer: $-2 \leq x < 6$

If you answered **correctly**, go to problem 103.
If you answered **incorrectly**, review "Solving Inequalities by Graphing on a Number Line" on page 189.

103. Solve for y: $2|y + 7| - 3 = 3$.

Answer: $-4, -10$

If you answered **correctly**, go to problem 105.
If you answered **incorrectly**, go to problem 104.

104. Solve for y: $|3y - 1| = 7$.

Answer: $\frac{8}{3}$ or -2

If you answered **correctly**, go to problem 105.
If you answered **incorrectly**, review "Absolute Value Equations" on page 192.

105. Solve for w: $|3w - 1| \geq 5$.

Answer: $w \leq -\frac{4}{3}$ or $w \geq 2$

If you answered **correctly**, go to problem 107.
If you answered **incorrectly**, go to problem 106.

106. Solve for t: $|5 - t| < 13$.

Answer: $-8 < t < 18$

If you answered **correctly**, go to problem 107.
If you answered **incorrectly**, review "Absolute Value Inequalities" on page 199.

107. Simplify $\sqrt{72x^5 y^2 z^7}$.

Answer: $6x^2 yz^3 \sqrt{2xz}$

If you answered **correctly**, go to problem 109.
If you answered **incorrectly**, go to problem 108.

108. Simplify $\sqrt{200xy^2z^3}$.

Answer: $10yz\sqrt{2xz}$

If you answered **correctly**, go to problem 109.
If you answered **incorrectly**, review "Simplifying Square Roots" on page 203.

109. Simplify and add $3\sqrt{18xy^4} + 5y\sqrt{8xy^2}$.

Answer: $19y^2\sqrt{2x}$

If you answered **correctly**, go to problem 111.
If you answered **incorrectly**, go to problem 110.

110. Simplify and subtract $\sqrt{20ab^2} - b\sqrt{5a}$.

Answer: $b\sqrt{5a}$

If you answered **correctly**, go to problem 111.
If you answered **incorrectly**, review "Simplifying Square Roots" on page 203.

111. Simplify $\sqrt[3]{8xy^3} - y\sqrt[3]{x}$.

Answer: $y\sqrt[3]{x}$

If you answered **correctly**, go to problem 113.
If you answered **incorrectly**, go to problem 112.

112. Simplify $\sqrt[3]{54t^4}$.

Answer: $3t\sqrt[3]{2t}$

If you answered **correctly**, go to problem 111.
If you answered **incorrectly**, review "Simplifying Other Roots" on page 206.

113. Solve for x: $\sqrt{3x-2} = 4$.

Answer: 6

If you answered **correctly**, go to problem 115.
If you answered **incorrectly**, go to problem 114.

114. Solve for x: $\sqrt{6-x} = 5$.

Answer: -19

If you answered **correctly**, go to problem 115.
If you answered **incorrectly**, review "Radical Equations" on page 208.

For problems 115 through 124, solve for the value(s) of the variable.

115. $4x^2 - 17x - 15 = 0$

Answer: $-\dfrac{3}{4}$, 5

If you answered **correctly**, go to problem 117.
If you answered **incorrectly**, go to problem 116.

116. $3x^2 + 20x - 7 = 0$

Answer: $7, \dfrac{1}{3}$

If you answered **correctly**, go to problem 117.
If you answered **incorrectly**, review "Solving Quadratic Equations by Factoring" on page 224.

117. $2x^2 + 7x + 1 = 0$

Answer: $\dfrac{-7 \pm \sqrt{41}}{4}$

If you answered **correctly**, go to problem 119.
If you answered **incorrectly**, go to problem 118.

118. $x^2 + 5x - 1 = 0$

Answer: $\dfrac{-5 \pm \sqrt{29}}{2}$

If you answered **correctly**, go to problem 119.
If you answered **incorrectly**, review "Solving Quadratic Equations with the Quadratic Formula" on page 228.

119. $z^6 + 7z^3 - 8 = 0$

Answer: $1, -2$

If you answered **correctly**, go to problem 121.
If you answered **incorrectly**, go to problem 120.

120. $z^4 - 13z^2 + 36 = 0$

Answer: $2, -2, 3, -3$

If you answered **correctly**, go to problem 121.
If you answered **incorrectly**, review "Solving Quadratic-Like Equations" on page 237.

121. $y(y + 7)(y - 9) \geq 0$

Answer: $-7 \leq y \leq 0$ or $y \geq 9$

If you answered **correctly**, go to problem 123.
If you answered **incorrectly**, go to problem 122.

122. $(x + 1)(x - 4) < 0$

Answer: $-1 < x < 4$

If you answered **correctly**, go to problem 123.
If you answered **incorrectly**, review "Quadratic and Other Inequalities" on page 241.

123. $\sqrt{y + 31} = 11 - y$

Answer: 5 only

If you answered **correctly**, go to problem 125.
If you answered **incorrectly**, go to problem 124.

124. $\sqrt{3x-5}=x-3$

Answer: 7

If you answered **correctly**, go to problem 125.
If you answered **incorrectly**, review "Radical Equations and Quadratics" on page 245.

125. What is the vertex of the parabola $y=-4(x+1)^2-5$?

Answer: $(-1, -5)$

If you answered **correctly**, go to problem 127.
If you answered **incorrectly**, go to problem 126.

126. The graph of the parabola $y=-4(x+1)^2-5$

    A. opens upward

    B. opens downward

    C. crosses the y-axis in two places

Answer: B

If you answered **correctly**, go to problem 127.
If you answered **incorrectly**, review "Graphing Other Curves" on page 273.

127. Where does the graph of $y = (x + 1)(x - 2)(x + 3)$ cross the x-axis?

Answer: $-1, 2, -3$

If you answered **correctly**, go to problem 129.
If you answered **incorrectly**, go to problem 128.

128. Where does the graph of $y = (x - 4)(x + 5)(x - 6)$ cross the x-axis?

Answer: $4, -5, 6$

If you answered **correctly**, go to problem 129.
If you answered **incorrectly**, review "Graphing Other Curves" on page 273.

129. Where is the equation of the line that has a slope of $-\frac{2}{3}$ and goes through the point $(-6,1)$?

Answer: $y=-\frac{2}{3}x-3$

If you answered **correctly**, go to problem 131.
If you answered **incorrectly**, go to problem 130.

130. What is the equation of the line that has a slope of 7 and goes through the point $(2,-3)$?

Answer: $y = 7x - 17$

If you answered **correctly**, go to problem 131.
If you answered **incorrectly**, review "Finding the Equation of a Line" on page 283.

131. What is a point in the solution of $2x - 4y < 7$?

    A. (4,0)

    B. (1,0)

    C. (−2,−3)

Answer: B

If you answered **correctly**, go to problem 133.
If you answered **incorrectly**, go to problem 132.

132. What is a point in the solution of $x + 8 \geq 7 + y$?

    A. (0,2)

    B. (5,3)

    C. (−6,−4)

Answer: B

If you answered **correctly**, go to problem 133.
If you answered **incorrectly**, review "Graphing Inequalities" on page 286.

133. What is the common solution of the lines $y = 3x - 7$ and $2x - y = 5$?

Answer: (2,−1)

If you answered **correctly**, go to problem 135.
If you answered **incorrectly**, go to problem 134.

134. What is the common solution of the lines $x + y = 18$ and $y = -2x + 26$?

Answer: (8,10)

If you answered **correctly**, go to problem 135.
If you answered **incorrectly**, review "Solving Systems of Equations by Graphing" on page 290.

135. Which of the following is a function of x?

    A. $y = \pm\sqrt{x^2 - 4}$

    B. $y = x^2 - 4$

    C. $y^2 = x^3 - 4$

Answer: B

If you answered **correctly**, go to problem 137.
If you answered **incorrectly**, go to problem 136.

136. Which of the following is a function of x?

    A. $y = 3x - 7$

    B. $x^2 + y^2 = 7$

    C. $|y| = 3x - 1$

Answer: A

If you answered **correctly**, go to problem 137.
If you answered **incorrectly**, review "Functions and Relations" on page 313.

137. What is the domain of the function $y = \dfrac{3}{x-6}$?

Answer: $x \neq 6$

If you answered **correctly**, go to problem 139.
If you answered **incorrectly**, go to problem 138.

138. What is the domain of the function $y = \sqrt{x+2}$?

Answer: $x \geq -2$

If you answered **correctly**, go to problem 139.
If you answered **incorrectly**, review "Domain and Range" on page 316.

139. What is the range of the function $y = \sqrt{4-x}$?

Answer: $y \geq 0$

If you answered **correctly**, go to problem 141.
If you answered **incorrectly**, go to problem 140.

140. What is the range of the function $y = x^2$?

Answer: $y \geq 0$

If you answered **correctly**, go to problem 141.
If you answered **incorrectly**, review "Domain and Range" on page 316.

141. Which is the inverse of the function $f(x) = \dfrac{2}{x+3}$?
    A. $f^{-1}(x) = \dfrac{x+3}{2}$
    B. $f^{-1}(x) = \dfrac{2-3x}{x}$
    C. $f^{-1}(x) = \dfrac{2}{x-3}$

Answer: B

If you answered **correctly**, go to problem 143.
If you answered **incorrectly**, go to problem 142.

142. Which is the inverse of the function $f(x) = 8x - 1$?
    A. $f^{-1}(x) = \dfrac{x+1}{8}$
    B. $f^{-1}(x) = 8x + 1$
    C. $f^{-1}(x) = \dfrac{1}{8x-1}$

Answer: A

If you answered **correctly**, go to problem 143.
If you answered **incorrectly**, review "Inverse Functions and Function Notation" on page 325.

143. One number is 2 less than 5 times another. Their sum is 16. What is their product?

Answer: 39

If you answered **correctly**, go to problem 145.
If you answered **incorrectly**, go to problem 144.

144. One number is 6 greater than another. Their sum is 194. What are the numbers?

Answer: 94 and 100

If you answered **correctly**, go to problem 145.
If you answered **incorrectly**, review "Number Problems" on page 333.

145. The sum of 4 consecutive multiples of 4 is 184. What is the smallest of them?

Answer: 40

If you answered **correctly**, go to problem 147.
If you answered **incorrectly**, go to problem 146.

146. The sum of 3 consecutive odd integers is −39. What are they?

Answer: −15, −13, −11

If you answered **correctly**, go to problem 147.
If you answered **incorrectly**, review "Consecutive Integer Problems" on page 335.

147. Sam is twice as old as Don. Twenty years ago, he was 4 times as old as Don. How old is Sam now?

Answer: 60

If you answered **correctly**, go to problem 149.
If you answered **incorrectly**, go to problem 148.

148. Ted is 8 years older than Jeff. Thirty years ago, Ted was 5 times as old as Jeff. How old is Jeff now?

Answer: 32

If you answered **correctly**, go to problem 149.
If you answered **incorrectly**, review "Age Problems" on page 337.

149. A rectangle is twice as long as it is wide. If the length is decreased by 1 inch and the width is increased by 1 inch, the area increases by 4 square inches. What is the original width of the rectangle?

Answer: 5

If you answered **correctly**, go to problem 151.
If you answered **incorrectly**, go to problem 150.

150. A rectangle is 14 inches longer than it is wide. If the area is 240 square inches, then what is the length?

Answer: 24 inches

If you answered **correctly**, go to problem 151.
If you answered **incorrectly**, review "Geometric Problems" on page 339.

151. Sidney left for school at 8:00 AM traveling at 35 mph. Ron left the same place at 8:30 AM, following Sidney's path. He was traveling at 40 mph. How far had Sidney gone before Ron caught up with him?

Answer: 140 miles

If you answered **correctly**, go to problem 153.
If you answered **incorrectly**, go to problem 152.

152. Mary and Carrie left their campsite at the same time. Mary headed north walking at the rate of 4 mph, and Carrie headed south walking at the rate of 6 mph. How long did it take before they were 40 miles apart?

Answer: 4 hours

If you answered **correctly**, go to problem 153.
If you answered **incorrectly**, review "Distance Problems" on page 344.

153. How many quarts of 40% solution and how many quarts of 90% solution should be mixed to create a mixture of 20 quarts of 80% solution?

Answer: 4 of 40% and 16 of 90%

If you answered **correctly**, go to problem 155.
If you answered **incorrectly**, go to problem 154.

154. How many quarts of 20% solution should be mixed with 6 quarts of 60% solution to produce a 50% solution?

Answer: 2 quarts

If you answered **correctly**, go to problem 155.
If you answered **incorrectly**, review "Mixture Problems" on page 347.

155. Charlie can do a job in 6 hours, and Dan can do the same job in 8 hours. How long will it take the two of them to do the job if they work together?

Answer: $3\frac{3}{7}$ hours

If you answered **correctly**, you are finished! Congratulations.
If you answered **incorrectly**, go to problem 156.

156. Rick can do a job in 30 minutes, and Tim can do the same job in 36 minutes. How long will it take for the two of them to do the job if they work together?

Answer: $16\frac{4}{11}$ minutes

If you answered **correctly**, you are finished! Congratulations.
If you answered **incorrectly**, review "Working Together Problems" on page 353.

Index